Physiological Consequences of Brain Insulin Action

The brain is crucial for the regulation of whole-body metabolism and behavior. The pancreas-derived hormone insulin modulates brain function and affects energy metabolism as well as cognition. This is partially achieved by modulating several types of brain cell populations including those relevant to satiety and reward. Importantly, brains of Alzheimer's disease patients exhibit a signature of brain insulin resistance with perturbed brain metabolism. This book will cover the basics of insulin signaling in the brain and will describe concepts of insulin resistance, a feature of type 2 diabetes. Moreover, insulin's effect on cognitive function will be pointed out and the association between brain insulin resistance and neurodegenerative diseases discussed. Additionally, potential behavioral and pharmacological concepts which can affect brain insulin signaling will be summarized.

Key Features:

- Summarizes insulin and the closely related IGF-1 receptor signaling
- Depicts concepts of insulin resistance
- Highlights the importance of conserved brain insulin signaling for brain function, metabolism, and behavior
- Describes potential behavioral and pharmacological approaches to support brain insulin signaling

Oxidative Stress and Disease

Series Editor:
Enrique Cadenas, MD, PhD
Helmut Sies, MD
University of Southern California School of Pharmacy
Los Angeles, California

PUBLISHED TITLES

Mitochondria in Liver Disease
edited by Derick Han and Neil Kaplowitz

Fetal and Early Postnatal Programming and its Influence on Adult Health
edited by Mulchand S. Patel and Jens H. Nielsen

Biomedical Application of Nanoparticles
edited by Bertrand Rihn

The Biology of the First 1,000 Days
edited by Crystal D. Karakochuk, Kyly C. Whitfield, Tim J. Green, and Klaus Kraemer

Hydrogen Peroxide Metabolism in Health and Disease
edited by Margreet C M Vissers, Mark Hampton, and Anthony J. Kettle

Glutathione
edited by Leopold Flohé

Vitamin C: Biochemistry and Function
edited by Margreet C M Vissers and Qi Chen

Cancer and Vitamin C
edited by Margreet C M Vissers and Qi Chen

Mammalian Heme Peroxidases: Diverse Roles in Health and Disease
edited by Clare Hawkins and William M. Nauseef

Redox Regulation of Differentiation and De-differentiation
edited by Carsten Berndt and Christopher Horst Lillig

Proteostasis and Proteolysis
edited by Niki Chondrogianni, Elah Pick and Anna Gioran

Oxidative Eustress in Exercise Physiology
edited by James N. Cobley and Gareth W. Davison

Physiological Consequences of Brain Insulin Action
edited by André Kleinridders

For more information about this series, please visit:
https://www.crcpress.com/Oxidative-Stress-and-Disease/book-series/CRCOXISTRDIS

Physiological Consequences of Brain Insulin Action

Edited by
André Kleinridders

CRC Press
Taylor & Francis Group
Boca Raton London New York

CRC Press is an imprint of the
Taylor & Francis Group, an **informa** business

First edition published 2023
by CRC Press
6000 Broken Sound Parkway NW, Suite 300, Boca Raton, FL 33487-2742

and by CRC Press
4 Park Square, Milton Park, Abingdon, Oxon, OX14 4RN

CRC Press is an imprint of Taylor & Francis Group, LLC

ISBN: 9780367529482 (hbk)
ISBN: 9780367529512 (pbk)
ISBN: 9781003079927 (ebk)

DOI: 10.1201/9781003079927

Typeset in Joanna
by codeMantra

CONTENTS

SERIES PREFACE

PHYSIOLOGICAL CONSEQUENCES OF BRAIN INSULIN ACTION

EDITOR: ANDRÉ KLEINRIDDERS

THE EDITOR

Dr. André Kleinridders is a Professor for Experimental and Molecular Nutritional Medicine at the Institute of Nutritional Science at the University of Potsdam. Prof. Kleinridders' research programs address the impact of brain insulin signaling on whole body pathophysiology and behavior with implications for diabetes, obesity and neurodegeneration. His studies focus on understanding the mechanisms of central insulin resistance (involving impairment of mitochondrial function) and its consequences on energy homeostasis and neurological alterations associated with aging. Dr. Kleinridders' expertise in endocrinology and genetics expands to areas of high significance for human pathophysiology, including metabolism, insulin signaling, and insulin resistance, the latter an umbrella term with implications for not only diabetes and obesity but also for neurodegenerative disorders, such as Alzheimer's disease. His research focuses on insulin action in the brain and its effect on mitochondrial stress responses and reduction of diet-induced weight gain. His recent publications establish Dr. Kleinridders' unique research focus on the role of brain insulin in human pathophysiology.

PHYSIOLOGICAL CONSEQUENCES OF BRAIN INSULIN ACTION

PHYSIOLOGICAL CONSEQUENCES OF BRAIN INSULIN ACTION, focuses on different aspects of brain insulin and its regulation of a myriad of systemic effects ranging from starve-feed cycles to behavioral patterns. This book is particularly attractive because it places brain insulin signaling in the physiologic maintenance of whole-body homeostasis as well as impairment of brain insulin signaling in the pathology of diabetes and obesity. Moreover, brain insulin resistance is a critical factor in the pathogenesis of Alzheimer's disease, sometimes termed diabetes type III, hence the significance of brain insulin in modulation of cognitive functions. The book entails mechanistic aspects of brain insulin signaling and that triggered by activation of the insulin-like growth factor receptor (IGF1R), the regulation of neuronal circuits and how insulin modulates metabolism in different brain cell populations. How hypothalamic insulin signaling controls energy metabolism, food intake, and behavior, is viewed among the systemic effects of brain insulin. These chapters expand the effects of brain insulin in terms of insulin resistance as it occurs in type 2 diabetes and Alzheimer's disease as well as diseases associated with protein misfolding. Pharmacological and dietary interventions that have the potential to overcome insulin resistance are also addressed.

Overall, the major thrust of this book is the recognition of the multifaceted effects of brain insulin and its signaling pathways not only in the brain and in different brain cell populations but how brain insulin modulates systemic tissue functions. This acquires further relevance in terms of brain insulin resistance and its impairment of whole-body homeostasis and association with cognitive deficits.

ENRIQUE CADENAS
HELMUT SIES

Series Editors
Oxidative Stress and Disease

PREFACE

The discovery of insulin represents a landmark achievement in medical sciences. Insulin was discovered by Frederick Grant Banting and Charles Herbert Best in the laboratory of John James Rickard McLeod more than 100 years ago and saved millions of lives from a formerly non-curable disease. While type 1 diabetes is characterized by insulin deficiency, which can be treated by the administration of insulin, the majority of patients suffer from type 2 diabetes. These patients exhibit initially reduced sensitivity to insulin, known as insulin resistance. Insulin resistance causes reduced glucose uptake in skeletal muscle and adipose tissue. In the liver, insulin resistance causes deteriorated hepatic glucose production, leading to elevated blood glucose levels overall. Although glucose uptake in the brain is mostly independent of insulin, insulin and its receptor have already been detected in the late 1960s and 1970s in the brain, and research over the last decades has clearly demonstrated that the brain is an insulin-sensitive organ.

Insulin in the brain regulates food intake, affects energy metabolism and glucose handling in peripheral tissues and even modulates fertility. In addition, there is ample evidence that insulin exhibits neuroprotective function. The presence of brain insulin resistance is not only a feature of unhealthy obesity but is also observed in the brains of patients suffering from Alzheimer's disease. Together, these observations highlight the importance of the precise regulation of insulin action in the central nervous system for a healthy metabolism and proper cognitive function.

This book will cover the molecular mechanisms of insulin signaling and the establishment of insulin resistance. As the brain shows region-specific differences in terms of function and cell composition, cell- and region-specific effects of insulin action are discussed. Diabetes-associated disorders including mood disorders and neurodegenerative diseases will be described and associated with brain insulin resistance. The complex interaction of peripheral organs with the brain is characterized to a great extent in this book, as is the hormonal crosstalk of insulin with leptin in the central nervous system. Importantly, novel insights about brain insulin signaling in humans will be explained in addition to behavioral and pharmacological treatment options to counteract insulin resistance. In summary, this book will provide a comprehensive overview about the physiological consequences of brain insulin action.

ACKNOWLEDGEMENTS

I would like to thank all the people who helped with the preparation of this book. A special thanks goes to all authors and co-authors who dedicated their time and work to contribute to important chapters. Without their efforts, this book would not have been possible.

EDITOR

Prof. Dr. André Kleinridders is a Professor for Experimental and Molecular Nutritional Medicine at the Institute of Nutritional Science at the University of Potsdam, Germany. Dr. Kleinridders' research programs aim at understanding causes and consequences of brain insulin resistance with implications for diabetes, obesity and neurological disorders. His studies focus on the interaction of mitochondrial function and insulin signaling and its consequences on energy homeostasis and behavior. Several publications establish Dr. Kleinridders' unique research focus on the role of brain insulin in metabolic disorders.

CONTRIBUTORS

CARME AULADELL
Department of Cellular Biology, Physiology and Immunology, Faculty of Biology
University of Barcelona
Barcelona, Spain

KELLY A. BORGES
Department of Biomedical Sciences
New York Institute of Technology College of Osteopathic Medicine
Old Westbury, New York, USA

SEBASTIAN BULLICH
Centre de Recherches sur la Cognition Animale (CRCA), Centre de Biologie Intégrative (CBI)
Toulouse University
Toulouse, France

WEIKANG CAI
Department of Biomedical Sciences
New York Institute of Technology College of Osteopathic Medicine
Old Westbury, New York, USA

ANTONI CAMINS
Departament de Farmacologia, Toxicologia i Química Terapèutica, Facultat de Farmàcia i Ciències de l'Alimentació
Universitat de Barcelona
Barcelona, Spain
and
Biomedical Research Networking Centre
Neurodegenerative Diseases (CIBERNED)
Madrid, Spain
and
Institut de Neurociències
Universitat de Barcelona
Barcelona, Spain

AMANDA CANO
Biomedical Research Networking Centre
Neurodegenerative Diseases (CIBERNED)
Madrid, Spain
and
Institute of Nanoscience and Nanotechnology (IN2UB)
University of Barcelona
Barcelona, Spain
and
Unitat de Farmàcia, Tecnologia Farmacèutica i Fisico-química, Facultat de Farmàcia i Ciències de l'Alimentació
Universitat de Barcelona
Barcelona, Spain
and
Research Center and Memory Clinic, Fundació
Ace Alzheimer Center Barcelona
Barcelona, Spain

SABELA CASADO
Departamento de Fisioloxía and Centro de Investigación en Medicina Molecular (CIMUS)
Universidade de Santiago de Compostela, Instituto de Investigaciones Sanitarias de Santiago de Compostela (IDIS)
Santiago de Compostela, Spain
and
CIBER Fisiopatología de la Obesidad y Nutrición (CIBERobn)
Madrid, Spain

SIMRAN CHOPRA
Institute of Nutritional Science, Molecular and Experimental Nutritional Medicine
University of Potsdam
Nuthetal, Germany

MIREN ETTCHETO
Departament de Farmacologia, Toxicologia i Química Terapèutica, Facultat de Farmàcia i Ciències de l'Alimentació
Universitat de Barcelona
Barcelona, Spain
and
Departament de Bioquímica i Biotecnologia, Facultat de Medicina i Ciències de la Salut
Universitat Rovira i Virgili
Reus, Spain
and
Biomedical Research Networking Centre
Neurodegenerative Diseases (CIBERNED)
Madrid, Spain
and
Institut de Neurociències
Universitat de Barcelona
Barcelona, Spain

ANA M. FERNANDEZ
Department of Functional and Systems Neuroscience
Cajal Institute, CSIC
Madrid, Spain
and
Ciberned
Madrid, Spain

XAVIER FIORAMONTI
Univ. Bordeaux
Bordeaux INP
NutriNeuro,
Bordeaux, France

JAUME FOLCH
Departament de Bioquímica i Biotecnologia, Facultat de Medicina i Ciències de la Salut
Universitat Rovira i Virgili
Reus, Spain
and
Biomedical Research Networking Centre
Neurodegenerative Diseases (CIBERNED)
Madrid, Spain

JIALIN FU
Dianne Nunnally Hoppes Laboratory for Diabetes Complications, Section of Vascular Cell Biology, Joslin Diabetes Center
Harvard Medical School
Boston, Massachusetts, USA

BRUNO P. GUIARD
Centre de Recherches sur la Cognition Animale (CRCA), Centre de Biologie Intégrative (CBI)
Toulouse University
Toulouse, France

LINUS HABERBOSCH
Department of Endocrinology and Metabolic Diseases
Charité University Medicine
Berlin, Germany

ROBERT HAUFFE
Institute of Nutritional Science, Molecular and Experimental Nutritional Medicine
University of Potsdam
Nuthetal, Germany

QIAN HUANG
Department of Biomedical Sciences
New York Institute of Technology College of Osteopathic Medicine
Old Westbury, New York, USA

REINER JUMPERTZ-VON SCHWARTZENBERG
German Center for Diabetes Research (DZD)
Neuherberg, Germany
and
Department of Internal Medicine IV, Division of Diabetology, Endocrinology and Nephrology
Eberhard-Karls University Tübingen
Tübingen, Germany

STEFAN KABISCH
Clinic of Endocrinology and Metabolic Medicine
Campus Benjamin Franklin
Charité University Medicine
Berlin, Germany
and
German Center for Diabetes Research (DZD)
München-Neuherberg, Germany

KAJ KAMSTRA
Centre for Neuroendocrinology and Brain Health Research Centre
University Otago
Dunedin, New Zealand
and
Department of Physiology
Dunedin School of Medicine, University of Otago
Dunedin, New Zealand

ANDRÉ KLEINRIDDERS
Institute of Nutritional Science, Molecular and Experimental Nutritional Medicine
University of Potsdam
Nuthetal, Germany

STEPHANIE KULLMANN
Institute for Diabetes Research and Metabolic Diseases of the Helmholtz Center Munich
University of Tübingen
Tübingen, Germany
and
German Center for Diabetes Research (DZD)
Tübingen, Germany

THOMAS LAEGER
Physiology and Pathophysiology of Nutrition
University of Potsdam
Potsdam, Germany

RACHEL N. LIPPERT
'Neurocircuit Development and Function' Junior Research Group
German Institute of Human Nutrition
Nuthetal, Germany
and
German Center for Diabetes Research (DZD)
Düsseldorf, Germany
and
NeuroCure Cluster of Excellence
Charité – Universitätsmedizin
Berlin, Germany

BRIAN C.LIU
Department of Immunology (former Integrative Physiology and Pathobiology), Program in Pharmacology and Experimental Therapeutics and Pharmacology and Drug Development, Graduate School of Biomedical Sciences
Tufts University School of Medicine
Boston, Massachusetts, USA

HUGO MARTIN
Univ. Bordeaux
Bordeaux INP
NutriNeuro,
Bordeaux, France

LAURA MARTINEZ-RACHADELL
Department of Functional and Systems Neuroscience
Cajal Institute, CSIC
Madrid, Spain
and
Ciberned
Madrid, Spain

LUKAS MAURER
Department of Endocrinology and Metabolic Diseases
Charité University Medicine
Berlin, Germany

PATRICIA MIRANDA-AZPIAZU
Laboratory of Neurobiology of Insulin Peptides
Achucarro Basque Center for Neuroscience
Leioa, Spain

RUBÉN NOGUEIRAS
Departamento de Fisioloxía and Centro de Investigación en Medicina Molecular (CIMUS)
Universidade de Santiago de Compostela, Instituto de Investigaciones Sanitarias de Santiago de Compostela (IDIS)
Santiago de Compostela, Spain
and
CIBER Fisiopatología de la Obesidad y Nutrición (CIBERobn)
Madrid, Spain

TADEU DE OLIVEIRA DIZ
Departamento de Fisioloxía and Centro de Investigación en Medicina Molecular (CIMUS)
Universidade de Santiago de Compostela, Instituto de Investigaciones Sanitarias de Santiago de Compostela (IDIS)
Santiago de Compostela, Spain

EMMANUEL N. POTHOS
Department of Immunology (former Integrative Physiology and Pathobiology), Program in Pharmacology and Experimental Therapeutics and Pharmacology and Drug Development, Graduate School of Biomedical Sciences
Tufts University School of Medicine
Boston, Massachusetts, USA

ELENA SANCHEZ-LOPEZ
Biomedical Research Networking Centre Neurodegenerative Diseases (CIBERNED)
Madrid, Spain
and
Institute of Nanoscience and Nanotechnology (IN2UB)
University of Barcelona
Barcelona, Spain
and
Unitat de Farmàcia, Tecnologia Farmacèutica i Fisico-química, Facultat de Farmàcia i Ciències de l'Alimentació
Universitat de Barcelona
Barcelona, Spain

IGNACIO TORRES ALEMAN
Ciberned
Madrid, Spain
and
Laboratory of Neurobiology of insulin peptides
Achucarro Basque Center for Neuroscience
Leioa, Spain
and
Ikerbasque Foundation for Science
Bilbao, Spain

SULAY TOVAR
Departamento de Fisioloxía and Centro de Investigación en Medicina Molecular (CIMUS)
Universidade de Santiago de Compostela, Instituto de Investigaciones Sanitarias de Santiago de Compostela (IDIS)
Santiago de Compostela, Spain
and
CIBER Fisiopatología de la Obesidad y Nutrición (CIBERobn)
Madrid, Spain

ALEXANDER TUPS
Centre for Neuroendocrinology and Brain Health Research Centre
University Otago
Dunedin, New Zealand
and
Department of Physiology
Dunedin School of Medicine, University of Otago
Dunedin, New Zealand

SELMA YAGOUB
'Neurocircuit Development and Function' Junior Research Group
German Institute of Human Nutrition
Nuthetal, Germany

QINGCHEN ZHANG
Department of Immunology (former Integrative Physiology and Pathobiology), Program in Pharmacology and Experimental Therapeutics and Pharmacology and Drug Development, Graduate School of Biomedical Sciences
Tufts University School of Medicine
Boston, Massachusetts, USA

NOTICE

Scientific knowledge constantly advances. The authors and editor reassessed their sources of information, which they considered as solid to give detailed information about the respective scientific topics. Yet due to possible human error and/or scientific progression, neither the editor nor the authors nor any other party warrants that the presented information of this book is in every aspect accurate and thus disclaims responsibility for any errors or omissions of the provided information in this work.

ABBREVIATIONS

[Ca2+]i	Cytosolic Ca^{2+} Concentration
5-HT	Serotonin
AARE	Amino Acid Response Element
ACE	Angiotensin-Converting Enzyme
Acrp30	Adipocyte Complement-Related Protein of 30 kDa
AD	Alzheimer's Disease
ADCY	Adenylate Cyclase
AdipoR	Adiponectin Receptor
ADP/ATP	Adenosine Di-/Triphosphate
Ads	Antidepressants
AGE	Advanced Glycation End-Products
AgRP	Agouti-Related Protein
AKT/PKB	Ak Strain Transforming Kinase/Protein Kinase B
AMPA	Alpha-Amino-3-Hydroxy-5-Methyl-4-Isoxazolepropionic Acid
AMPK	5′AMP-Activated Protein Kinase
ANGPTL6	Angiopoietin-Related Protein 6
ANLS	Astrocyte-Neuron Lactate Shuttle
APOE	Apolipoprotein E
ARC	Arcuate Nucleus
ASD	Autism Spectrum Disorder
ATF4	Activating Transcription Factor 4
Aβ	Amyloid Beta (Aβ)
BACE1	Beta-Secretase 1
Bad	BCL2 Associated Agonist Of Cell Death
BAT	Brown Adipose Tissue
BBB	Blood–Brain Barrier
BCAA	Branched-Chain Amino Acids
Bcl-2	B-cell Lymphoma 2
BCSF	Blood–Cerebrospinal Fluid Barrier
BDI	Beck's Depression Inventory
BDNF	Brain-Derived Neurotrophic Factor
BHB	Beta-Hydroxybutyrate
BIR	Brain Insulin Resistance
BMI	Body Mass Index
BOLD	Blood-Oxygen-Level-Dependent Contrast
CART	Cocaine and Amphetamine-Related Transcript
CBF	Imaging or Cerebral Blood Flow
CDKAL1	Cyclin-Dependent Kinase 5 Regulatory Subunit-Associated Protein 1-Like 1
ChREBP	Carbohydrate-Response Element-Binding Protein
CNO	Clozapine-N-Oxyde
CNS	Central Nervous System
CRE	Camp Response Element
CREB	Camp Response Element Binding Protein
CRP	C-Reactive Peptide
CSF	Cerebrospinal Fluid
CVD	Cardiovascular Disease
DA3–CH	GLP-1/GIP Dual Receptor Agonist

DAG	Diacyl Glycerol
DASH	Dietary Approaches to Stop Hypertension
DAT	Dopamine Transporter
db/db	Leptin Receptor-Deficient
DCX	Doublin or Lissencephalin-X
DHA	Docosahexaenoic Acid
DIO	Diet-Induced Obesity
DKK1	Dickkopf-Related Protein 1
DM2	See T2D, T2DM
DMH	Dorso-Medial Hypothalamus
DPP-4	Dipeptidyl Peptidase-4
DRD2	Dopamin Receptor D2
DREADD	Designer Receptor Exclusively Activated by Designer Drug
DRN	Dorsal Raphe Nucleus
DRP1	Dynamin-Related Protein 1
Dvl	Dishevelled
EEG	Electroencephalography
EGF	Epidermal Growth Factor
eIF2α	Eukaryotic Initiation Factor 2 Alpha
Emx-Cre	Emx Locus Cre Recombinase
EOAD	Early-Onset AD
EPA	Eicosapentaenoic Acid
EPAC	Exchange Protein Activated by Camp
ER	Endoplasmic Reticulum
ERK	Extracellular Signal-Regulated Kinase
ETC	Electron Transport Chain
F2,6P2	Fructose-2,6-Bisphosphate
FA	Friedreich's Ataxia
fa/fa	Zucker Fatty Rat
FABP2	Fatty-Acid Binding Protein 2
FDA	Food And Drug Administration
FDG	^{18}F Fluorodeoxyglucose
FDG-PET	2-Fluoro-2-Deoxy-D-Glucose Positron Emission Tomography
FFA	Free Fatty Acid
FGF-2	Fibroblast Growth Factor 2
FGF21	Fibroblast Growth Factor 21
FGFR	Fibroblast Growth Factor Receptor
fMRI	Functional Magnetic Resonance Imaging
FOXO	Forkhead Box O
FTO	Fused-Toes and Obesity
Fzd	Frizzled
G6P	Glucose-6-Phosphate
GABA	Gamma-Amino Butyric Acid
GAD	Generalized Anxiety Disorder

GCN2	General Control Nonderepressible 2
GDF15	Growth/Differentiation Factor 15
GDP/GTP	Guanosine Di-/Triphosphate
GH	Growth Hormone
GHRH	Growth Hormone Releasing Hormone
GHSR	Growth Hormone Secretagogue Receptor
GI	Glycemic Index
GIP	Glucose-Dependent Insulinotropic Polypeptide; Gastric Inhibitory Peptide
GIRKO	Glial Insulin Receptor Knock-Out
GLP1	Glucagon-Like Peptide 1
GluA1	Glutamate Receptor 1
GLUT	Glucose Transporter
GPCR	G-Protein-Coupled Receptor
GRB2, Grb-2	Growth Factor Receptor-Bound Protein 2
GRP94	Endoplasmin
GS	Glycogen Synthase
GSK	Glycogen Synthase Kinase
GSK3β	Glycogen Synthase Kinase-3β
GWAS	Genome Wide Association Study
Hb	Habenula
HFD	High-Fat Diet
HGP	Hepatic Glucose Production
HMG	Hemimegalencephaly
HMG-CoA	B-Hydroxy β-Methylglutaryl-CoA
HMW	High-Molecular-Weight
HOMA-IR	Homeostasis Model Assessment of Insulin Resistance
HPA axis	Hypothalamic–Pituitary–Adrenal Axis
HSGAG	Heparan Sulfate Glycosaminoglycan
ICV	Intracerebroventricular
IDE	Insulin Degrading Enzyme
IDI	Intranasal Delivery of Insulin
IGF-1, IGF1	Insulin-Like Growth Factor 1
IGF-1R, IGF1R	Insulin-Like Growth Factor 1 Receptor
IGF2	Insulin like Growth Factor 2
IGFBP	Insulin-Like Growth Factor Binding Protein
IKK	Inhibitor of Nuclear Factor Kappa-B Kinase

IL	Interleukin
IL-6	Interleukin 6
ILP	Insulin Peptides
IP3	Inositol-1,4,5-Triphosphate
IP3R	Inositol-1,4,5-Triphosphate (IP3) Receptor
IQ	Intelligence Quotient
IR, InsR, INSR	Insulin Receptor
IRE1	Serine/Threonine-Protein Kinase/Endoribonuclease
IRS	Insulin Receptor Substrate
IRX3	Iroquois Homeobox Protein 3
iv	Intravenous
iv GTT	Intravenous Glucose Tolerance Test
JAK2	Janus Kinase 2
JNK	C-Jun N-Terminal Kinase
K-ATP	ATP-Sensitive Potassium Channels
KIR6x	Inward Rectifier K(+) Channel
Klb-KO	Bklotho Knockout
KO	Knockout
LAT1	L-type Amino Acid Transporter 1
LDH1/5	Lactate Dehydrogenase 1/5
LEF/TCF	Lymphoid Enhancer Factor/T Cell Factor
LepR b	Leptin Receptor Isoform b
LH	Lateral Hypothalamus
LHA	Lateral Hypothalamic Area
LMW	Low-Molecular-Weight
LOAD	Late Onset AD
LPS	Lipopolysaccharide
LRP6	Low-Density Lipoprotein Receptor-Related Protein 6
LTD	Long-Term (Synaptic) Depression
LTP	Long Term Potentiation
MAM	Mitochondrial Associated Membrane
MAO	Monoamine Oxidase
MAP	Mitogen-Activated Protein
MAPK	Mitogen-Activated Protein Kinase
MBH	Medio Basal Hypothalamus
MC3R	Melanocortin-3 Receptor
MC4R	Melanocortin 4 Receptor
MCH	Melanin Concentrating Hormone
MCI	Mild Cognitive Impairment
MCRs	Melanocortin Receptors
MCU	Mitochondrial Ca^{2+} Uniporter
ME	Median Eminence
MEG	Magnetoencephalography
MEK	Mitogen-Activated Protein Kinase
MFN1/2	Mitofusin 1/2
MMSE	Mini-Mental State Examination
MMW	Middle-Molecular-Weight
MRI	Magnetic Resonance Imaging
MRN	Median Raphe Nucleus
MSR	Mitochondrial Stress Response
mTNS	Medial Nucleus of the Solitary Tract
mTOR	Mechanistic Target of Rapamycin
mTORC	Mammalian Target of Rapamycin Complex
Munc-18	Mammalian Uncoordinated 18 Proteins
NAC	N-Acetylcysteine
NAcc	Nucleus Accumbens
NAFLD	Non-alcoholic Fatty Liver Disease
NFT	Neurofibrillary Tangles
NF-κB	Nuclear Factor Kappa B
NHID	National Health Information Database
NIRKO	Neuronal Insulin Receptor Knock-Out
NLRP3	Nucleotide-Binding Oligomerization Domain (NOD)-Like Receptor Protein 3
NMDA	N-Methyl-d-Aspartate
NPY	Neuropeptide Y
NRF1/2	Nuclear Respiratory Factor 1/2
NRIs	Norepinephrine Reuptake Inhibitors
NSB	Non-caloric Sweetened Beverages
NSC	Neural Stem Cells
NTS	Nucleus of the Solitary Tract
NTS	Nucleus Tractus Solitarius
NZO	New Zealand Obese
ob/ob	Leptin-Deficient
oGTT	Oral Glucose Tolerance Test
OPA1	Optic Atrophy 1
P	Postnatal Day
PA	Physical Activity
PC2	Prohormone Convertase 2
PCK-1	Phosphoenolpyruvate Carboxykinase 1
PCOS	Polycystic Ovarian Syndrome
PD	Parkinson's Disease

PDK, PDPK	Phosphatidylinositol Dependent Protein Kinase
PDX-1	Pancreatic and Duodenal Homeobox 1
PET	Positron Emission Tomography
PFC	Prefrontal Cortex
PFK	Phosphofructokinase
PFKFB3	6-Phosphofructo-2-Kinase/ Fructose-2,6-Bisphosphatase-3
PGC1α/β	Peroxisome Proliferator-Activated Receptor Gamma Co-activator 1-α/β
PH	Pleckstrin Homology Domain
PI3K, PI-3K	Phosphoinositide 3-Kinase
PiD	Pick's Disease
PIP2	Phosphatidylinositol 4,5-Bisphosphate
PIP3	Phosphatidylinositol 3,4,5-Trisphosphate
PKA	Protein Kinase A
PKC	Protein Kinase C
PKR	Protein kinase
PLCβ	Phospholipase C Beta
POMC	Pro-opiomelanocortin
PPAR	Peroxisome Proliferator-Activated Receptor
PPP	Pentose Phosphate Pathway
PREDIMED	Prevention with Mediterranean Diet
PRK	Interferon-Inducible Double-Stranded RNA-Dependent Protein Kinase Activator A
PSD-95	Postsynaptic Density Protein 95
p-Tau	Hyperphosphorylated Tau
PTB	Phosphotyrosine-Binding Domain
PTEN	Dual-Specificity Protein Phosphatase
PTG	Protein Targeting to Glycogen
PTP1B	Tyrosine-Protein Phosphatase Non-Receptor Type 1
PTPN11	Tyrosine-Protein Phosphatase Non-Receptor Type 11
PVH	Para-Ventricular Hypothalamus
PVN	Paraventricular Nucleus
RA	Receptor Agonists
Rac	Ras-related C3 Botulinum Toxin Substrate
RAF	Rapidly Accelerated Fibrosarcoma
Ras	Rat Sarcoma Virus
rb-2	Retinoblastoma Protein 2
RCT	Randomized Controlled Trial
Rheb-GTP	Ras Homolog Enriched in Brain-GTP
ROS	Reactive Oxygen Species
RPS6	40S Ribosomal Protein S6
RSPO3	R-Spondin-3
RTKs	Receptor Tyrosine Kinases
RYGB	Roux-en Y Gastric Bypass
RyR	Ryanodine Receptor
S6K	Ribosomal S6 Kinase
SERCA	Sarco-Endoplasmic Reticulum Ca^{2+}-ATPase
SERT	Serotonin Transporter
SES	Socioeconomic Status
SF-1	Steroidogenic Factor 1
SGLT2	Sodium/Glucose Cotransporter 2
SH2B1	SH2 Domain-Containing Adaptor Protein
SHC	SHC-Transforming Protein
SIDD	Severe Insulin-Deficient Diabetes Mellitus Type 2
SIRD	Severe Insulin-Resistant Diabetes Mellitus Type 2
si-RNA	Small Interfering RNA
SIRT1	Sirtuin 1
SIX3	Six Homeobox 3
SMOC1	SPARC-Related Modular Calcium-Binding Protein 1
SNARE	N-Ethylmaleimide-Sensitive Factor Attachment Protein Receptors
SNP	Single Nucleotid Polymorphism
SNpc	Substantia Nigra Pars Compacta
SOCS	Suppressor of Cytokine Signalling
SOD	Super Oxide Dismutase
SOS	Son of Sevenless Homolog
SOX1	Sry-Box Transcription Factor 1
SPECT	Single Photon Emission Computed Tomography
SSB	Sugar-Sweetened Beverages
SSRIs	Selective Serotonin Reuptake Inhibitors
STAT3	Signal Transducer and Activator of Transcription 3
STZ	Streptozotocin
SUR	Sulfonylurea Receptors
T1D, T1DM	Type 1 Diabetes Mellitus
T2D, T2DM	Type 2 Diabetes Mellitus
T3DM	Type 3 Diabetes
TCAs	Tricyclic Antidepressants

TCF7l2	Transcription Factor 7-Like 2
TCPTP	T cell Protein Tyrosine Phosphatase
TFAM	Mitochondrial Transcriptional Factor A
TLR4	Toll-Like Receptor 4
TNF	Tumor Necrosis Factor
TrkB	Tropomyosin Receptor Kinase B
TSC	Tuberous Sclerosis
TXNIP	Thioredoxin-Interacting Protein
TyG	Triglyceride Glucose
UPR	Unfolded Protein Reponse
VEGF	Vascular Endothelial Growth Factor
VMH	Ventro-Medial Hypothalamus
VMN	Ventromedial Nucleus
VTA	Ventral Tegmental Area
WAT	White Adipose Tissue
WHO	World Health Organization
WNT	Wingless-Integration1
αMSH	Alpha-Melanocyte stimulating hormone

CHAPTER ONE

Molecular Mechanisms of Brain Insulin Signaling[1]

Simran Chopra, Robert Hauffe, and André Kleinridders

CONTENTS

INTERACTION BETWEEN METABOLISM AND INSULIN

With global type 2 diabetes (T2D) rates on an exponential rise, a major hallmark feature of this metabolic condition is insulin resistance. As a result, the physiological functions and downstream mechanistic actions of insulin are, to this date, being predominantly studied in its main target tissues, including the liver, muscle, and white adipocytes. Although several studies have also demonstrated a correlation between metabolic and neurodegenerative diseases, a higher risk of cognitive decline has been observed in diabetic patients (1–3). This suggests that alterations in brain insulin action can be observed as a pathological feature in both conditions, although other common signaling pathways might additionally be responsible for this, such as inflammation and apoptosis. However, the brain was not considered an insulin-sensitive organ until the late 1970s

[1] Simran Chopra and Robert Hauffe have contributed equally.

DOI: 10.1201/9781003079927-1

when the insulin receptor (IR) was discovered to be abundantly distributed in the brain (4, 5). Ever since then, metabolic and cognitive outcomes that influence memory, olfactory perception, emotional regulation, eating behavior, and peripheral metabolism, have been demonstrated to be affected by insulin signaling in the central nervous system (CNS) (3, 5). Nevertheless, to appreciate these effects of insulin action in the CNS, it is critical to understand the molecular nature of this signaling cascade with its expression patterns in the brain.

After digestion of food in the gastrointestinal tract, carbohydrates, mainly in the form of glucose, are released into the bloodstream. This leads to an increase in blood glucose levels which induces the release of the hormone insulin from β-cells located in the islands of Langerhans of the pancreas. The insulin gene on the human chromosome 11 encodes a 110 amino acid precursor called preproinsulin, which gets directed to the endoplasmic reticulum for folding into proinsulin, consisting of three polypeptide chains A, B, and C. After further transportation into the trans-Golgi network, the C-peptide is cleaved off, leaving behind the functional insulin as a 51 amino acid protein with a molecular weight of 5.8 kDa. This final product is stored in granules within the β-cells to be released in response to rising circulating glucose levels. Thus, glucose from the bloodstream is taken up by pancreatic β-cells via the glucose transporter 2 (GLUT 2) and enters the classical adenosine triphosphate (ATP) generating pathway of mitochondrial oxidative phosphorylation. Here, the conversion of adenosine diphosphate (ADP) to ATP increases the ATP/ADP ratio, which deactivates the ATP-dependent potassium channel K_{ATP}, leading to the depolarization of the β-cell membrane. This causes voltage-gated calcium channels (predominantly L-type Ca^{2+} channels) to open, and the resulting calcium influx triggers exocytosis of insulin into circulation (6, 7). The release of insulin subsequently decreases blood glucose levels through the insulin receptor-mediated increase in glucose uptake in the peripheral tissues. This thereby decreases hepatic glucose production and also regulates food intake.

To exert its effects in the CNS, insulin is transported across the blood-brain barrier by saturable insulin transporters (8–10) and is released into the cerebrospinal fluid (CSF) where it is distributed to insulin-sensitive brain regions. Upon insulin binding to the IR, the IR changes its conformation to bring kinase domains of the IR in proximity to tyrosine phosphorylation sites on the β-chain, allowing the autophosphorylation of at least eight tyrosine residues (11). This then leads to the activation of insulin receptor substrate (IRS) proteins and subsequently of two general downstream signaling arms through (i) the phosphoinositide 3-kinase (PI3K) and protein kinase B (PKB, also called AKT) axis, and (ii) the mitogen-activated protein kinase (MAPK), discussed in detail below (Figure 1.1)

INSULIN AND INSULIN GROWTH FACTOR 1 RECEPTOR

In vertebrates, there are two other structurally related peptide hormones to insulin, insulin growth factor (IGF)-1 and -2, which are responsible for mitogenic growth, while insulin plays a predominant role of being a metabolic regulatory hormone (12, 13). The biological effects of these peptide hormones are mediated via plasma membrane-bound receptors in target cells which are the IR and the IGF-1 receptor (IGF-1R), respectively. Both IR and IGF-1R have highly homologous structures where both have an extracellular α subunit that harbors the ligand-binding site and a β subunit that contains the tyrosine kinase domain (14, 15). As a result of having high structural homology, the IR and IGF-1R can not only bind each other's ligands with a lower affinity as homodimers, but they can also heterodimerize to form IR/IGF-1R hybrids if both receptors are expressed in the same region (13, 16). Different forms of dimers, as well as their occurrence and known or putative functions, are discussed below in the "Insulin and Insulin Growth Factor 1 Receptor Heterodimers" section.

Furthermore, IR has two distinct isoforms formed through alternative splicing: IR-A that excludes exon 11, and IR-B that includes exon 11. These isoforms further differ in receptor activation and signaling, where IR-A has a higher affinity for IGF-2 and is expressed predominantly during prenatal life and contributes toward embryogenesis and fetal development. However, IR-B is primarily expressed in insulin-sensitive tissues to enhance the metabolic actions of insulin. Moreover, as the IR has two isoforms, this can further lead to the formation of two variants of the IR/IGF-1R hybrid receptors which tend to have a higher affinity toward IGF-1 and IGF-2 than insulin (16–18) (Figure 1.2).

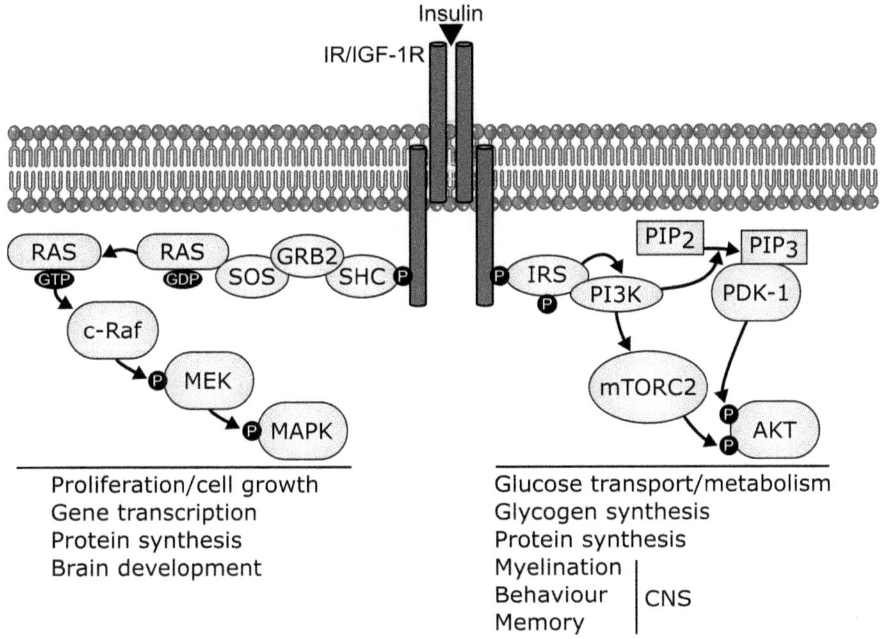

Figure 1.1 Overview of the insulin signaling pathway and associated downstream effects. Insulin affects a variety of catabolic cellular processes through two major avenues. (I) Activation of the MAPK signaling cascade through with involvement of SHC, GRB2, SOS, and the GTPase RAS. (II) Activation of AKT through the interaction of IRS, PI3K, PDK-1, and mTORC2.

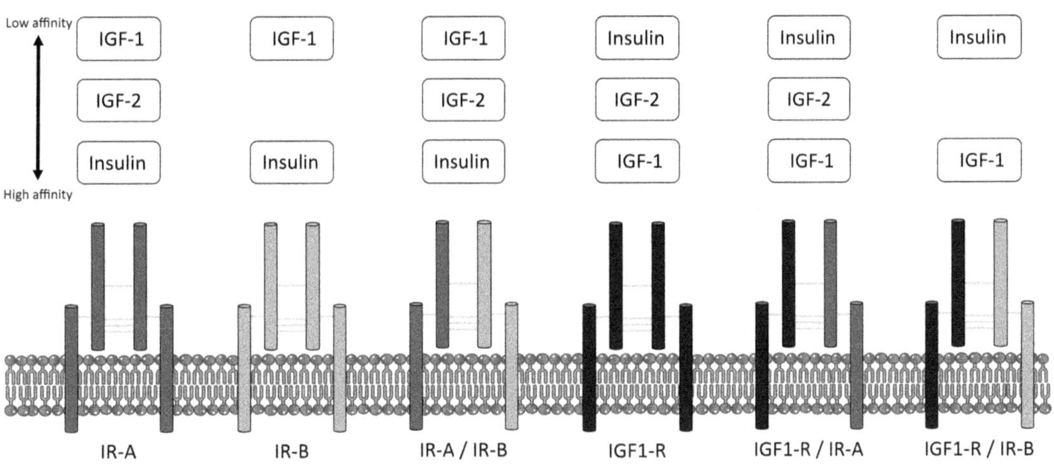

Figure 1.2 Activation of the IR and IGF1R through insulin and IGF molecule binding. The insulin receptor has two isoforms (IR-A and IR-B), while IGF1-R exists as a single isoform. Both receptors can be expressed as homodimers or heterodimers depending on the expression levels of each receptor. Insulin primarily binds to IR-B and the heterodimer of IR-A/IR-B, but has a weaker affinity for the IGF-1R and IGF-1R/IR heterodimers. IGF-1 and IGF-2 ligands have the greatest affinity for the IGF-1R and IGF-1R/IR heterodimers as well as IR-A.

INSULIN SIGNALING CASCADE IN THE BRAIN

The main goal of insulin signaling after feeding is the reduction of circulating glucose. This is achieved via (i) an increase in glucose uptake in peripheral tissues, such as skeletal muscle and adipose tissue, (ii) the suppression of hepatic glucose production (HGP), and (iii) limiting nutrient influx through a reduction of food intake. Interestingly, while peripheral insulin directly affects glucose uptake in muscle and adipose tissue, insulin signaling in the CNS controls whole-body energy homeostasis via control of food intake, HGP, lipogenesis, and thermogenesis. The control of food intake is mainly regulated through the activity of two clusters of hypothalamic neurons, namely orexigenic (appetite-inducing) NPY/AgRP neurons and anorexigenic (appetite-reducing) POMC neurons. Leading on from this, after having passed the blood-brain barrier, insulin binds the IR in these areas to exert its effects on whole-body metabolism (Figure 1.1).

Interestingly, a neuronal IR KO mouse model (NIRKO mouse) displayed increased food intake (19). Following on from these findings, it could be shown that insulin administration alters neuronal activation of both POMC and AgRP neurons (20–22). Thus, insulin stimulation modulates the activity of both of these neuronal populations to reduce food intake (23–26). Moreover, insulin also affects HGP at least partially via AgRP neuronal activity, as deletion of IR in AgRP, but not in POMC neurons, led to impaired suppression of HGP (27). Furthermore, mice lacking the IR in the hypothalamus have impaired HGP, but re-expressing the IR specifically in AgRP neurons reconstituted the suppression of HGP, while re-expression in POMC neurons conversely increased HGP and modestly food intake (28). Ablation of the IR specifically in POMC neurons impairs the inhibition of adipose tissue lipolysis (29). Additionally, NIRKO mice also show signs of anxiety and depressive-like behavior through altered dopamine turnover, highlighting an intriguing link between central insulin resistance and behavioral disorders as well (30).

DOWNSTREAM SIGNALING FROM THE INSULIN AND INSULIN GROWTH FACTOR 1 RECEPTOR

The IRS protein family in mammals consists of IRS 1–4 where IRS-1 and IRS-2 are ubiquitously expressed and mediate the actions of insulin/IGF-1 in most tissues, IRS-3 is present in adipose tissue of rodents but in humans is a pseudogene and IRS-4 is expressed primarily in the hypothalamus. Each member of the IRS family shares a common structure and demonstrates a unique physiological function which may be due to sequence divergence or difference in expression patterns (17, 31, 32).

IRS proteins contain an N-terminal pleckstrin homology (PH) domain adjacent to a phosphotyrosine-binding (PTB) domain, which is followed by several regulatory tyrosine and serine/threonine phosphorylation sites. Here, the PTB domain represents the binding site of IRS proteins to phosphotyrosine residues on the insulin receptor after insulin activation. The PH domain can bind phosphatidylinositol lipids such as phosphatidylinositol 4,5-bisphosphate (PIP_2) as a substrate of PI3K. Furthermore, the tyrosine phosphorylation sites coordinate downstream signaling cascades by also binding effector proteins like the above-mentioned PI3K or growth factor receptor-bound protein 2 (GRB2). By contrast, the specific serine phosphorylation sites, phosphorylated by the c-Jun N-terminal kinase (JNK1, also known as MAPK8) and other protein kinases, inhibit insulin-stimulated tyrosine phosphorylation. First, this is a negative feedback loop to limit over-stimulation of downstream pathways by insulin (33); second, it also represents one mechanism of insulin resistance if it persists. Upon recruitment of IRS-1 to the insulin receptor by binding of the PTB domain to phosphorylated IR tyrosine residue Tyr999 (note: original publications refer to this tyrosine residue as Tyr960 which shifted due to subsequent sequence updates), the kinase domain of the IR can then phosphorylate tyrosine residues on IRS-1 (34, 35). While the human and murine IRS-1 amino acid sequences are well conserved, the exact position of the tyrosine (Y) residues differ slightly, denoted by either hY or mY for the human or mouse residue, respectively. Thus, the phosphorylated tyrosine residues on IRS-1 are hY465/mY460, hY612/mY608, hY632/mY628, hY662/mY658, hY896/mY891, hY941/mY935, hY989/mY983, hY1179/mY1173, and hY1229/mY1220 (36–39). From here the downstream signaling branches out to either engage the PI3K-AKT or the MAPK pathways.

The insulin signaling pathway contains a negative feedback loop to limit over-stimulation of downstream pathways by insulin, where specific serine phosphorylation sites, phosphorylated by the JNK1 and other protein kinases, inhibit insulin-stimulated tyrosine phosphorylation. However, this also represents a common feature of insulin resistance. Phosphorylation of hSer312 (mSer307) is commonly associated with insulin resistance (40). Interestingly, a mSer307Ala substitution on the other hand renders mice insulin resistant, presumably because serine phosphorylations serve to modulate the severity of insulin resistance through maintaining insulin signaling in metabolic diseases (41). Furthermore, evidence suggests there is a cluster of inhibitory serine phosphorylations in the IRS-1 C-terminal region consisting of hS312, hS616, hS636, hS891, hS1101 and hS1223 (mS307, mS612, mS632, mS887, mS891, mS1097, and mS1214) (42). This demonstrates that IRS signaling is essential in propagating insulin-mediated processes.

Insulin Signaling through the AKT Pathway

One of the most well-described downstream pathways of the IR signaling cascade is the PI3K-AKT signaling pathway. The serine/threonine kinase PI3K is a heterodimer consisting of a regulatory (p85) and a catalytic subunit (p110). PI3K is recruited to the insulin receptor through an SH2 domain on p85, a process facilitated by activated IRS-1 (43). The formation of this complex is thought to release the inhibition of the catalytic p110 subunit by p85 (44). The p110 subunit, now in proximity to its lipid substrate, is free to catalyze the reaction of PIP2 to PIP3. PIP3 acts as a secondary messenger that recruits phosphatidylinositol-dependent protein kinase (PDK1) to the cell membrane. PDK1 subsequently binds AKT via its PH domain and activates it by phosphorylation at Thr308 (45, 46). However, for full AKT activation, further phosphorylation of AKT at Ser473 via mammalian target of rapamycin complex (mTORC) 2-itself activated and bound by PI3K- is necessary (47). AKT itself is known to act on a variety of pathways, though this is dependent on the affected tissue, including the translocation of GLUT4 vesicles to facilitate glucose uptake, the induction of glycogen synthesis, protein synthesis (via mTOR1), and the suppression of gluconeogenesis in peripheral tissues. In the CNS specifically, AKT is known to induce myelination (48), thereby affecting neuronal function, such as affective behavior and spatial memory (49) (Figure 1.1).

Moreover, insulin-induced AKT activation is crucial to modulate gene regulation via blocking the activity of FOXO transcription factors. There are four members of the FOXO family which include FOXO1, FOXO3, FOXO4, and FOXO6, where FOXO1 and FOXO3 are expressed in almost all tissues. AKT phosphorylates FOXO proteins at several serine residues (50) which facilitates the binding of FOXOs to proteins of the 14-3-3 family. This interaction leads to the nuclear exclusion of FOXO proteins and inhibition of their transcriptional activity (50). It has been suggested that the FOXO1 function is an important regulator of food intake and energy balance (51). While insulin inhibits FOXO1 activity and reduces food intake, active FOXO1 enhances food intake. FOXO6 is primarily expressed in the hippocampus of the brain and liver. It has been observed that upon whole-body deletion of FOXO6, although the mice exhibited normal learning behavior, they displayed impaired memory consolidation. In addition, through the process of stereotactic injection of viruses into the hippocampus of wild-type adult mice, it was revealed that FOXO6 activity in the hippocampus was important to promote memory consolidation (52, 53), indicating that insulin action can affect memory via FOXO6 regulation.

Furthermore, mTOR is a serine/threonine kinase known to exist in the form of two complexes, mTORC1 and mTORC2, although more is known about the mTORC1 signaling pathway. Upon AKT activation, mTORC1 inhibits autophagy and promotes protein synthesis in neurons (54). In addition, activated hypothalamic mTOR signaling reduces food intake (55). Furthermore, in neurons, it has been established that both insulin and IGF-1 are able to increase mRNA translation which is thought to occur via mTORC1 phosphorylation (56).

EXTRACELLULAR SIGNAL-REGULATED PROTEIN KINASES ISOFORMS

The second largest insulin signaling pathway similarly starts with autophosphorylation sites in SH domains on the IR, which then binds and phosphorylates the protein SHC. Following the binding of Growth factor receptor-bound protein 2 (GRB2), GRB2 then recruits and activates the guanine nucleotide exchange factor Son of sevenless homolog (SOS). Activated SOS promotes the dissociation of guanosine diphosphate (GDP) from the Ras GTPase which allows Ras to bind guanosine triphosphate (GTP) and become activated. GTP-bound Ras can then interact with the proto-oncogene serine/threonine-protein kinase c-Raf, which finally activates the serine/tyrosine/threonine Mitogen-activated protein kinase (MEK, also known as MAP2K or MAPKK). MEK in turn phosphorylates and activates the serine/threonine kinases mitogen-activated kinase (MAPKs;

originally named "extracellular signal-regulated kinase" (ERKs)). This sequence of kinases is a classic example of a protein kinase cascade, where the numerous constituents allow for feedback regulation and signal amplification.

Activated MAPKs affect a multitude of downstream processes, such as the translation of mRNA to proteins. This is influenced by MAPK phosphorylation of 40S ribosomal protein S6 kinase (RPS6) to activate ribosomal protein S6 and thereby increase translation (57). MAPKs also regulate the levels and activities of several transcription factors, leading to altered transcription of genes which is consequently important for the cell cycle. Thus, MAPK stimulates cell growth and mitochondrial function in insulin-sensitive cells (58, 59) (Figure 1.1).

Negative Modulation of the Insulin Signaling Pathway

The insulin signaling pathway is a tightly regulated molecular mechanism that controls a range of metabolically important functions in cells and generally induces energy-consuming processes, such as protein synthesis, growth, and proliferation. Moreover, it also controls both peripheral metabolic functions (e.g., glucose uptake or hepatic glucose production), as well as central metabolic functions (e.g., control of appetite and thus food intake). As such, a hyperactivation could lead to detrimental metabolic outcomes. To counteract the hyperactivation of the insulin response, the signaling pathway is subjected to a variety of negative feedback modulations on most of its protein components via posttranslational modifications, transcriptional control, and targeted protein degradation. However, if these feedback loops are continually activated, a chronic insulin-resistant state can occur.

As outlined above, the insulin signaling pathway is activated through a series of phosphorylation events. As protein phosphorylation is a reversible posttranslational modification, it represents a readily available site for fine-tuning the amplitude of the signaling response as well as a site for a negative feedback loop. Enzymatic dephosphorylation is carried out by phosphatases. In the case of the insulin signaling pathway, we can differentiate between two cases: (i) dephosphorylation of proteins in a negative feedback loop to stop the signaling cascade and (ii) inhibitory phosphorylation of proteins to limit the intensity of the cellular response to insulin (Figure 1.3).

Genetic Mutations Associated with Insulin Resistance

The importance of the insulin signaling pathway is further highlighted by inactivating mutations of its components that are associated with insulin resistance. For example, patients with missense mutations in the IR gene that lead to reduced expression or the expression of a truncated protein show severe insulin resistance and growth retardation (60, 61). Treatment of affected patients is exceedingly difficult, as the lack of a functional IR can as of yet not be adequately compensated. One investigatory treatment approach involves the application of recombinant IGF-1, which can improve symptoms and affect longevity, although only in a subset of patients (concisely reviewed by Harouch et al. (62)). Following the signaling pathway downstream, a polymorphism in the human IRS-1 gene that leads to a Gly972Arg substitution is also associated with insulin resistance (63).

Dephosphorylation of Proteins and Metabolites Involved in the Insulin Signaling Pathway

The main negative modulator responsible for fine-tuning insulin signaling is the tyrosine-protein phosphotase non-receptor type 1 (PTP1B). PTP1B directly dephosphorylates tyrosine residues of the insulin receptor, thereby stopping the downstream activation of IRS proteins. IRS itself can be dephosphorylated by PTPN11 (also known as SHP-2) (Figure 1.3). Furthermore, the metabolite PIP_3 is subject to dephosphorylation by Phosphatidylinositol 3,4,5-trisphosphate 3-phosphatase and dual-specificity protein phosphatase (PTEN), and Phosphatidylinositol 3,4,5-trisphosphate 5-phosphatase 1 (SHIP), that subsequently stops the signaling cascade (Figure 1.3). Downstream of the insulin signaling cascade, AKT can be deactivated by serine/threonine-protein phosphatase 2A catalytic subunit alpha (PP2a). Furthermore, AKT can be dephosphorylated by PH domain and Leucine-rich repeat Protein Phosphatase (PHLPP), which dephosphorylate AKT on Ser473, which is a phosphorylation site needed for the full activation of AKT (Figure 1.3). However, it needs to be stated that these feedback systems, that act on all levels of the signaling cascade, serve a vital role in limiting the cellular response to insulin (64). It is readily imaginable that unchecked activation of insulin signaling and the subsequently engaged energy-consuming processes, such as protein synthesis, growth, and proliferation could quickly lead to detrimental outcomes for the cells. Indeed, loss of function mutations in PTEN for example lead to a loss of feedback inhibition, which is the underlying cause of a variety of diseases collectively called PTEN hamartoma tumor syndromes (PHTS). These syndromes are characterized by the inability of cells to stop cell division, which leads to the growth of non-cancerous tumors (hamartoma) and a substantial life-long risk of cancer, particularly in the breast, thyroid, or uterus (65). Moreover, in the Cowden syndrome, which is part of the PHT syndrome, over 80% of patients show neurodevelopmental disorders, such as macrocephaly or hemimegalencephaly (66). Furthermore, an estimated 70% of prostate cancer patients have lost at least one gene copy of PTEN at the time of diagnosis (67). Similarly, hyperactivation of AKT, or gene amplification leading to higher total AKT functionality, has been described in a variety of cancers, such as gastric, breast, ovarian, pancreatic, colon, esophageal, and thyroid cancers (comprehensively reviewed by Mundi et al. (68)). Additionally, it has been found that gain of function mutations of AKT that result in its hyperactivation lead to forms of Juvenile Granulosa Cell tumors (JGCT). Mechanistically, in-frame duplications within the PH domain of AKT lead to higher-than-normal distribution at the cell membrane and an increased degree of activating phosphorylation (69, 70).

Inhibitory Phosphorylations of Proteins in the Insulin Signaling Pathway

Inhibitory phosphorylation events, *i.e.*, phosphorylation of protein residues which then prevent activation of the respective enzyme, are mediated mainly by serine-threonine kinases. Again, IRS-1 plays a crucial role in this process. Both JNK and Inhibitor of nuclear factor kappa-B kinase (IKK) phosphorylate IRS on serine residues within the phosphotyrosine-binding domain, thereby inhibiting the binding of IRS to the insulin receptor, and effectively stopping the signaling cascade (Figure 1.3). Insulin desensitization through increased serine phosphorylation of IRS proteins is a response present in all insulin-sensitive tissues and is closely linked to the degree of cellular stress, through *e.g.* inflammation, or oxidative, mitochondrial, and endoplasmic reticulum (ER) stress. Generally, inflammation is a response to the presence of cellular stressors and would be terminated after the causative factor is eliminated. The various serine phosphorylations result from different stressors. Here, hS312 (mS307) as stated above, is targeted by JNK and IKK, which in turn can be activated due to different cellular stress responses, most notably inflammation and oxidative stress (71, 72). The PKC and mTOR pathways are responsible for phosphorylation of IRS-1 at hS616 (mS612) and hS636 (mS632), respectively (73, 74). Finally, hS1101 (mS1097) is phosphorylated by PKC-θ, which is activated specifically by free fatty acids (75). Another regulatory pathway is present through the action of Suppressor of Cytokine Signaling 3 (SOCS 3). This protein acts negatively on the insulin signaling pathway in a two-fold manner: First, it has been demonstrated to bind the insulin receptor and block IRS-1 recruitment (76, 77), and second, it can direct IRS-1 to ubiquitin-mediated degradation (78) (Figure 1.3). Thus, if a stressor persists, as is the case in *e.g.*, diet-induced over-stimulation of the insulin signaling cascade, the inflammatory response becomes chronic. This can lead to chronic deactivation of IRS proteins and thus chronic insulin resistance. In mouse models of diet-induced cognitive impairment, phosphorylation of inhibitory serine residues was seen to be increased in the brain (79). Similarly, transgenic mice overexpressing a chimeric mouse/human amyloid precursor protein and a mutant human presenilin 1 (APP/PS1 mice), which is a model of AD, have elevated phosphorylation of IRS-1 at hSer312(mSer307) and hSer636(mSer632) residues (80).

Another potent inducer of brain insulin resistance is an oversupply of free fatty acids through a mechanism called lipotoxicity. Saturated fatty acids, such as palmitate, are known to cause and worsen insulin-resistant states. They are also recognized to be present in the CSF of humans with obesity (81), where it was observed that palmitate increased activation of JNK, IKK, and PKC *in vitro* and in rodent models, thereby increasing inhibitory serine phosphorylation of IRS proteins (82–86). In the CNS specifically, palmitate administration also led to the translocation of PKC-θ in NPY/AgRP neurons to the cell membrane and was associated with insulin resistance. This was seen to be reversed by the knockdown of PKC-θ (87).

Furthermore, it has been shown that both mitochondrial and ER stress contributes to brain insulin resistance. Here, mitochondrial stress is commonly associated with a deteriorated mitochondrial respiration, leading to increased production of reactive oxygen species (ROS) and subsequent oxidative stress, JNK activation, and ultimately insulin resistance (88). As an example, inducing mitochondrial dysfunction through interference with the mitochondrial protein folding machinery leads to neuronal insulin resistance through JNK activation *in vitro and in vivo* (89, 90). Mitochondrial stress is closely linked to ER stress. Generally, ER stress occurs when the capacity of the ER to fold proteins becomes saturated, which is when there is an increased need for *de novo* synthesis of proteins in response to cellular demands (*e.g.* stress responses), or alterations in the protein synthesis machinery. The ER stress signaling network negatively impacts insulin signaling by inducing serine phosphorylation of insulin signaling molecules, but also intersecting with the insulin signaling pathway via Interferon-inducible double-stranded RNA-dependent protein kinase activator A (PRK) and Serine/threonine-protein kinase/endoribonuclease (IRE1), which act in concert to activate JNK, ultimately inducing inhibitory IRS phosphorylation (38, 91) (succinctly reviewed in Flamment et al. (92)).

Overall, it can be seen that through the activation of the downstream signaling components of the insulin and IGF signaling pathways these hormones are critical in regulating normal brain function. The different sensitivities of this pathway or different region-specific outcomes might be also caused by differences in expression patterns in various brain regions, which will be described in the following section.

INSULIN AND INSULIN GROWTH FACTOR 1 RECEPTOR

The IR was initially known as a 'housekeeping gene' due to its wide expression patterns and promoter characteristics which are typical of a housekeeping gene (93). In mammalian tissues, IR protein expression is seen to be the highest in adipose tissue. Relative to adipose tissue, IR expression in the liver and heart was approximately 30% and 10% in the muscle, spleen, placenta, and kidney. On the contrary, IGF-1R exists as a single isoform and is barely expressed in the liver and adipose tissue but has a similar expression to IR in muscle (94). Moreover, apart from IR and IGF-1R subunits only being present in peripheral tissues, they are also widely expressed throughout the brain with a distinct regional distribution. Although, the kinetics and pharmacological properties of the IR in the brain are similar to the IR found in peripheral tissues there are structural differences that make them distinct from one another. This includes the α subunit of the IR in the brain is smaller (predominantly IR-A expression) than the α subunit of the IR in peripheral tissues owing to a difference in molecular size and the degree of glycosylation in brain IRs tends to be lower than in peripheral tissues (16, 95, 96).

IR EXPRESSION IN THE BRAIN

Currently, there is not one brain region known which lacks either the IR or IGF1-R, further emphasizing how crucial they are for maintaining normal brain function. IR was first shown to be present through techniques of immunocytochemistry and radioreceptor assays in adult rat brains with the highest densities observed predominantly in the olfactory bulb, which was then followed by the hippocampus, hypothalamus, and cerebellum. Whereas moderate to low densities of the IR were observed in regions of the thalamus, the caudate-putamen in the striatum, the mesencephalic nuclei in the midbrain, and the brain stem (4, 97, 98). However, it is important to note that there are marked differences in the expression levels of the IR during different stages of development. In rats, during neonatal and early postnatal stages of development the caudate-putamen neuroepithelium, amygdaloid body, and some brainstem nuclei showed higher densities of IR for a brief duration of time. Over the course of brain maturation into the adult rat brain, the same regions showed a lower IR density (99, 100). Furthermore, human brains obtained from patient postmortem samples have also demonstrated that the IR expression is highest in the hypothalamus, cerebellum, and cortical and subcortical regions (3).

At a cellular level, it has been observed that the predominant isoform in the brain is the IR-A isoform which has a lower molecular weight, differential glycosylation, and has the greatest density in neurons (16). While the IR-B isoform was seen to be mostly expressed in the glial cells (particularly astrocytes). More specifically, IR-A expression is more evident in neural progenitor cells. Thus, the expression pattern of IR-A in the brain, which is involved in mitogenic effects, may be indicative of the importance of central insulin/IGF signaling toward brain development and function (15, 18, 101). On the other hand, the physiological function of the IR has been determined through genetic knockout (KO) mice where neuron-specific deletion of IR showed phenotypes of normal brain size and development while also showing metabolic phenotypes of mild obesity, and insulin resistance (19).

INSULIN GROWTH FACTOR 1 RECEPTOR

The IGF-1R expression also follows a regional distribution with developmental regulation. Unlike IR, IGF-1R appears to be present in the ventral floor plate even before the formation of the brain (97, 102). During later embryonic stages and early postnatal stages, the highest densities of the IGF-1R are seen in most hypothalamic and thalamic nuclei, the midbrain, and the brainstem, which was then shown to transiently decrease through later postnatal stages and eventually only low levels were present in the adult brain. On the contrary, in the adult brain, the highest expression was perceived in the cortex, hippocampus, and the dorsal thalamus, while moderate expression was seen in the olfactory bulb the hypothalamus and the cerebellum. Although the olfactory bulb, cortex, hippocampus, choroid plexus, and cerebellum showed a decrease in IGF-1R expression as brain development progressed. Although, relative to earlier developmental stages, these brain regions still demonstrated high expression levels at any given stage of development (100, 103). In postmortem tissue from human brain samples, IGF-1R expression was highest in the hippocampus and amygdala, while moderate expression levels were observed in the cerebellum, cerebral cortex, and caudate nucleus (104). IGF-1R transcripts in the rat brain have been expressed largely in developing astrocytes and oligodendrocytes, which coincides with the role of IGF-1 seen previously to promote proliferation and differentiation of astrocytes and oligodendrocytes during early development (105). Genetic KOs of IGF-1R in the brain have further verified a reduction in brain size, growth retardation, and behavioral changes (106).

INSULIN AND INSULIN GROWTH FACTOR 1 RECEPTOR HETERODIMERS

Moreover, apart from existing as homodimers in the brain, IR and IGF-1R can also heterodimerize with one another, as seen in peripheral tissues (16). The IR and IGF-1R heterodimers (of both IR splice variants) intriguingly demonstrated lower ligand affinity toward insulin but showed similar ligand affinity to IGF-1 as the IGF-1R homodimer, suggesting that in the brain a response toward IGF-1 over insulin may be favored. Furthermore, the IGF-2 ligand binds to the IR-A homodimer with the highest affinity which may enhance mitogenic signaling. As well as this, IGF-2 binds to the IGF-1R homodimer and IGF-1R/IR-A heterodimer with a higher affinity than insulin. As aforementioned, heterodimerization of these receptors depends on the ratio of the receptors being expressed in specific brain regions (16) (Figure 1.2). Although the expression patterns of these heterodimers in the different brain regions are still not entirely understood, a redundancy in IR and IGF-1R gene expression in many brain regions has been observed which contributes toward the formation of the heterodimers. Both IR and IGF-1R are ubiquitously expressed in the neuron-dense granule cell layers of the olfactory bulb, dentate gyrus, cerebellar cortex, and pyramidal cell layers of the piriform cortex (107). In the rabbit brain, it has been further shown that over half of the IGF-1Rs and a large proportion of IRs occur as heterodimers This suggests that the formation of hybrid receptors may be present throughout the entire CNS. However the formation of the heterodimers is also dependent on the relative expression of each receptor which further differs between cells (16). Moreover, the hybrid receptors are shown to behave similarly to the IGF1-R homodimers and thereby potentially cause more activation of the PI3K pathway (108).

INSULIN RECEPTOR SUBSTRATE 1 EXPRESSION IN THE BRAIN

Results of various biochemical assays have demonstrated that IRS-1 is widely distributed in the CNS of the rat brain. The presence of IRS-1 in rat brain has specifically been detected in all neurons of the cerebral cortex but prominently in the piriform cortex, the hippocampus where localization

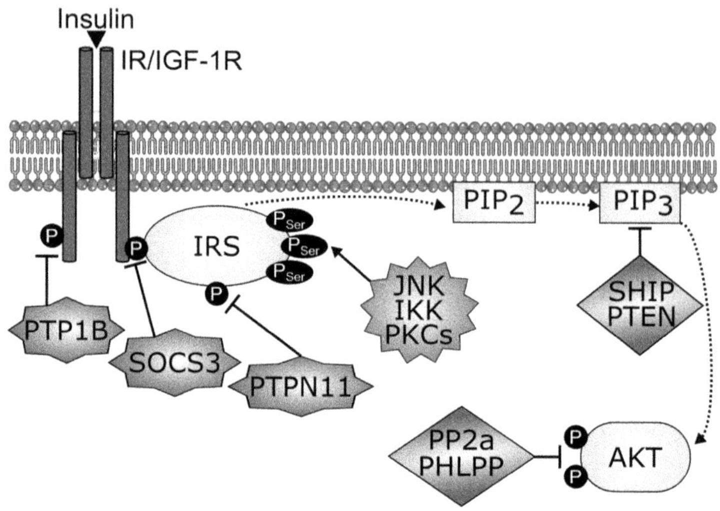

Figure 1.3 Overview of negative modulators of the insulin signaling pathway. Negative modulation of insulin signaling follows three major principles. (I) Serine/Threonine phosphorylation of e.g. IRS1 to inhibit its activation is mediated through kinases such as JNK, IKK, or PKCs. (II) Dephosphorylation of IR, IRS1, or AKT by PTP1B, PTPN11 (IRS1), or PP2a and PHLPP (AKT). (III) Inhibition of the conversion of PIP2 to PIP3, thereby inhibiting AKT activation and inhibition of binding to the IR (SOCS3).

of IRS-1 was seen to be mainly within cell bodies of the pyramidal neurons and dentate gyrus granules and in the brainstem IRS-1 was present in various brainstem nuclei. While in the cerebellum IRS-1 expression was limited to only the Purkinje neurons and in the hypothalamic and thalamic regions localization was specific to the habenular nuclei. On the contrary, IRS-1 was barely detected in the olfactory bulb. The distribution of IRS-1 in the different brain regions also shows a strong correlation with the distribution of IR where overlap in expression was seen in the cerebral cortex, hippocampus, and brainstem. While there was also an overlap in distribution seen between IRS-1 and IGF-1R specifically in the cerebellar cortex and the hippocampus (109, 110). Furthermore, stimulating the hippocampus and cerebellar cortex of human brain tissues from postmortem patients with 1nM insulin activated the IR in these regions but also observed the binding of IR to IRS-1 rather than IR to IRS-2 (111).

Moreover, mice lacking IRS-1 in the CNS were seen to have no noticeable changes in food intake, or expression of the appetite-related neuropeptides controlled *via* insulin signaling (POMC, AGRP, and NPY) (112). However, IRS-1 KO mice demonstrated a significant reduction in growth but also have an increase in the brain/body weight ratio due to the brain size remaining near normal. This could presumably also be through the decreased release of growth hormone releasing hormone in the hypothalamus (112). Nevertheless, it is plausible that other IRS proteins compensate for the lack of IRS in the brain as whole-body IRS-1 KO mice show a certain resistance against IGF-1 induced brain growth and myelination, they also react with a compensatory up-regulation of IRS-2 and IRS-4. This suggests some functional redundancy between IRS proteins (113).

INSULIN RECEPTOR SUBSTRATE 2 EXPRESSION IN THE BRAIN

Various immunohistochemical studies have demonstrated the neural distribution of IRS-2 in the rat brain. IRS-2 was shown to be highly expressed in various hypothalamic nuclei including the arcuate nucleus (ARC), and ventromedial nucleus (VMN) and paraventricular nucleus (PVN). IRS-2 was also seen to be highly expressed in the cerebellum (98). More specifically, IRS-2 was seen to be co-localized in alpha-melanocyte-stimulating hormone and neuropeptide Y containing neurons within the ARC. Although within the hypothalamus, the lateral hypothalamic area and dorsomedial nucleus seem not to express IRS-2. In the medial and periventricular parvocellular areas of the PVN and in all three subnuclei of the VMN IRS-2 was expressed. Various regions of the hindbrain such as the area postrema, medial nucleus of the solitary tract (mNTS) and the dorsal vagal complex also demonstrated IRS-2 expression. In the ventral tegmental area (VTA) of the midbrain, the dopamine neurons in the region also portrayed IRS-2 expression (114).

Additionally, stimulating the hippocampus and cerebellar cortex of human brain tissues from postmortem patients with 1nM IGF-1 activated IGF-1R and mediated IGF-1R binding to IRS-2 rather than IRS-1 (111). Furthermore, KO of IRS-2 in the hypothalamus of mice leads to increased food intake and obesity (114, 115). Overall, this suggests that IRS-2 may play a crucial role in central energy homeostasis (114, 116). IRS-2 KO mice also showed a decrease in neuronal proliferation thus leading to a reduced brain size (117, 118), and increased Tau phosphorylation, which is a known hallmark of AD progression (117, 119), while a brain-specific deletion of IRS2 disrupted synaptic plasticity, highlighting the neuronal importance of IRS proteins, in absence of metabolic disorders, such as hyperglycemia (120). On the other hand, IRS-2 serves as a prime example of how the insulin signaling cascade operates within a finely tuned equilibrium of stimulation and inhibition, as the hyperactivation, through overexpression of IRS-2 in neurons, led to impaired locomotor activity along with obesity and insulin resistance (121).

INSULIN RECEPTOR SUBSTRATE 4 EXPRESSION IN THE BRAIN

Unlike the other members of the IRS protein family, IRS-4 is seen to have a narrower distribution within the brain. IRS-4 has a wide expression range within the hypothalamus where it was highly dense in the medial preoptic nucleus, VMN and the ARC. Furthermore, IRS-4 has been detected in the amygdala and moderately expressed in the granular layer of the olfactory and accessory olfactory bulbs. In the midbrain, IRS-4 was identified in the VTA, the substantia nigra pars compacta and the raphe nucleus. On the contrary, IRS-4 was not detected in the regions of the cortex, hippocampus, caudate-putamen, thalamus, substantia nigra pars reticulata and the superior and inferior colliculi (31). An overlap between IRS-1, IRS-2, and IRS-4 was observed in the neurons of the ARC, and in the VMN suggesting that there is a partial functional redundancy by IRS-4 toward IRS-1 and IRS-2 (31, 110, 122). Although the KO models of IRS-4 only caused very mild effects on growth and energy homeostasis, this did not affect the expression levels of IRS-1 and IRS-2 in various insulin-sensitive tissues including the brain (123). Nevertheless, functional synergy has been observed between IRS-2 and IRS-4 in regulating central insulin action (122). Moreover, neurons within the periventricular nucleus of the hypothalamus that express IRS-4 have been seen to play a role in mediating normal feeding, body weight and energy expenditure, as inhibition of these neurons was seen to result in obesity and lower energy expenditure (124).

AKT ISOFORMS AND THEIR EXPRESSION IN THE BRAIN

AKT has three isoforms which include AKT 1–3 and all are known to be expressed in the brain (125). These isoforms are known to be closely related due to the similarities in their structural homology yet are thought to play a distinct role in regulating cell growth, proliferation and metabolism throughout the body (126). AKT 1 KO mice demonstrated overall growth impairment, while AKT 2 KO mice portrayed a diabetes-mellitus-like syndrome phenotype due to being insulin resistant and AKT 3 KO mice show a reduction in brain size (125, 127).

More specifically, in the brain the AKT pathway is central to a diverse array of functions where by mediating downstream substrates including forkhead box O (FOXO), mTORC1, and glycogen synthase kinase 3β it is able to play a critical role in controlling axon growth, gene transcription, protein synthesis stress responses, neurotransmission and synaptic plasticity (128–130). It is also essential for normal brain development and thus, its dysregulation has been recognized as a causative factor of neurodevelopmental, neurodegenerative, and neuropsychiatric diseases.

In adult mice brains it has been demonstrated that the relative expression of AKT 3 was the highest in the overall brain as it represented around 50% of all total AKT proteins, followed by AKT 1 which denoted approximately 30% of all total AKT proteins, while AKT 2 accounted for the remaining expression levels of all AKT proteins. While AKT 2 was predominantly expressed in the cerebellum and AKT 1 was chiefly expressed in the spinal cord (125), AKT 3 was observed to be expressed in all brain regions of the adult brain, primarily in the hippocampus, cortex and neural progenitors, supporting a crucial function in brain development (131).

Accordingly, activating mutations in AKT 3 are found in brain samples of patients displaying Hemimegalencephaly (HMG), which is a brain disorder characterized by an enlarged, malformed cerebral hemisphere leading to severe epilepsy (132, 133). Accordingly, a whole-body deletion of AKT 3 in mice specifically reduces brain size by 20%–25% (125, 134). Conversely, whole-body deletion of AKT 1 leads to a proportional size reduction of all organs (127, 135). Similarly, deletions in the human genome containing the AKT 3 gene (1q42-q44) have been associated with developmental deficiencies, such as agenesis of the corpus callosum (ACC) and microcephaly (136), whereas a duplication of the region encompassing *AKT* 3 (1q43q44) leads to macrocephaly (137). Thus, AKT 3 is an important regulator of brain growth.

EXTRACELLULAR SIGNAL-REGULATED PROTEIN KINASES ISOFORMS AND THEIR EXPRESSION IN THE BRAIN

The two main isoforms of ERK are ERK 1 and 2, which are known to share an amino acid sequence identity of approximately 85% yet have distinct biological roles (138, 139). Knockout studies have demonstrated that ERK 1 deficient mice portray a resistance toward obesity when challenged with a high-fat diet, show increased levels of activity within open-field tests and a striatum-dependent enhancement in long-term memory (140–142). While ERK 2 knockout mice showed lethality in offspring, conditional inactivation of the *ERK* 2 gene in murine neural progenitor cells of the developing cortex demonstrated reduced proliferation of neural progenitors, resulting in less neuron formation. These mice also portrayed impairments in associative learning (143, 144). Interestingly, a microdeletion on chromosome 22q11.2 containing the *ERK* 2 gene in humans was seen to decrease ERK 2 expression, leading to impairments in neurodevelopment (145).

Moreover, in mammals ERKs are known to be expressed ubiquitously throughout the body but are known to have the highest expression in the brain and spinal cord (138, 146). In relation to this, adequate brain development has been shown to be compromised upon ERK 1 and 2 deficiencies, ERK inhibition reduces insulin signaling cascade in regards to mitochondrial function suggesting a significant regulatory role of ERK signaling within the CNS (147). In rat brains, it has been demonstrated that there is a difference in regional distribution between ERK 1 and 2 where the frontal cortex had a greater distribution of ERK 2 in areas including the parietal cortex, nucleus accumbens, and caudate-putamen. While the brain stem and spinal cord portrayed a higher expression of ERK 1, especially in the hypothalamus, thalamus, VTA, midbrain, pons-medulla, and cerebellum (148). Moreover, primarily through ERK activation, it has been shown that insulin signaling in the hypothalamus improves mitochondrial function and as a result decreases starvation-induced autophagy (149).

CONCLUDING REMARKS

Insulin signaling in the brain has been demonstrated to affect metabolic and cognitive outcomes which include peripheral metabolism, eating behavior, memory, olfactory perception, and emotional regulation. As we have established in this chapter, this complex system is able to mediate insulin and IGF-1 dependent biological responses through the regulation of its downstream components and the integration of various signaling pathways.

Overall, it can be seen that the ubiquitous expression of the IR and IGF-1R, and their downstream signaling components in various regions of the brain, make these signaling pathways critical in regulating normal brain function. It has been observed that the IR-A isoform is the highest expressed in neurons while the IGF-1R in comparison to IR is highly expressed in the brain throughout development. Moreover, the IR and IGF-1R can also heterodimerize with one another which is highly reliant on the ratio of the receptors being expressed in specific brain regions. Investigating the mechanisms of the insulin signaling pathway in relation to the brain will be vital in further determining the correlations that exist between metabolic and neurodegenerative diseases and decipher potential therapeutic target sites.

ACKNOWLEDGMENTS

The research work of the group and AK has been supported by the Deutsche Forschungsgemeinschaft grant project KL 2399/7-1, Deutsche Diabetes Gesellschaft grants and by grants from the German Ministry of Education and Research (BMBF) and the State of Brandenburg (DZD grant 82DZD00302). We also thank previous group members for their work.

REFERENCES

1. G. J. Biessels, M. W. J. Strachan, F. L. J. Visseren, L. J. Kappelle, R. A. Whitmer, Dementia and cognitive decline in type 2 diabetes and prediabetic stages: Towards targeted interventions. *The Lancet Diabetes Endocrinol.* **2**, 246–255 (2014).
2. R. Ghasemi, L. Dargahi, A. Haeri, M. Moosavi, Z. Mohamed, A. Ahmadiani, Brain insulin dysregulation: Implication for neurological and neuropsychiatric disorders. *Mol Neurobiol.* **47**, 1045–1065 (2013).
3. S. Kullmann, A. Kleinridders, D. M. Small, A. Fritsche, H.-U. Häring, H. Preissl, M. Heni, Central nervous pathways of insulin action in the control of metabolism and food intake. *The Lancet Diabetes Endocrinol.* **8**, 524–534 (2020).
4. J. Havrankova, J. Roth, M. Brownstein, Insulin receptors are widely distributed in the central nervous system of the rat. *Nature.* **272**, 827–829 (1978).
5. S. Kullmann, M. Heni, M. Hallschmid, A. Fritsche, H. Preissl, H.-U. Häring, Brain insulin resistance at the crossroads of metabolic and cognitive disorders in humans. *Physiol. Rev.* **96**, 1169–1209 (2016).
6. F. M. Ashcroft, P. Rorsman, Electrophysiology of the pancreatic β-cell. *Prog. Biophys. Mol. Biol.* **54**, 87–143 (1989).
7. P. Rorsman, F. M. Ashcroft, Pancreatic β-cell electrical activity and insulin secretion: Of mice and men. *Physiol. Rev.* **98**, 117–214 (2018).
8. W. A. Banks, J. B. Jaspan, W. Huang, A. J. Kastin, Transport of insulin across the blood-brain barrier: Saturability at euglycemic doses of insulin. *Peptides.* **18**, 1423–1429 (1997).
9. W. A. Banks, A. J. Kastin, Differential permeability of the blood–brain barrier to two pancreatic peptides: Insulin and amylin. *Peptides.* **19**, 883–889 (1998).
10. W. A. Banks, J. B. Owen, M. A. Erickson, Insulin in the brain: There and back again. *Pharmacol. Ther.* **136**, 82–93 (2012).
11. J. M. Tavaré, R. M. Denton, Studies on the autophosphorylation of the insulin receptor from human placenta. Analysis of the sites phosphorylated by two-dimensional peptide mapping. *Biochem. J.* **252**, 607–615 (1988).
12. S. Jin Chan, D. F. Steiner, Insulin through the ages: Phylogeny of a growth promoting and metabolic regulatory hormonel. *Am. Zool.* **40**, 213–222 (2000).
13. C. Hernández-Sánchez, A. Mansilla, F. de Pablo, R. Zardoya, Evolution of the insulin receptor family and receptor isoform expression in vertebrates. *Mol. Biol. Evol.* **25**, 1043–1053 (2008).
14. M. F. White, C. R. Kahn, The insulin signaling system. *J. Biol. Chem.* **269**, 1–4 (1994).
15. I. Pomytkin, J. P. Costa-Nunes, V. Kasatkin, E. Veniaminova, A. Demchenko, A. Lyundup, K.-P. Lesch, E. D. Ponomarev, T. Strekalova, Insulin receptor in the brain: Mechanisms of activation and the role in the CNS pathology and treatment. *CNS Neurosci. Ther.* **24**, 763–774 (2018).
16. A. Belfiore, F. Frasca, G. Pandini, L. Sciacca, R. Vigneri, Insulin receptor isoforms and insulin receptor/insulin-like growth factor receptor hybrids in physiology and disease. *Endocr. Rev.* **30**, 586–623 (2009).
17. J. Boucher, A. Kleinridders, R. C. Kahn, Insulin receptor signaling in normal and insulin-resistant states. *Cold Spring Harbor Perspect. Biol.* **6**, 1–23 (2014).
18. A. Kleinridders, Deciphering brain insulin receptor and insulin-like growth factor 1 receptor signalling. *J. Neuroendocrinol.* **28** (2016), doi:10.1111/jne.12433.
19. J. C. Brüning, D. Gautam, D. J. Burks, J. Gillette, M. Schubert, P. C. Orban, R. Klein, W. Krone, D. Müller-Wieland, C. R. Kahn, Role of brain insulin receptor in control of body weight and reproduction. *Science.* **289**, 2122–2125 (2000).
20. M. Claret, M. A. Smith, R. L. Batterham, C. Selman, A. I. Choudhury, L. G. D. Fryer, M. Clements, H. Al-Qassab, H. Heffron, A. W. Xu, J. R. Speakman, G. S. Barsh, B. Viollet, S. Vaulont, M. L. J. Ashford, D. Carling, D. J. Withers, AMPK is essential for energy homeostasis regulation and glucose sensing by POMC and AgRP neurons. *J Clin Invest.* **117**, 2325–2336 (2007).
21. J. W. Hill, C. F. Elias, M. Fukuda, K. W. Williams, E. D. Berglund, W. L. Holland, Y.-R. Cho, J.-C.

Chuang, Y. Xu, M. Choi, D. Lauzon, C. E. Lee, R. Coppari, J. A. Richardson, J. M. Zigman, S. Chua, P. E. Scherer, B. B. Lowell, J. C. Brüning, J. K. Elmquist, Direct insulin and leptin action on pro-opiomelanocortin neurons is required for normal glucose homeostasis and fertility. *Cell Metab.* **11**, 286–297 (2010).

22. K. W. Williams, L. O. Margatho, C. E. Lee, M. Choi, S. Lee, M. M. Scott, C. F. Elias, J. K. Elmquist, Segregation of acute leptin and insulin effects in distinct populations of arcuate proopiomelanocortin neurons. *J. Neurosci.* **30**, 2472–9 (2010).
23. B. F. Belgardt, J. C. Brüning, CNS leptin and insulin action in the control of energy homeostasis. *Ann. N. Y. Acad. Sci.* **1212**, 97–113 (2010).
24. L. Varela, T. L. Horvath, Leptin and insulin pathways in POMC and AgRP neurons that modulate energy balance and glucose homeostasis. *EMBO Rep.* **13**, 1079–1086 (2012).
25. K. Loh, L. Zhang, A. Brandon, Q. Wang, D. Begg, Y. Qi, M. Fu, R. Kulkarni, J. Teo, P. Baldock, J. C. Brüning, G. Cooney, G. Neely, H. Herzog, Insulin controls food intake and energy balance via NPY neurons. *Mol. Metabol.* **6**, 574–584 (2017).
26. A. Kleinridders, E. N. Pothos, Impact of brain insulin signaling on dopamine function, food intake, reward, and emotional behavior. *Curr. Nutr. Rep.* **8**, 83–91 (2019).
27. A. C. Könner, R. Janoschek, L. Plum, S. D. Jordan, E. Rother, X. Ma, C. Xu, P. Enriori, B. Hampel, G. S. Barsh, C. R. Kahn, M. A. Cowley, F. M. Ashcroft, J. C. Brüning, Insulin action in AgRP-expressing neurons is required for suppression of hepatic glucose production. *Cell Metab.* **5**, 438–449 (2007).
28. H. V. Lin, L. Plum, H. Ono, R. Gutiérrez-Juárez, M. Shanabrough, E. Borok, T. L. Horvath, L. Rossetti, D. Accili, Divergent regulation of energy expenditure and hepatic glucose production by insulin receptor in agouti-related protein and POMC neurons. *Diabetes.* **59**, 337–346 (2010).
29. A. C. Shin, N. Filatova, C. Lindtner, T. Chi, S. Degann, D. Oberlin, C. Buettner, Insulin receptor signaling in POMC, but not AgRP, neurons controls adipose tissue insulin action. *Diabetes.* **66**, 1560–1571 (2017).
30. A. Kleinridders, W. Cai, L. Cappellucci, A. Ghazarian, W. R. Collins, S. G. Vienberg, E. N. Pothos, R. C. Kahn, Insulin resistance in brain alters dopamine turnover and causes behavioral disorders. *Proc. Natl. Acad. Sci.* **112**, 3463–3468 (2015).
31. S. Numan, D. S. Russell, Discrete expression of insulin receptor substrate-4 mRNA in adult rat brain. *Mol. Brain Res.* **72**, 97–102 (1999).
32. M. F. White, IRS proteins and the common path to diabetes. *Am. J. Physiol. Endocrinol. Metabol.* **283**, E413–E422 (2002).
33. Y. H. Lee, J. Giraud, R. J. Davis, M. F. White, c-Jun N-terminal Kinase (JNK) mediates feedback inhibition of the insulin signaling cascade*. *J. Biol. Chem.* **278**, 2896–2902 (2003).
34. D. Sawka-Verhelle, C. Filloux, S. Tartare-Deckert, I. Mothe, E. Van Obberghen, Identification of Stat 5B as a substrate of the insulin receptor. *Eur. J. Biochem.* **250**, 411–417 (1997).
35. M. F. White, J. N. Livingston, J. M. Backer, V. Lauris, T. J. Dull, A. Ullrich, C. R. Kahn, Mutation of the insulin receptor at tyrosine 960 inhibits signal transmission but does not affect its tyrosine kinase activity. *Cell.* **54**, 641–649 (1988).
36. X. J. Sun, D. L. Crimmins, M. G. Myers, M. Miralpeix, M. F. White, Pleiotropic insulin signals are engaged by multisite phosphorylation of IRS-1. *Mol. Cell. Biol.* **13**, 7418–7428 (1993).
37. J. Rush, A. Moritz, K. A. Lee, A. Guo, V. L. Goss, E. J. Spek, H. Zhang, X.-M. Zha, R. D. Polakiewicz, M. J. Comb, Immunoaffinity profiling of tyrosine phosphorylation in cancer cells. *Nat. Biotechnol.* **23**, 94–101 (2005).
38. X. Yang, A. Nath, M. J. Opperman, C. Chan, The double-stranded RNA–dependent protein kinase differentially regulates insulin receptor substrates 1 and 2 in HepG2 cells. *MBoC.* **21**, 3449–3458 (2010).
39. A. Koh, M. N. Lee, Y. R. Yang, H. Jeong, J. Ghim, J. Noh, J. Kim, D. Ryu, S. Park, P. Song, S.-H. Koo, N. R. Leslie, P.-O. Berggren, J. H. Choi, P.-G. Suh, S. H. Ryu, C1-Ten is a protein tyrosine phosphatase of insulin receptor substrate 1 (IRS-1), regulating IRS-1 stability and muscle atrophy. *Mol. Cell. Biol.* **33**, 1608–1620 (2013).
40. E. D. Werner, J. Lee, L. Hansen, M. Yuan, S. E. Shoelson, Insulin resistance due to phosphorylation of insulin receptor substrate-1 at serine 302*. *J. Biol. Chem.* **279**, 35298–35305 (2004).
41. K. D. Copps, N. J. Hancer, L. Opare-Ado, W. Qiu, C. Walsh, M. F. White, Irs1 serine 307 promotes insulin sensitivity in mice. *Cell Metab.* **11**, 84–92 (2010).

42. Z. Yi, P. Langlais, E. A. D. Filippis, M. Luo, C. R. Flynn, S. Schroeder, S. T. Weintraub, R. Mapes, L. J. Mandarino, Global assessment of regulation of phosphorylation of insulin receptor substrate-1 by insulin in vivo in human muscle. *Diabetes.* **56**, 1508–1516 (2007).
43. J. M. Backer, M. G. Myers, S. E. Shoelson, D. J. Chin, X. J. Sun, M. Miralpeix, P. Hu, B. Margolis, E. Y. Skolnik, J. Schlessinger, Phosphatidylinositol 3'-kinase is activated by association with IRS-1 during insulin stimulation. *EMBO J.* **11**, 3469–3479 (1992).
44. J. Yu, Y. Zhang, J. McIlroy, T. Rordorf-Nikolic, G. A. Orr, J. M. Backer, Regulation of the p85/p110 phosphatidylinositol 3′-kinase: Stabilization and inhibition of the p110α catalytic subunit by the p85 regulatory subunit. *Mol. Cell. Biol.* (1998), doi:10.1128/MCB.18.3.1379.
45. D. R. Alessi, M. Andjelkovic, B. Caudwell, P. Cron, N. Morrice, P. Cohen, B. A. Hemmings, Mechanism of activation of protein kinase B by insulin and IGF-1. *EMBO J.* **15**, 6541–6551 (1996).
46. D. R. Alessi, S. R. James, C. P. Downes, A. B. Holmes, P. R. J. Gaffney, C. B. Reese, P. Cohen, Characterization of a 3-phosphoinositide-dependent protein kinase which phosphorylates and activates protein kinase Bα. *Curr. Biol.* **7**, 261–269 (1997).
47. D. D. Sarbassov, D. A. Guertin, S. M. Ali, D. M. Sabatini, Phosphorylation and regulation of Akt/PKB by the rictor-mTOR complex. *Science.* **307**, 1098–1101 (2005).
48. S. P. Narayanan, A. I. Flores, F. Wang, W. B. Macklin, Akt signals through the mammalian target of rapamycin pathway to regulate CNS myelination. *J. Neurosci.* **29**, 6860–6870 (2009).
49. H. Wong, J. Levenga, L. LaPlante, B. Keller, A. Cooper-Sansone, C. Borski, R. Milstead, M. Ehringer, C. Hoeffer, Isoform-specific roles for AKT in affective behavior, spatial memory, and extinction related to psychiatric disorders. *eLife.* **9**, e56630 (2020).
50. G. Tzivion, M. Dobson, G. Ramakrishnan, FoxO transcription factors: Regulation by AKT and 14-3-3 proteins. *Biochimica et Biophysica Acta (BBA).* **1813**, 1938–1945 (2011).
51. M.-S. Kim, Y. K. Pak, P.-G. Jang, C. Namkoong, Y.-S. Choi, J.-C. Won, K.-S. Kim, S.-W. Kim, H.-S. Kim, J.-Y. Park, Y.-B. Kim, K.-U. Lee, Role of hypothalamic Foxo1 in the regulation of food intake and energy homeostasis. *Nat. Neurosci.* **9**, 901–906 (2006).
52. D. A. M. Salih, A. J. Rashid, D. Colas, L. de la Torre-Ubieta, R. P. Zhu, A. A. Morgan, E. E. Santo, D. Ucar, K. Devarajan, C. J. Cole, D. V. Madison, M. Shamloo, A. J. Butte, A. Bonni, S. A. Josselyn, A. Brunet, FoxO6 regulates memory consolidation and synaptic function. *Genes Dev.* **26**, 2780–2801 (2012).
53. Y. Wang, Y. Zhou, D. T. Graves, FOXO Transcription Factors: Their Clinical Significance and Regulation. *BioMed Res. Int.* **2014**, e925350 (2014).
54. S. Gabbouj, S. Ryhänen, M. Marttinen, R. Wittrahm, M. Takalo, S. Kemppainen, H. Martiskainen, H. Tanila, A. Haapasalo, M. Hiltunen, T. Natunen, Altered insulin signaling in Alzheimer's disease brain – special emphasis on PI3K-Akt pathway. *Front. Neurosci.* **13**, 629 (2019).
55. D. Cota, K. Proulx, K. A. B. Smith, S. C. Kozma, G. Thomas, S. C. Woods, R. J. Seeley, Hypothalamic mTOR signaling regulates food intake. *Science.* **312**, 927–930 (2006).
56. J. Chenal, K. Pierre, L. Pellerin, Insulin and IGF-1 enhance the expression of the neuronal monocarboxylate transporter MCT2 by translational activation via stimulation of the phosphoinositide 3-kinase–Akt–mammalian target of rapamycin pathway. *Eur. J. Neurosci.* **27**, 53–65 (2008).
57. M. Pende, S. H. Um, V. Mieulet, M. Sticker, V. L. Goss, J. Mestan, M. Mueller, S. Fumagalli, S. C. Kozma, G. Thomas, S6K1−/−/S6K2−/− mice exhibit perinatal lethality and rapamycin-sensitive 5′-terminal oligopyrimidine mRNA translation and reveal a mitogen-activated protein kinase-dependent S6 kinase pathway. *Mol. Cell. Biol.* (2004), doi:10.1128/MCB.24.8.3112-3124.2004.
58. M. Schell, K. Wardelmann, A. Kleinridders, Untangling the effect of insulin action on brain mitochondria and metabolism. *J. Neuroendocrinol.* **n/a**, e12932.
59. C. M. Taniguchi, B. Emanuelli, C. R. Kahn, Critical nodes in signalling pathways: Insights into insulin action. *Nat. Rev. Mol. Cell Biol.* **7**, 85–96 (2006).
60. S. I. Taylor, T. Kadowaki, H. Kadowaki, D. Accili, A. Cama, C. McKeon, Mutations in insulin-receptor gene in insulin-resistant patients. *Diabetes Care.* **13**, 257–279 (1990).
61. T. Kadowaki, H. Kadowaki, M. M. Rechler, M. Serrano-Rios, J. Roth, P. Gorden, S. I. Taylor, Five mutant alleles of the insulin receptor gene

in patients with genetic forms of insulin resistance. *J. Clin. Invest.* **86**, 254–264 (1990).
62. S. Ben Harouch, A. Klar, T. C. Falik Zaccai, *GeneReviews®*, M. P. Adam, H. H. Ardinger, R. A. Pagon, S. E. Wallace, L. J. Bean, G. Mirzaa, A. Amemiya, Eds. (University of Washington, Seattle, Seattle, WA, 1993; http://www.ncbi.nlm.nih.gov/books/NBK476444/).
63. M. Stumvoll, A. Fritsche, A. Volk, N. Stefan, A. Madaus, E. Maerker, A. Teigeler, M. Koch, F. Machicao, H. Häring, The Gly972Arg polymorphism in the insulin receptor substrate-1 gene contributes to the variation in insulin secretion in normal glucose-tolerant humans. *Diabetes.* **50**, 882–885 (2001).
64. R. A. Haeusler, T. E. McGraw, D. Accili, Biochemical and cellular properties of insulin receptor signalling. *Nat. Rev. Mol. Cell Biol.* **19**, 31–44 (2018).
65. R. Pilarski, C. Eng, Will the real Cowden syndrome please stand up (again)? Expanding mutational and clinical spectra of the PTEN hamartoma tumour syndrome. *J. Med. Genet.* **41**, 323–326 (2004).
66. A. M. Jelsig, N. Qvist, K. Brusgaard, C. B. Nielsen, T. P. Hansen, L. B. Ousager, Hamartomatous polyposis syndromes: A review. *Orphanet J. Rare Dis.* **9**, 101 (2014).
67. Z. Chen, L. C. Trotman, D. Shaffer, H.-K. Lin, Z. A. Dotan, M. Niki, J. A. Koutcher, H. I. Scher, T. Ludwig, W. Gerald, C. Cordon-Cardo, P. Paolo Pandolfi, Crucial role of p53-dependent cellular senescence in suppression of Pten-deficient tumorigenesis. *Nature.* **436**, 725–730 (2005).
68. P. S. Mundi, J. Sachdev, C. McCourt, K. Kalinsky, AKT in cancer: New molecular insights and advances in drug development. *Br. J. Clin. Pharmacol.* **82**, 943–956 (2016).
69. L. Bessière, A.-L. Todeschini, A. Auguste, S. Sarnacki, D. Flatters, B. Legois, C. Sultan, N. Kalfa, L. Galmiche, R. A. Veitia, A hot-spot of in-frame duplications activates the oncoprotein AKT1 in Juvenile Granulosa cell tumors. *EBioMedicine.* **2**, 421–431 (2015).
70. A. Auguste, L. Bessière, A.-L. Todeschini, S. Caburet, S. Sarnacki, J. Prat, E. D'angelo, P. De La Grange, O. Ariste, F. Lemoine, B. Legois, C. Sultan, A. Zider, L. Galmiche, N. Kalfa, R. A. Veitia, Molecular analyses of juvenile granulosa cell tumors bearing AKT1 mutations provide insights into tumor biology and therapeutic leads. *Human Mol. Genet.* **24**, 6687–6698 (2015).
71. V. Aguirre, E. D. Werner, J. Giraud, Y. H. Lee, S. E. Shoelson, M. F. White, Phosphorylation of Ser307 in insulin receptor substrate-1 blocks interactions with the insulin receptor and inhibits insulin action*. *J. Biol. Chem.* **277**, 1531–1537 (2002).
72. J. Wu, C. Mei, H. Vlassara, G. E. Striker, F. Zheng, Oxidative stress-induced JNK activation contributes to proinflammatory phenotype of aging diabetic mesangial cells. *Am. J. Physiol. Renal Physiol.* **297**, F1622–F1631 (2009).
73. K. De Fea, R. A. Roth, Protein kinase C modulation of insulin receptor substrate-1 tyrosine phosphorylation requires serine 612. *Biochemistry.* **36**, 12939–12947 (1997).
74. O. N. Ozes, H. Akca, L. D. Mayo, J. A. Gustin, T. Maehama, J. E. Dixon, D. B. Donner, A phosphatidylinositol 3-kinase/Akt/mTOR pathway mediates and PTEN antagonizes tumor necrosis factor inhibition of insulin signaling through insulin receptor substrate-1. *PNAS.* **98**, 4640–4645 (2001).
75. Y. Li, T. J. Soos, X. Li, J. Wu, M. DeGennaro, X. Sun, D. R. Littman, M. J. Birnbaum, R. D. Polakiewicz, Protein kinase C θ inhibits insulin signaling by phosphorylating IRS1 at Ser1101*. *J. Biol. Chem.* **279**, 45304–45307 (2004).
76. B. Emanuelli, P. Peraldi, C. Filloux, D. Sawka-Verhelle, D. Hilton, E. V. Obberghen, SOCS-3 is an insulin-induced negative regulator of insulin signaling*. *J. Biol. Chem.* **275**, 15985–15991 (2000).
77. B. Emanuelli, P. Peraldi, C. Filloux, C. Chavey, K. Freidinger, D. J. Hilton, G. S. Hotamisligil, E. Van Obberghen, SOCS-3 inhibits insulin signaling and is up-regulated in response to tumor necrosis factor-α in the adipose tissue of obese mice*. *J. Biol. Chem.* **276**, 47944–47949 (2001).
78. L. Rui, M. Yuan, D. Frantz, S. Shoelson, M. F. White, SOCS-1 and SOCS-3 block insulin signaling by ubiquitin-mediated degradation of IRS1 and IRS2 *. *J. Biol. Chem.* **277**, 42394–42398 (2002).
79. V. Kothari, Y. Luo, T. Tornabene, A. M. O'Neill, M. W. Greene, T. Geetha, J. R. Babu, High fat diet induces brain insulin resistance and cognitive impairment in mice. *Biochimica et Biophysica Acta (BBA).* **1863**, 499–508 (2017).
80. T. R. Bomfim, L. Forny-Germano, L. B. Sathler, J. Brito-Moreira, J.-C. Houzel, H. Decker, M. A. Silverman, H. Kazi, H. M. Melo, P. L. McClean, C. Holscher, S. E. Arnold, K. Talbot, W. L. Klein, D. P. Munoz, S. T. Ferreira, F. G. D. Felice, An

anti-diabetes agent protects the mouse brain from defective insulin signaling caused by Alzheimer's disease–associated Aβ oligomers. *J. Clin. Invest.* **122**, 1339–1353 (2012).
81. H. M. Melo, G. da S. S. da Silva, M. R. Sant'Ana, C. V. L. Teixeira, J. R. Clarke, V. S. M. Coreixas, B. C. de Melo, J. T. S. Fortuna, L. Forny-Germano, J. H. Ledo, M. S. Oliveira, C. P. Figueiredo, R. Pardossi-Piquard, F. Checler, J. M. Delgado-García, A. Gruart, L. A. Velloso, M. L. F. Balthazar, D. E. Cintra, S. T. Ferreira, F. G. D. Felice, Palmitate is increased in the cerebrospinal fluid of humans with obesity and induces memory impairment in mice via pro-inflammatory TNF-α. *Cell Reps.* **30**, 2180–2194.e8 (2020).
82. J. K. Kim, Y.-J. Kim, J. J. Fillmore, Y. Chen, I. Moore, J. Lee, M. Yuan, Z. W. Li, M. Karin, P. Perret, S. E. Shoelson, G. I. Shulman, Prevention of fat-induced insulin resistance by salicylate. *J. Clin. Invest.* **108**, 437–446 (2001).
83. U. Özcan, Q. Cao, E. Yilmaz, A.-H. Lee, N. N. Iwakoshi, E. Özdelen, G. Tuncman, C. Görgün, L. H. Glimcher, G. S. Hotamisligil, Endoplasmic reticulum stress links obesity, insulin action, and type 2 diabetes. *Science.* **306**, 457–461 (2004).
84. S. Sinha, G. Perdomo, N. F. Brown, R. M. O'Doherty, Fatty acid-induced insulin resistance in L6 myotubes is prevented by inhibition of activation and nuclear localization of nuclear factor κB *. *J. Biol. Chem.* **279**, 41294–41301 (2004).
85. H. Shi, M. V. Kokoeva, K. Inouye, I. Tzameli, H. Yin, J. S. Flier, TLR4 links innate immunity and fatty acid–induced insulin resistance. *J. Clin. Invest.* **116**, 3015–3025 (2006).
86. S. Schenk, M. Saberi, J. M. Olefsky, Insulin sensitivity: Modulation by nutrients and inflammation. *J. Clin. Invest.* **118**, 2992–3002 (2008).
87. S. C. Benoit, C. J. Kemp, C. F. Elias, W. Abplanalp, J. P. Herman, S. Migrenne, A.-L. Lefevre, C. Cruciani-Guglielmacci, C. Magnan, F. Yu, K. Niswender, B. G. Irani, W. L. Holland, D. J. Clegg, Palmitic acid mediates hypothalamic insulin resistance by altering PKC-θ subcellular localization in rodents. *J. Clin. Invest.* **119**, 2577–2589 (2009).
88. T. Tiganis, Reactive oxygen species and insulin resistance: the good, the bad and the ugly. *Trends Pharmacol. Sci.* **32**, 82–89 (2011).
89. A. Kleinridders, H. P. M. M. Lauritzen, S. Ussar, J. H. Christensen, M. A. Mori, P. Bross, R. C. Kahn, Leptin regulation of Hsp60 impacts hypothalamic insulin signaling. *J. Clin. Invest.* **123**, 4667–80 (2013).
90. K. Wardelmann, M. Rath, J. P. Castro, S. Blümel, M. Schell, R. Hauffe, F. Schumacher, T. Flore, K. Ritter, A. Wernitz, T. Hosoi, K. Ozawa, B. Kleuser, J. Weiß, A. Schürmann, A. Kleinridders, Central acting Hsp10 regulates mitochondrial function, fatty acid metabolism, and insulin sensitivity in the Hypothalamus. *Antioxidants.* **10**, 711 (2021).
91. T. Nakamura, M. Furuhashi, P. Li, H. Cao, G. Tuncman, N. Sonenberg, C. Z. Gorgun, G. S. Hotamisligil, Double-stranded RNA-dependent protein kinase links pathogen sensing with stress and metabolic homeostasis. *Cell.* **140**, 338–348 (2010).
92. M. Flamment, E. Hajduch, P. Ferré, F. Foufelle, New insights into ER stress-induced insulin resistance. *Trends Endocrinol. Metab.* **23**, 381–390 (2012).
93. C. McKeon, V. Moncada, T. Pham, P. Salvatore, T. Kadowaki, D. Accili, S. I. Taylor, Structural and functional analysis of the insulin receptor promoter. *Mol. Endocrinol.* **4**, 647–656 (1990).
94. E. M. Bailyes, B. T. Navé, M. A. Soos, S. R. Orr, A. C. Hayward, K. Siddle, Insulin receptor/IGF-I receptor hybrids are widely distributed in mammalian tissues: Quantification of individual receptor species by selective immunoprecipitation and immunoblotting. *Biochem. J.* **327**, 209–215 (1997).
95. E. Blázquez, E. Velázquez, V. Hurtado-Carneiro, J. M. Ruiz-Albusac, Insulin in the brain: Its pathophysiological implications for states related with central insulin resistance, type 2 diabetes and Alzheimer's disease. *Front. Endocrinol.* **5**, 161 (2014).
96. K. A. Heidenreich, N. R. Zahniser, P. Berhanu, D. Brandenburg, J. M. Olefsky, Structural differences between insulin receptors in the brain and peripheral target tissues. *J. Biol. Chem.* **258**, 8527–8530 (1983).
97. A. M. Fernandez, I. Torres-Alemán, The many faces of insulin-like peptide signalling in the brain. *Nat. Rev. Neurosci.* **13**, 225–239 (2012).
98. A. Kleinridders, H. Ferris, W. Cai, R. C. Kahn, Insulin action in brain regulates systemic metabolism and brain function. *Diabetes.* **63**, 2232–2243 (2014).
99. S.-L. Chiu, H. T. Cline, Insulin receptor signaling in the development of neuronal structure and function. *Neural Dev.* **5**, 7 (2010).

100. S. Kar, J.-G. Chabot, R. Quirion, Quantitative autoradiographic localization of [125I]insulin-like growth factor I, [125I]insulin-like growth factor II, and [125I]insulin receptor binding sites in developing and adult rat brain. *J. Comp. Neurol.* **333**, 375–397 (1993).
101. C. J. Garwood, L. E. Ratcliffe, S. V. Morgan, J. E. Simpson, H. Owens, I. Vazquez-Villaseñor, P. R. Heath, I. A. Romero, P. G. Ince, S. B. Wharton, Insulin and IGF1 signalling pathways in human astrocytes in vitro and in vivo; characterisation, subcellular localisation and modulation of the receptors. *Mol. Brain.* **8**, 51 (2015).
102. C. A. Bondy, H. Werner, C. T. Roberts Jr., D. LeRoith, Cellular pattern of insulin-like growth factor-I (IGF-I) and type I IGF receptor gene expression in early organogenesis: Comparison with IGF-II gene expression. *Mol. Endocrinol.* **4**, 1386–1398 (1990).
103. A. Baron-Van Evercooren, C. Olichon-Berthe, A. Kowalski, G. Visciano, E. Van Obberghen, Expression of IGF-I and insulin receptor genes in the rat central nervous system: A developmental, regional, and cellular analysis. *J. Neurosci. Res.* **28**, 244–253 (1991).
104. A. Adem, S. S. Jossan, R. d'Argy, P. G. Gillberg, A. Nordberg, B. Winblad, V. Sara, Insulin-like growth factor 1 (IGF-1) receptors in the human brain: Quantitative autoradiographic localization. *Brain Res.* **503**, 299–303 (1989).
105. L. M. Garcia-Segura, J. R. Rodriguez, I. Torres-Aleman, Localization of the insulin-like growth factor I receptor in the cerebellum and hypothalamus of adult rats: An electron microscopic study. *J. Neurocytol.* **26**, 479–490 (1997).
106. L. Kappeler, C. D. M. Filho, J. Dupont, P. Leneuve, P. Cervera, L. Périn, C. Loudes, A. Blaise, R. Klein, J. Epelbaum, Y. L. Bouc, M. Holzenberger, Brain IGF-1 receptors control mammalian growth and lifespan through a neuroendocrine mechanism. *PLOS Biol.* **6**, e254 (2008).
107. C. A. Bondy, M. A. Bach, W.-H. Lee, Mapping of brain insulin and insulin-like growth factor receptor gene expression by in situ hybridization. *Neuroprotocols.* **1**, 240–249 (1992).
108. J. Chen, A. M. Nagle, Y.-F. Wang, D. N. Boone, A. V. Lee, Controlled dimerization of insulin-like growth factor-1 and insulin receptors reveals shared and distinct activities of holo and hybrid receptors. *J. Biol. Chem.* **293**, 3700–3709 (2018).
109. D. G. Baskin, A. J. Sipols, M. W. Schwartz, M. F. White, Immunocytochemical detection of insulin receptor substrate-1 (IRS-1) in rat brain: colocalization with phosphotyrosine. *Regul. Pept.* **48**, 257–266 (1993).
110. F. Folli, L. Bonfanti, E. Renard, C. R. Kahn, A. Merighi, Insulin receptor substrate-1 (IRS-1) distribution in the rat central nervous system. *J. Neurosci.* **14**, 6412–6422 (1994).
111. K. Talbot, H.-Y. Wang, H. Kazi, L.-Y. Han, K. P. Bakshi, A. Stucky, R. L. Fuino, K. R. Kawaguchi, A. J. Samoyedny, R. S. Wilson, Z. Arvanitakis, J. A. Schneider, B. A. Wolf, D. A. Bennett, J. Q. Trojanowski, S. E. Arnold, Demonstrated brain insulin resistance in Alzheimer's disease patients is associated with IGF-1 resistance, IRS-1 dysregulation, and cognitive decline. *J. Clin. Invest.* **122**, 1316–1338 (2012).
112. T. Hayashi, T. Kubota, I. Mariko, I. Takamoto, M. Aihara, Y. Sakurai, N. Wada, T. Miki, T. Yamauchi, N. Kubota, T. Kadowaki, Lack of brain insulin receptor substrate-1 causes growth retardation, with decreased expression of growth hormone-releasing hormone in the hypothalamus. *Diabetes.* **70**, 1640–1653 (2021).
113. P. Ye, L. Li, P. K. Lund, A. J. D'Ercole, Deficient expression of insulin receptor substrate-1 (IRS-1) fails to block insulin-like growth factor-I (IGF-I) stimulation of brain growth and myelination. *Brain Res. Dev. Brain Res.* **136**, 111–121 (2002).
114. A. W. Pardini, H. T. Nguyen, D. P. Figlewicz, D. G. Baskin, D. L. Williams, F. Kim, M. W. Schwartz, Distribution of insulin receptor substrate-2 in brain areas involved in energy homeostasis. *Brain Res.* **1112**, 169–178 (2006).
115. X. Lin, A. Taguchi, S. Park, J. A. Kushner, F. Li, Y. Li, M. F. White, Dysregulation of insulin receptor substrate 2 in β cells and brain causes obesity and diabetes. *J. Clin. Invest.* **114**, 908–916 (2004).
116. D. J. Withers, Insulin receptor substrate proteins and neuroendocrine function. *Biochem. Soc. Trans.* **29**, 525–529 (2001).
117. M. Schubert, D. P. Brazil, D. J. Burks, J. A. Kushner, J. Ye, C. L. Flint, J. Farhang-Fallah, P. Dikkes, X. M. Warot, C. Rio, G. Corfas, M. F. White, Insulin receptor substrate-2 deficiency impairs brain growth and promotes Tau phosphorylation. *J. Neurosci.* **23**, 7084–7092 (2003).
118. M. Udelhoven, M. Pasieka, U. Leeser, W. Krone, M. Schubert, Neuronal insulin receptor substrate 2 (IRS2) expression is regulated by ZBP89 and SP1 binding to the IRS2 promoter. *J. Endocrinol.* **204**, 199–208 (2010).

119. R. Killick, G. Scales, K. Leroy, M. Causevic, C. Hooper, E. E. Irvine, A. I. Choudhury, L. Drinkwater, F. Kerr, H. Al-Qassab, J. Stephenson, Z. Yilmaz, K. P. Giese, J.-P. Brion, D. J. Withers, S. Lovestone, Deletion of Irs2 reduces amyloid deposition and rescues behavioural deficits in APP transgenic mice. *Biochem. Biophys. Res. Commun.* **386**, 257–262 (2009).
120. D. A. Costello, M. Claret, H. Al-Qassab, F. Plattner, E. E. Irvine, A. I. Choudhury, K. P. Giese, D. J. Withers, P. Pedarzani, Brain deletion of insulin receptor substrate 2 disrupts hippocampal synaptic plasticity and metaplasticity. *PLOS ONE.* **7**, e31124 (2012).
121. J. Zemva, M. Udelhoven, L. Moll, S. Freude, O. Stöhr, H. S. Brönneke, R. B. Drake, W. Krone, M. Schubert, Neuronal overexpression of insulin receptor substrate 2 leads to increased fat mass, insulin resistance, and glucose intolerance during aging. *AGE.* **35**, 1881–1897 (2013).
122. M. Sadagurski, X. C. Dong, M. G. Myers, M. F. White, Irs2 and Irs4 synergize in non-LepRb neurons to control energy balance and glucose homeostasis. *Mol. Metab.* **3**, 55–63 (2014).
123. V. R. Fantin, Q. Wang, G. E. Lienhard, S. R. Keller, Mice lacking insulin receptor substrate 4 exhibit mild defects in growth, reproduction, and glucose homeostasis. *Am. J. Physiol. Endocrinol. Metab.* **278**, E127–E133 (2000).
124. A. K. Sutton, I. E. Gonzalez, M. Sadagurski, M. Rajala, C. Lu, M. B. Allison, J. M. Adams, M. G. Myers, M. F. White, D. P. Olson, Paraventricular, subparaventricular and periventricular hypothalamic IRS4-expressing neurons are required for normal energy balance. *Sci. Rep.* **10**, 5546 (2020).
125. R. M. Easton, H. Cho, K. Roovers, D. W. Shineman, M. Mizrahi, M. S. Forman, V. M.-Y. Lee, M. Szabolcs, R. de Jong, T. Oltersdorf, T. Ludwig, A. Efstratiadis, M. J. Birnbaum, Role for Akt3/protein kinase Bγ in attainment of normal brain size. *Mol. Cell. Biol.* **25**, 1869–1878 (2005).
126. C. C. Kumar, V. Madison, AKT crystal structure and AKT-specific inhibitors. *Oncogene.* **24**, 7493–7501 (2005).
127. W. S. Chen, P.-Z. Xu, K. Gottlob, M.-L. Chen, K. Sokol, T. Shiyanova, I. Roninson, W. Weng, R. Suzuki, K. Tobe, T. Kadowaki, N. Hay, Growth retardation and increased apoptosis in mice with homozygous disruption of the akt1 gene. *Genes Dev.* **15**, 2203–2208 (2001).
128. C. O' Neill, PI3-kinase/Akt/mTOR signaling: Impaired on/off switches in aging, cognitive decline and Alzheimer's disease. *Experi. Gerontol.* **48**, 647–653 (2013).
129. P. Salcedo-Tello, A. Ortiz-Matamoros, C. Arias, GSK3 function in the brain during development, neuronal plasticity, and neurodegeneration. *Int. J. Alzheimer's Dis.* **2011**, e189728 (2011).
130. L. Stoica, P. J. Zhu, W. Huang, H. Zhou, S. C. Kozma, M. Costa-Mattioli, Selective pharmacogenetic inhibition of mammalian target of Rapamycin complex I (mTORC1) blocks long-term synaptic plasticity and memory storage. *PNAS.* **108**, 3791–3796 (2011).
131. A. Poduri, G. D. Evrony, X. Cai, P. C. Elhosary, R. Beroukhim, M. K. Lehtinen, L. B. Hills, E. L. Heinzen, A. Hill, R. S. Hill, B. J. Barry, B. F. D. Bourgeois, J. J. Riviello, A. J. Barkovich, P. M. Black, K. L. Ligon, C. A. Walsh, Somatic activation of AKT3 causes hemispheric developmental brain malformations. *Neuron.* **74**, 41–48 (2012).
132. L. Flores-Sarnat, H. B. Sarnat, G. Dávila-Gutiérrez, A. Álvarez, Hemimegalencephaly: Part 2. Neuropathology suggests a disorder of cellular lineage. *J. Child Neurol.* **18**, 776–785 (2003).
133. S. Gowda, F. Salazar, W. E. Bingaman, P. Kotagal, D. L. Lachhwani, A. Gupta, S. Davis, J. Niezgoda, E. Wyllie, Surgery for catastrophic epilepsy in infants 6 months of age and younger: Clinical article. *J. Neurosur.* **5**, 603–607 (2010).
134. O. Tschopp, Z.-Z. Yang, D. Brodbeck, B. A. Dummler, M. Hemmings-Mieszczak, T. Watanabe, T. Michaelis, J. Frahm, B. A. Hemmings, Essential role of protein kinase Bγ (PKBγ/Akt3) in postnatal brain development but not in glucose homeostasis. *Development.* **132**, 2943–2954 (2005).
135. H. Cho, J. L. Thorvaldsen, Q. Chu, F. Feng, M. J. Birnbaum, Akt1/PKBα is required for normal growth but dispensable for maintenance of glucose homeostasis in mice*. *J. Biol. Chem.* **276**, 38349–38352 (2001).
136. E. Boland, J. Clayton-Smith, V. G. Woo, S. McKee, F. D. C. Manson, L. Medne, E. Zackai, E. A. Swanson, D. Fitzpatrick, K. J. Millen, E. H. Sherr, W. B. Dobyns, G. C. M. Black, Mapping of deletion and translocation breakpoints in 1q44 implicates the serine/threonine kinase AKT3 in postnatal microcephaly and agenesis of the corpus callosum. *Am. J. Human Genet.* **81**, 292–303 (2007).
137. D. Wang, S. Zeesman, M. A. Tarnopolsky, M. J. M. Nowaczyk, Duplication of AKT3 as a cause of macrocephaly in duplication 1q43q44. *Am. J. Med. Genet. A.* **161**, 2016–2019 (2013).

138. T. G. Boulton, S. H. Nye, D. J. Robbins, N. Y. Ip, E. Radzlejewska, S. D. Morgenbesser, R. A. DePinho, N. Panayotatos, M. H. Cobb, G. D. Yancopoulos, ERKs: A family of protein-serine/threonine kinases that are activated and tyrosine phosphorylated in response to insulin and NGF. *Cell.* **65**, 663–675 (1991).
139. Y. Mebratu, Y. Tesfaigzi, How ERK1/2 activation controls cell proliferation and cell death: Is subcellular localization the answer? *Cell Cycle.* **8**, 1168–1175 (2009).
140. F. Bost, M. Aouadi, L. Caron, P. Even, N. Belmonte, M. Prot, C. Dani, P. Hofman, G. Pagès, J. Pouysségur, Y. L. Marchand-Brustel, B. Binétruy, The extracellular signal–regulated kinase isoform ERK1 is specifically required for in vitro and in vivo adipogenesis. *Diabetes.* **54**, 402–411 (2005).
141. C. Mazzucchelli, C. Vantaggiato, A. Ciamei, S. Fasano, P. Pakhotin, W. Krezel, H. Welzl, D. P. Wolfer, G. Pagès, O. Valverde, A. Marowsky, A. Porrazzo, P. C. Orban, R. Maldonado, M. U. Ehrengruber, V. Cestari, H.-P. Lipp, P. F. Chapman, J. Pouysségur, R. Brambilla, Knockout of ERK1 MAP kinase enhances synaptic plasticity in the striatum and facilitates striatal-mediated learning and memory. *Neuron.* **34**, 807–820 (2002).
142. J. C. Selcher, T. Nekrasova, R. Paylor, G. E. Landreth, J. D. Sweatt, Mice lacking the ERK1 isoform of MAP kinase are unimpaired in emotional learning. *Learn. Mem.* **8**, 11–19 (2001).
143. M. Aouadi, B. Binetruy, L. Caron, Y. Le Marchand-Brustel, F. Bost, Role of MAPKs in development and differentiation: Lessons from knockout mice. *Biochimie.* **88**, 1091–1098 (2006).
144. I. S. Samuels, J. C. Karlo, A. N. Faruzzi, K. Pickering, K. Herrup, J. D. Sweatt, S. C. Saitta, G. E. Landreth, Deletion of ERK2 mitogen-activated protein kinase identifies its key roles in cortical neurogenesis and cognitive function. *J. Neurosci.* **28**, 6983–6995 (2008).
145. T. H. Shaikh, R. J. O'Connor, M. E. Pierpont, J. McGrath, A. M. Hacker, M. Nimmakayalu, E. Geiger, B. S. Emanuel, S. C. Saitta, Low copy repeats mediate distal chromosome 22q11.2 deletions: Sequence analysis predicts breakpoint mechanisms. *Genome Res.* **17**, 482–491 (2007).
146. T. G. Boulton, M. H. Cobb, Identification of multiple extracellular signal-regulated kinases (ERKs) with antipeptide antibodies. *Cell Regul.* **2**, 357–371 (1991).
147. Y. Satoh, S. Endo, T. Nakata, Y. Kobayashi, K. Yamada, T. Ikeda, A. Takeuchi, T. Hiramoto, Y. Watanabe, T. Kazama, ERK2 contributes to the control of social behaviors in mice. *J. Neurosci.* **31**, 11953–11967 (2011).
148. J. Ortiz, H. W. Harris, X. Guitart, R. Z. Terwilliger, J. W. Haycock, E. J. Nestler, Extracellular signal-regulated protein kinases (ERKs) and ERK kinase (MEK) in brain: Regional distribution and regulation by chronic morphine. *J. Neurosci.* **15**, 1285–1297 (1995).
149. K. Wardelmann, S. Blümel, M. Rath, E. Alfine, C. Chudoba, M. Schell, W. Cai, R. Hauffe, K. Warnke, T. Flore, K. Ritter, J. Weiß, C. R. Kahn, A. Kleinridders, Insulin action in the brain regulates mitochondrial stress responses and reduces diet-induced weight gain. *Mol. Metab.* **21**, 68–81 (2019).

CHAPTER TWO

Insulin/IGF Signaling in Early Brain Development

Selma Yagoub
Rachel N. Lippert

CONTENTS

EFFECTS ON THE DEVELOPING CENTRAL NERVOUS SYSTEM

> One cannot help getting the impression that these structural changes, at any rate, are rather unspecific results of gene action. Some more fundamental change in development seems to upset those developmental processes, which at a given time happen to be in a critical and hence susceptible stage
>
> – H Grüneberg 1943 'The Genetics of the Mouse'

A century ago, insulin was initially described as a peripheral hormone, unable to cross the **blood-brain barrier** (1), which is the structure maintaining a protective and only semi-permeable divide between peripheral circulation and the central nervous system.

In 1945, in a series of experiments using chick embryos, Landauer made the first report on the morphological effects of excess insulin treatment in embryonic stages (2). Here, injection of insulin directly into the yolk resulted in a 'rumpless' phenotype among other developmental abnormalities. P.K. Duraiswami went on to show that insulin introduction into the chicken embryo at multiple stages post-fertilization resulted in predominantly skeletal defects (3).

DOI: 10.1201/9781003079927-2

Later, Margolis and Altszuler, in 1967, were the first to show the presence of insulin in the *cerebrospinal fluid*, the clear, colorless fluid found circulating throughout the brain (4). Only in the 1980s, evidence showed that peripheral insulin can cross the blood-brain barrier and the hormone was detected in the brain in both humans and rodents (5). However, at this time, the brain was considered to be "insensitive to insulin" since insulin was shown to not affect *glucose transporters*, the specialized proteins which are trafficked to the cellular membrane to increase glucose uptake into the respective cells in response to increased insulin signaling. Specifically, this insensitivity was suggested due to a lack of stimulation of glucose transporters located at the blood-brain barrier and cerebral glucose metabolism in humans (6).

Interestingly, it was already suggested that insulin and affiliated proteins and their receptors could play an important role in the central nervous system during neural growth and development (7). This role in central nervous system development highlights the diverse function of insulin and associated hormone signaling pathways in mediating early events of nervous system formation, independent of later metabolic and glucoregulatory functions of insulin signaling in adult organisms.

FIRST STEPS IN CENTRAL NERVOUS SYSTEM DEVELOPMENT: NEURULATION

In the initial phases of nervous system development in humans and in animals, the developing neural tube undergoes closure during a process called *neurulation* (8). During this phase, elevated proinsulin, insulin, or blockade of insulin receptor (InsR) signaling either promote cell survival or increase apoptotic events, respectively (9). The most commonly used model to study neurulation has historically been the chick embryo which only has a single preproinsulin gene similar to humans, whereas rodents express two preproinsulin genes (10). This developmental step precedes the formation of the pancreas; however, preproinsulin expression is found in early neurulation in *Xenopus laevis* (11). The source of this preproinsulin is in part due to the activity of promoting factors, such as Insulin-promoting-factor-1, Ipf1 whose expression and activity precede the formation of the pancreas but supports early detection of insulin in the dorsal foregut endoderm at the 20 somite stage in rodents (12–14). However, other studies show that neuronal subpopulations may also be able to produce and secrete insulin either via detection in subregions of the brain using *in situ* hybridization (15) or through primary neuronal cultures at various time points in early development (16, 17).

Preproinsulin is found by embryonic day 1.5–2 in chick embryos and is detected as early as embryonic day 8.5 in mice (the preproinsulin II gene, specifically) (12, 18). In this model, preproinsulin produced by prepancreatic cells was sufficient to increase cell survival in embryonic day 1.5 embryos, a time point before *prohormone convertase* **2** (PC2) is expressed. PC2 is a serine protease and is the enzyme that further processes the preproinsulin gene to result in the final mature insulin form commonly studied in adult humans and animal models (Figure 2.1). This effect on cell survival in early embryonic phases was also confirmed in cultured developing embryos treated directly with insulin (19). Further studies showed insulin action at its receptor in membrane preparations from prepancreatic embryos at the onset of neurulation, via phosphorylation of an exogenous substrate, further confirming the activity of insulin in this early developmental phase (20). Although the relative binding of radiolabeled insulin was reduced in comparison to similar hormones, the relative stimulation of phosphorylation was similar, further underscoring the role of minimal changes in insulin in the development of profound effects in the central nervous system cellular signaling cascades. The insulin receptor is found in two isoforms, InsR-A and InsR-B. In early development, a predominance of InsR-A is noted in the developing central nervous system (21). This dominant expression of InsR-A isoform in the brain exists into adulthood (22). In human-derived neurons or brain tissue, InsR-A predominates in neurons but the InsR-B isoform is highly expressed in microglia. In adult tissues, however, both InsR-A and InsR-B can be detected in mature neurons (23, 24), Thus the predominance of InsR-A during development in the brain is likely due to the high metabolic demand during development, and the shifting ratio of InsR-A/InsR-B from fetal development to adulthood may reflect the changing metabolic demand of the tissue.

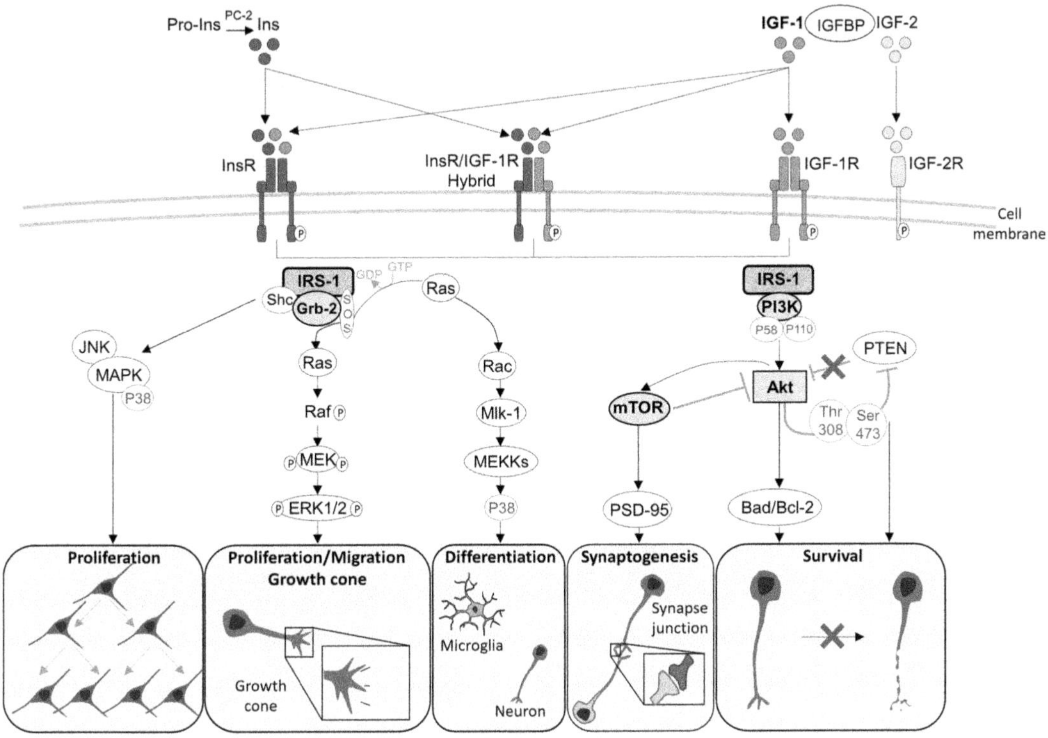

Figure 2.1 Insulin and IGF signaling pathways in brain development.

Functional *Insulin-like Growth Factor* (IGF) is also expressed early in development and plays a significant role in the proper formation of the central nervous system (20). Both isoforms of IGF, referred to as IGF-1 and IGF-2, have varying roles in development and adulthood, promoting the growth and survival of neuronal cells through their respective receptors, IGF-1R and IGF-2R and the hybrid Insulin/IGF-1 receptor. Using bovine embryos, a protective effect- of IGF-1 was noted in the early developing embryo. However, due to the anti-apoptotic effects of IGF-1, it was indicated that IGF-1 may inhibit neurulation via the down-regulation of specific genes involved in this process (25, 26). On the other hand, additional studies in vertebrates have yielded insights pertaining to IGF-2. Interestingly, IGF-2 is shown to have a high affinity and ability to signal through the IR-A isoform and points toward dynamic crosstalk of the two signaling pathways in early development (21, 27). In zebrafish, there are two co-orthologs of IGF2, *igf2a* and *igf2b*. Found primarily in and near the notochord, *igf2a* and *igf2b*, play a role in neurulation. The use of an oligomer molecule called a *morpholino* can decrease gene expression and morpholino-mediated knockdown of either *igf2a* or *igf2b* results in defects in dorsal midline development. This is highlighted by changes in the rate of segmentation, notochord undulations, ectopic fusion of the nephron primordia, defects in ventral forebrain development and increased ventral curvature.

While normal neurulation and effects of both insulin and IGF have been studied in the normal developmental context, it is critical to point out here that in states of elevated hormone signaling, defects to this phase of central nervous system formation are seen. Specifically, in conditions of elevated blood glucose of the mother, such as in gestational diabetes mellitus or generalized Type II diabetic mothers, an increase in neural tube defects is noted (28). Insulin itself cannot cross the placental barrier but the elevated levels of blood glucose of the mother are sensed by the developing fetus. This elevated blood glucose results in elevated insulin production in the fetus. As excessive transport of glucose across the placenta in this disease state is linked to neural tube defects in children, and given the above-mentioned roles of insulin and IGF signaling in the effective closure of the neural cord, this disruption in the levels of these hormones may result in these detrimental effects in early development.

EXPANDING CENTRAL NERVOUS SYSTEM CELLS: NEUROGENESIS

After the closure of the neural chord, the proliferation of neuronal cells occurs. This neurogenesis is mediated in part by insulin but has primarily been studied in the context of IGF signaling. Seminal studies have concentrated on the role of these peptides in neurogenesis in the chick embryonic retina. Further studies using *in vitro* models have confirmed the role of insulin, IGF-1 and IGF-2 as trophic factors for neurons in culture. Using cultured sympathetic neurons from 7-day-old chick embryos, insulin, IGF-1 and IGF-2 induced a significant increase in proliferation of this neuronal cell type (29). The specific action of IGF1 and IGF2 on this process was via the *Insulin-like Growth Factor Binding Proteins* (IGFBPs) rather than directly via receptor-mediated signaling. In primary fetal rat neuronal cells, IGF-1 increases the survival and expansion of neuronal and non-neuronal cell populations (30). Further, stimulation of growth of neural stem cells (NSCs) by other growth factors, such as EGF and FGF-2 requires the presence of IGF-1 (31). This effect is mediated by the phosphorylation of Akt/Bad/Bcl-2 signaling and the phosphorylation of Akt at Thr308 and Ser473 and is inhibited in the presence of excess PTEN, a protein phosphatase, as shown in Figure 2.1 (32, 33). The derivation of neural tissues from embryonic stem cells necessitates IGF-2 signaling via the IGF1-R and subsequent modulation of neural markers, SOX1, IRX3, and SIX3, further indicating a role of the IGF axis in neurogenesis and differentiation (34).

Further studies supplementing a long-lasting insulin analog to neonatal rats showed significant reductions in neuronal number in specific regions of the brain which ultimately correlated with later metabolic dysfunction (35). This reduction in neuronal number was region-specific and implicates elevated insulin signaling in site-specific neurogenesis in the neonatal brain which will be discussed later in this chapter.

NEURONAL SURVIVAL AND DEVELOPMENT

The *growth cone* is the actively extending portion of the developing neuron, providing a leading point for emerging neurites searching for the proper synaptic targets. While this is an extremely complex system, utilizing gradients of molecular cues and both presynaptic and postsynaptic signaling, the expression and activity of receptor tyrosine kinases in the developing extension are critical for this process. The IGF-1R is found localized in the developing growth cone, underscoring the necessary balance of insulin and IGF-1 hormones in development (36, 37). Enrichment specifically of the beta subunit of the IGF-1R is found at the developing growth cone in cultured hippocampal pyramidal neurons (38, 39). Additionally, this region within the developing growth cone is critical for glial cell interactions, necessary for proper brain structure organization (40). Studies using cell culture of PC12 cells overexpressing the insulin receptor also indicate that insulin itself plays a role in neurite outgrowth of cultured neuronal cells (41, 42). This action is via MAP kinase activation, where translocation of MAP kinase to the nucleus is correlated with increased expression of insulin receptors on the cell surface. Insulin acts on the cone growth also by activating the Shc/-Grb-2/MAPK signaling pathways (43).

Both IGF-1 and IGF-2 stimulate neurite outgrowth and promote increased connectivity of sympathetic, sensory, and motor neuronal populations (29). IGFs are also known to influence not only growth but also the maturation of neuronal cells. This is distinct via the promotion of dendritic and axonal growth, synaptic spine formation, general synaptogenesis, and the promotion of myelination (44). However, despite the primary role of IGF in mediated much of the cell survival phenotypes, in disease states where elevated insulin is present, such as in gestational diabetes mellitus, the action of insulin can also play a role in these processes. In rodent models, si-RNA mediated knockdown of the InsR, IGF-1R or IGF-2R leads to significant loss of cerebellar structure, showing the redundant role of this receptor family in the maintenance and health of brain regions (45). More recent studies explored this hypothesis and revealed that insulin induces neurite outgrowth by regulating SIRT1, a gene expressed principally in developing nervous systems (46, 47). These effects on outgrowth can continue from generalized projections to the level of individual synapses. Specifically in *Xenopus laevis* a role of insulin receptor signaling in synaptic density is shown (48). In addition, in the somatosensory cortex, the application of IGF-1 promotes the expression of synaptic proteins such as PSD-95 and increases ribosomal protein S6 phosphorylation and overall spine density (49).

Many aspects of cell survival are thought to be mediated by IGF-1 of neuronal and non-neuronal cells of the central nervous system. In culture, IGF-1 promotes the survival of pro-oligodendrocytes through activation of PI3K/mTOR/Akt and MEK/ERK pathways (Figure 2.1) and combined inhibition of glutamate-mediated caspase3 activation (50, 51). This action of IGF-1 is not limited to oligodendrocytes, in fact, studies show the increased survival of a number of neuronal populations including neural progenitor-like cells (52).

CELLULAR DIFFERENTIATION

In refined studies closely assessing embryonic development of specific subtypes of neurons in the brain, effects of insulin and IGF signaling can be seen. Using cultured avian embryos, a restricted time window occurring at embryonic day 3 was shown to lead to significant increases in insulin and IGF specifically in catecholaminergic neurons (53). This effect is not noted if the cells are cultured at embryonic day 2.5, underscoring the precise timing regulation of these hormones in the development and differentiation of neuronal cell types. Further studies in *Drosophila* analyzed this temporal effect of insulin signaling via the Ras/Rac/MEKK pathway (Figure 2.1) on the differentiation of neuronal subtypes (48). Through activation of these pathways by loss of TSC1 or PTEN, which act as repressor elements, or inhibition via loss of InsR expression, or downstream signaling components results in precocious or delayed differentiation respectively of neuronal cells while having no effect on the ultimate cell fate (54). Timing of insulin and related hormone signaling is critical for normal developmental trajectories and cascades of subsequent cues for central nervous system formation.

As described above, insulin and IGF are key actors in a myriad of important neurodevelopmental events. Table 2.1 summarizes the action of insulin and IGF-1/IGF-2 signalization pathways that promote neurogenesis, growth, and survival.

BRAIN REGION SPECIFIC EFFECTS IN DEVELOPMENT

While the role of insulin and IGF and their respective receptors have been studied across developmental processes, further studies pertaining to the roles of these factors have centered on specific subregions of the brain. The function of insulin, IGF, and their respective receptors on local development and function of neuronal circuits have been studied. Whole brain deletion of either IGF-1R or IR results in profound effects in animals. Specifically, using Nestin-Cre, a Cre recombinase targeted only to neuronal and glia precursors, IGF-1R deletion results in decreased bodyweight, subsequent decreases in growth hormone (GH) levels and a lengthening of the life span (55). Deletion of the InsR using the same Nestin-Cre mediated approach resulted in no change in brain weight, development or overall generalized morphology but did result in peripheral

TABLE 2.1
Summary of the Effect of Insulin and IGF on Brain Development Mechanisms

Neurodevelopmental Events	Insulin	IGF-1/IGF-2
Neurulation	Cell survival	Apoptotic signal
Neurogenesis	Neuronal proliferation	Neuronal proliferation in presence of IGFBP Phosphorylation of Akt/Bad/Blc-2 pathway and blocking PTEN signaling
Neuronal survival and development	Development of growth cones via InsR through MAP kinase and/or Shc/Grb-2/MAPK signaling pathways	Dendritic and axonal growth Formation of synaptic spine Synaptogenesis Myelinization
Cell differentiation	Ras/Rac/MEKK signaling pathway stimulation	Apoptotic signal Proliferation

changes in metabolic parameters (56). To understand the effects of these receptors in development, key regions involved in controlling metabolism, memory, and other functions, which are targeted by insulin and IGF signaling throughout their development, have been further studied in depth. Below, the primary regions targeted are described.

THE CORTEX

This integrating center of the brain is involved in higher cognitive functions such as attention and memory. It is well known today that IGF-1 and the two InsR isoforms, InsR-A and InsR-B, are expressed in the cortex (23). Niblock and colleagues demonstrated that IGF-1-treated brain slices obtained from postnatal day 10 (P10) rats stimulate dendritic growth in the primary somatosensory cortex and that IGFs influence dendritic branching and total dendritic length (57). Supporting this finding, IGF-1 brain-specific deletion in mice induces a disruption in lipid and microtubule metabolism, leading to impaired neuronal somatic and dendritic growth in the frontoparietal cortex (58). These results show that insulin/IGF receptor signaling is essential for normal cortical dendritic growth during the postnatal brain development period. However, due to the dramatic phenotype of IGF-1 knockouts and almost no gross anatomical phenotype in InsR knockout models, changes to neuronal outgrowth are likely driven by altered IGF binding. Throughout the development, dynamic changes in both IGF-1 and IGF-2 are noted, and varying expression in different subregions of the CNS suggests varying roles of both factors in development (59). Cortex specific deletion of both *Igf1r* and *InsR* using an Emx-Cre mouse model, which deletes floxed *Insr* and *Igf1r* at embryonic day 10.5, demonstrates that signaling via these receptors in cortical regions is critical for the axonal outgrowth of these neuronal subsets but is not required for proper migration and polarization of neurons, albeit in very defined subregions of the cortex (60).

In humans, gestational diabetes mellitus has been correlated to deficits in fine and gross motor functions, low verbal IQ, language impairment, inattention and hyperactivity, and poor general cognitive function in offspring (61–63). For this reason, many studies focus on understanding the impact of maternal diabetes or obesity on the development of the cortex.

THE HIPPOCAMPUS

In this region, insulin signaling plays a modulatory role in learning and memory processing. It has been shown that insulin is necessary for the postnatal maturation of glutamatergic and GABAergic synapses of the hippocampus and the mediation of interactions of different neurotransmission systems during neuronal activation (64, 65).

As in the cortex, the hippocampus also expresses InsR-A and B and IGF-1R (23). Popken and colleagues found that the initiated overexpression of IGF-1 stimulates the overall growth of the brain and increases neuronal number beginning prenatally and persisting in postnatal animals compared to wild-type littermates (66). In mice, the dentate gyrus is the most affected hippocampal region by either an overexpression of IGF-1 (44, 66, 67) or when IGF-1 action or expression is blunted (68–70). Zhang and colleagues bring further evidence showing that the actions of IGF-1 in the development of the hippocampus are complex and highly related (71). Further, dual deletion of *Igf1r* and *Insr* only in a subset of cortical neurons is sufficient to dramatically decrease hippocampal formation (60).

Dou and colleagues showed that an induced insulin deficiency by the injection of *streptozotocin*, a pancreatic beta-cell toxin, induced minor memory impairments that might be compensated by IGF-1 signaling pathways (72). Interestingly, recent studies revealed that maternal high-fat diet-induced insulin resistance impairs synaptic plasticity, learning, and memory in mothers (73, 74) and their offspring (73). Insulin treatment in the streptozotocin model is sufficient to protect neurogenesis, neuronal maturation (as indicated by a reduction in DCX) and neuronal survival in the dentate gyrus of the hippocampus in rats (75). Moreover, it has been suggested that an altered insulin signaling might play a key role in the appearance of Alzheimer's disease (76–78). This indicates that insulin is also implicated in maintaining "healthy" aging. This evidence reveals that insulin is a major factor in the modification of synaptic connections required for the development and maintenance of higher brain functions such as learning and memory. Collectively, these data indicate that insulin and IGF-1 through IGF-1R play a critical role in the development of the hippocampus during the early postnatal mice.

THE HYPOTHALAMUS

Within the brain, a region called the hypothalamus contains neurons which have been studied extensively for their role in energy homeostasis. Interestingly, in this region of the brain the *median eminence*, a circumventricular organ, is known to express high levels of the IGF-1R (79). Further, expression of the InsR on specific subpopulations of metabolic neurons located here is known to correlate to their neuronal projection development (80).

Maternal obesity and gestational diabetes are both associated with hyperinsulinemia and insulin resistance. In rodents, early studies showed that the injection of insulin between 15 and 20 days of gestation caused obesity in adult offspring and an increase of noradrenaline and noradrenergic neurons in the paraventricular nucleus of the hypothalamus (PVH) (81, 82). Moreover, an induced maternal insulin deficiency caused by the injection of streptozotocin results in insulin resistance and increased food intake in adult offspring that is associated with a marked reduction of arcuate nucleus specific projections to the PVH (83). The same observations were reported when pups were directly injected with streptozotocin (83). These findings indicate that maternal insulin plays a role in the development of intra-hypothalamic neuronal projections and the metabolic programming in offspring.

Interestingly, it has been revealed in rats that the bilateral injection of insulin at P2 and P8 induce morphological impairment of specific hypothalamic nuclei including the arcuate nucleus and the ventromedial nucleus of the hypothalamus in addition to lifelong metabolic disorders (84). The same phenotype was observed in rats that received chronic subcutaneous injections of insulin from P8 to P11, as well as a reduced volume of the VMH in adulthood (35). It is important to mention that because it is not possible to dissociate insulin injection from hypoglycemia, the role of insulin in the development of hypothalamic nuclei and neuronal projections *in vivo* is less studied than other metabolic hormones such as leptin. However, evidence from Vogt and colleagues using a genetically modified mouse line that lacks expression of the InsR specifically in the pro-opiomelanocortin (POMC)-expressing neurons of the arcuate nucleus in the hypothalamus results in a restoration of a subset of neuronal projections, indicating a negative role of elevated InsR and subsequent InsR signaling in neuronal development. With this model, they showed that the absence of InsR alters the normal development of intra-hypothalamic POMC projections to the PVH as well as in the dorsomedial nulceus of the hypothalamus (DMH) and lateral hypothalamus (LH) in offspring of mothers exposed to a high-fat diet during the lactation period (80).

INSULIN DEPENDENT DISEASES AND BRAIN DEVELOPMENT EFFECTS

In humans, the third trimester of pregnancy is known to be a critical period for brain development. Interestingly, insulin levels have been shown to increase during the third trimester of gestation in humans (85). In mouse models, a rise in insulin and expression of the InsR/IGF-1R is noted in the postnatal period (86, 87). This fluctuation of insulin levels in circulation matches an increase of InsR expression in the brain during the neuronal growth period reported in humans and rabbits (88, 89). These findings support the involvement of insulin/IGF signaling in brain growth and development.

GESTATIONAL DIABETES MELLITUS AND DECREASED MATERNAL INSULIN SENSITIVITY

In recent years, much of the focus on insulin signaling in brain development has centered on the role of excess insulin exposure. Specifically, this has centered on insulin resistance and hyperglycemia, hallmarks of gestational diabetes mellitus (90). While insulin itself cannot cross the placental barrier, the maternal inability to regulate blood glucose levels due to insulin resistance, results in an exposure of the developing fetus to excessive blood glucose, resulting in fetal insulin overproduction. This consequently results in babies born large for their gestational age, due to excessive growth of the fetus. Furthermore, consequent immediate hypoglycemia upon birth is evident due to the source of elevated blood glucose from the mother no longer being present but elevated fetal insulin level is still able to exert its blood glucose-lowering effects. However, the effects of gestational diabetes mellitus go further than those seen in just the growth of the fetus and circulating insulin and blood glucose levels. At the level of the central nervous system, gestational diabetes mellitus results in decreased fetal postprandial brain responses after an oral glucose tolerance test to the mother (91, 92). In animal models of overnutrition mediated hyperglycemia as well as gestational diabetes mellitus, a number of dramatic effects on neuronal properties are being studied in more detail. From effects on specific neuronal populations in the hypothalamus in the context of maternal overnutrition (80, 83, 93) to induction of insulin insufficiency using the streptozotocin model, changes in brain development and resulting overall offspring metabolism and behavior are affected by changes to insulin.

MALNUTRITION AND MATERNAL UNDERNUTRITION

While gestational diabetes mellitus often results in chronic hyperglycemia and subsequent overproduction of insulin by the fetus, states of chronic hypoglycemia can also result in dramatic effects to brain development. In conditions of chronic undernutrition, levels of IGF are significantly reduced in the developing neonate. Within the hypothalamus this decrease in circulating IGF is implicated in a disturbance to the development of the *growth hormone-releasing hormone-expressing* (GHRH) neurons, resulting in decreased growth hormone section from the pituitary gland and overall reduced development of the offspring (79).

Maternal malnutrition affects the brain growth of the fetus (94, 95). As mentioned previously, IGF-1 is implicated in normal brain development and plays a role in neurogenesis and neuronal survival through different kinase signaling pathways such as PI3K/Akt (33, 96, 97). In the context of maternal undernutrition, IGF-1 plasma levels are affected in both mothers and the fetus (98). Moreover, levels of IGFBP-1, a transporter of IGF-1 are increased in the plasma (98). These changes in circulating levels of IGFBP-1 (increased) and IGF-1 (decreased) may act in concert to reduce the availability of IGF-1 thus limiting fetal brain cell growth during maternal nutrient restriction (98). Overall, these studies demonstrated that IGF-1 and its transporter, IGFBP-1, are key modulators of the fetal brain cells growth in a context of maternal starvation.

POLYCYSTIC OVARIAN SYNDROME (PCOS)

PCOS is one of the most common disorders of premenopausal women. Endocrine and metabolic disorders are characterized by a combination of symptoms and signs of androgen excess and ovarian dysfunctions (99, 100). PCOS is frequently associated with obesity, diabetes, and cardiovascular risk factors (101–103). Recently, maternal hyperandrogenemia in PCOS has been involved in neurodevelopment disorders, such as autism spectrum disorder (ASD) and attention deficit hyperactivity disorder (ADHD) (100, 104–106). Interestingly, PCOS patients have shown an increased level of IGF-1 due to an insulin-induced decrease in the production of the binding proteins, IGFBP-1 and IGFBP-3 (107–109). Although there are growing evidence correlating neurodevelopmental disorders to PCOS and to dysfunctional IGF-1 signaling pathway, it is not shown yet the causality of IGF-1 signaling in the presence of neurodevelopmental disorders in offspring from mothers with PCOS.

CONCLUDING REMARKS

Generally, precisely timed regulation of levels of circulating hormones is necessary to result in the proper formation of the central nervous system. Perturbations through maternal environment, or loss of gene function within the insulin and IGF family of genes, can result in detrimental effects to neuronal circuitry. Insulin and IGF play a role in a variety of neurodevelopmental processes starting already at neurulation and continuing through the life of the organism. Through understanding the role of these hormones, their role in development can be seen as dramatically different than their function in the adult organism, further signifying the complex nature of hormonal signaling and interactions with the developing brain.

REFERENCES

1. N. Haugaard, M. Vaughan, E. S. Haugaard, W. C. Stadie, Studies of radioactive injected labeled insulin. *J Biol Chem* **208**, 549–563 (1954).
2. W. Landauer, Rumplessness of chicken embryos produced by the injection of insulin and other chemicals. *J Exp Zool*, 65–77 (1945).
3. P. K. Duraiswami, Insulin-induced skeletal abnormalities in developing chickens. *Br Med J* **2**, 384–390 (1950).
4. R. U. Margolis, N. Altszuler, Insulin in the cerebrospinal fluid. *Nature* **215**, 1375–1376 (1967).
5. A. Dorn et al., Insulin- and glucagonlike peptides in the brain. *Anat Rec* **207**, 69–77 (1983).

6. S. G. Hasselbalch et al., No effect of insulin on glucose blood-brain barrier transport and cerebral metabolism in humans. *Diabetes* **48**, 1915–1921 (1999).
7. N. R. Bhat, Insulin dependent neurite outgrowth in cultured embryonic mouse brain cells. *Brain Res* **313**, 315–318 (1983).
8. Y. Sakai, Neurulation in the mouse: manner and timing of neural tube closure. *Anat Rec* **223**, 194–203 (1989).
9. C. Hernandez-Sanchez, E. Rubio, J. Serna, E. J. de la Rosa, F. de Pablo, Unprocessed proinsulin promotes cell survival during neurulation in the chick embryo. *Diabetes* **51**, 770–777 (2002).
10. M. B. Soares et al., RNA-mediated gene duplication: The rat preproinsulin I gene is a functional retroposon. *Mol Cell Biol* **5**, 2090–2103 (1985).
11. A. R. Shuldiner, F. de Pablo, C. A. Moore, J. Roth, Two nonallelic insulin genes in Xenopus laevis are expressed differentially during neurulation in prepancreatic embryos. *Proc Natl Acad Sci U S A* **88**, 7679–7683 (1991).
12. G. K. Gittes, W. J. Rutter, Onset of cell-specific gene expression in the developing mouse pancreas. *Proc Natl Acad Sci U S A* **89**, 1128–1132 (1992).
13. J. Jonsson, L. Carlsson, T. Edlund, H. Edlund, Insulin-promoter-factor 1 is required for pancreas development in mice. *Nature* **371**, 606–609 (1994).
14. G. Teitelman, S. Alpert, J. M. Polak, A. Martinez, D. Hanahan, Precursor cells of mouse endocrine pancreas coexpress insulin, glucagon and the neuronal proteins tyrosine hydroxylase and neuropeptide Y, but not pancreatic polypeptide. *Development* **118**, 1031–1039 (1993).
15. S. U. Devaskar et al., Insulin gene expression and insulin synthesis in mammalian neuronal cells. *J Biol Chem* **269**, 8445–8454 (1994).
16. N. P. Birch, D. L. Christie, A. G. Renwick, Proinsulin-like material in mouse foetal brain cell cultures. *FEBS Lett* **168**, 299–302 (1984).
17. R. Schechter et al., Immunohistochemical and in situ hybridization study of an insulin-like substance in fetal neuron cell cultures. *Brain Res* **636**, 9–27 (1994).
18. L. Deltour et al., Differential expression of the two nonallelic proinsulin genes in the developing mouse embryo. *Proc Natl Acad Sci U S A* **90**, 527–531 (1993).
19. A. V. Morales, J. Serna, C. Alarcón, E. J. de la Rosa, F. de Pablo, Role of prepancreatic (pro)insulin and the insulin receptor in prevention of embryonic apoptosis. *Endocrinology* **138**, 3967–3975 (1997).
20. M. Girbau, L. Bassas, J. Alemany, F. de Pablo, In situ autoradiography and ligand-dependent tyrosine kinase activity reveal insulin receptors and insulin-like growth factor I receptors in prepancreatic chicken embryos. *Proc Natl Acad Sci U S A* **86**, 5868–5872 (1989).
21. F. Frasca et al., Insulin receptor isoform A, a newly recognized, high-affinity insulin-like growth factor II receptor in fetal and cancer cells. *Mol Cell Biol* **19**, 3278–3288 (1999).
22. I. Pomytkin et al., Insulin receptor in the brain: Mechanisms of activation and the role in the CNS pathology and treatment. *CNS Neurosci Ther* **24**, 763–774 (2018).
23. A. Kleinridders, H. A. Ferris, W. Cai, C. R. Kahn, Insulin action in brain regulates systemic metabolism and brain function. *Diabetes* **63**, 2232–2243 (2014).
24. B. Spencer, L. Rank, J. Metcalf, P. Desplats, Identification of insulin receptor splice variant B in neurons by in situ detection in human brain samples. *Sci Rep* **8**, 4070 (2018).
25. D. Sacks et al., Multisociety consensus quality improvement revised consensus statement for endovascular therapy of acute ischemic stroke. *Int J Stroke* **13**, 612–632 (2018).
26. A. Q. Bonilla et al., Developmental changes in thermoprotective actions of insulin-like growth factor-1 on the preimplantation bovine embryo. *Mol Cell Endocrinol* **332**, 170–179 (2011).
27. A. N. Ziegler et al., IGF-II promotes stemness of neural restricted precursors. *Stem Cells* **30**, 1265–1276 (2012).
28. M. F. Greene, Spontaneous abortions and major malformations in women with diabetes mellitus. *Semin Reprod Endocrinol* **17**, 127–136 (1999).
29. K. Zackenfels, R. W. Oppenheim, H. Rohrer, Evidence for an important role of IGF-I and IGF-II for the early development of chick sympathetic neurons. *Neuron* **14**, 731–741 (1995).
30. I. Torres-Aleman, F. Naftolin, R. J. Robbins, Trophic effects of insulin-like growth factor-I on fetal rat hypothalamic cells in culture. *Neuroscience* **35**, 601–608 (1990).
31. Y. Arsenijevic, S. Weiss, B. Schneider, P. Aebischer, Insulin-like growth factor-I is necessary for neural stem cell proliferation and demonstrates distinct actions of epidermal growth factor and fibroblast growth factor-2. *J Neurosci* **21**, 7194–7202 (2001).

32. S. R. Datta et al., Akt phosphorylation of BAD couples survival signals to the cell-intrinsic death machinery. *Cell* **91**, 231–241 (1997).
33. G. Otaegi et al., Modulation of the PI 3-kinase-Akt signalling pathway by IGF-I and PTEN regulates the differentiation of neural stem/precursor cells. *J Cell Sci* **119**, 2739–2748 (2006).
34. N. Takata, E. Sakakura, Y. Sasai, IGF-2/IGF-1R signaling has distinct effects on Sox1, Irx3, and Six3 expressions during ES cell derived-neuroectoderm development in vitro. *In Vitro Cell Dev Biol Anim* **52**, 607–615 (2016).
35. A. Plagemann et al., Hypothalamic insulin and neuropeptide Y in the offspring of gestational diabetic mother rats. *Neuroreport* **9**, 4069–4073 (1998).
36. S. Quiroga, Q. Santiago, R. S. Garofalo, S. G. Robert, H. P. Karl, Insulin-like growth factor I receptors of fetal brain are enriched in nerve growth cones and contain a beta-subunit variant. *Proc Natl Acad Sci USA* (1995).
37. B. Kim, H. L. Cheng, B. Margolis, E. L. Feldman, Insulin receptor substrate 2 and Shc play different roles in insulin-like growth factor I signaling. *J Biol Chem* **273**, 34543–34550 (1998).
38. F. Mascotti, A. Cáceres, K. H. Pfenninger, S. Quiroga, Expression and distribution of IGF-1 receptors containing a beta-subunit variant (betagc) in developing neurons. *J Neurosci* **17**, 1447–1459 (1997).
39. S. Quiroga, R. S. Garofalo, K. H. Pfenninger, Insulin-like growth factor I receptors of fetal brain are enriched in nerve growth cones and contain a beta-subunit variant. *Proc Natl Acad Sci U S A* **92**, 4309–4312 (1995).
40. M. J. Rigby, T. M. Gomez, L. Puglielli, Glial cell-axonal growth cone interactions in neurodevelopment and regeneration. *Front Neurosci* **14**, 203 (2020).
41. I. Dikic, J. Schlessinger, I. Lax, PC12 cells overexpressing the insulin receptor undergo insulin-dependent neuronal differentiation. *Curr Biol* **4**, 702–708 (1994).
42. L. A. Greene, A. S. Tischler, Establishment of a noradrenergic clonal line of rat adrenal pheochromocytoma cells which respond to nerve growth factor. *Proc Natl Acad Sci U S A* **73**, 2424–2428 (1976).
43. M. G. Myers, Jr. et al., Role of IRS-1-GRB-2 complexes in insulin signaling. *Mol Cell Biol* **14**, 3577–3587 (1994).
44. J. O'Kusky, P. Ye, Neurodevelopmental effects of insulin-like growth factor signaling. *Front Neuroendocrinol* **33**, 230–251 (2012).
45. S. M. De la Monte, M. Tong, N. Bowling, P. Moskal, si-RNA inhibition of brain insulin or insulin-like growth factor receptors causes developmental cerebellar abnormalities: Relevance to fetal alcohol spectrum disorder. *Mol Brain* **4**, 13 (2011).
46. J. Liu, P. Speder, A. H. Brand, Control of brain development and homeostasis by local and systemic insulin signalling. *Diabetes Obes Metab* **16** (Suppl 1), 16–20 (2014).
47. A. Niyomchan et al., Insulin attenuates arsenic-induced neurite outgrowth impairments by activating the PI3K/Akt/SIRT1 signaling pathway. *Toxicol Lett* **236**, 138–144 (2015).
48. S. L. Chiu, H. T. Cline, Insulin receptor signaling in the development of neuronal structure and function. *Neural Dev* **5**, 7 (2010).
49. G. Della Sala et al., Dendritic spine instability in a mouse model of CDKL5 disorder is rescued by insulin-like growth factor 1. *Biol Psychiatry* **80**, 302–311 (2016).
50. J. K. Ness, N. E. Mitchell, T. L. Wood, IGF-I and NT-3 signaling pathways in developing oligodendrocytes: differential regulation and activation of receptors and the downstream effector Akt. *Dev Neurosci* **24**, 437–445 (2002).
51. O. Bibollet-Bahena, G. Almazan, IGF-1-stimulated protein synthesis in oligodendrocyte progenitors requires PI3K/mTOR/Akt and MEK/ERK pathways. *J Neurochem* **109**, 1440–1451 (2009).
52. T. J. Huat et al., IGF-1 enhances cell proliferation and survival during early differentiation of mesenchymal stem cells to neural progenitor-like cells. *BMC Neurosci* **15**, 91 (2014).
53. V. Nataf, S. Monier, Effect of insulin and insulin-like growth factor I on the expression of the catecholaminergic phenotype by neural crest cells. *Brain Res Dev Brain Res* **69**, 59–66 (1992).
54. J. M. Bateman, H. McNeill, Temporal control of differentiation by the insulin receptor/tor pathway in Drosophila. *Cell* **119**, 87–96 (2004).
55. L. Kappeler et al., Brain IGF-1 receptors control mammalian growth and lifespan through a neuroendocrine mechanism. *PLoS Biol* **6**, e254 (2008).
56. J. C. Brüning et al., Role of brain insulin receptor in control of body weight and reproduction. *Science* **289**, 2122–2125 (2000).
57. M. M. Niblock, J. K. Brunso-Bechtold, D. R. Riddle, Insulin-like growth factor I stimulates dendritic growth in primary somatosensory cortex. *J Neurosci* **20**, 4165–4176 (2000).

58. C. M. Cheng et al., Insulin-like growth factor 1 is essential for normal dendritic growth. *J Neurosci Res* **73**, 1–9 (2003).
59. P. Rotwein, S. K. Burgess, J. D. Milbrandt, J. E. Krause, Differential expression of insulin-like growth factor genes in rat central nervous system. *Proc Natl Acad Sci U S A* **85**, 265–269 (1988).
60. J. Jin, P. Ravindran, D. Di Meo, A. W. Püschel, Igf1R/InsR function is required for axon extension and corpus callosum formation. *PLoS One* **14**, e0219362 (2019).
61. T. D. Clausen et al., Cognitive function in adult offspring of women with Type 1 diabetes. *Diabet Med* **28**, 838–844 (2011).
62. G. Dionne, M. Boivin, J. R. Seguin, D. Perusse, R. E. Tremblay, Gestational diabetes hinders language development in offspring. *Pediatrics* **122**, e1073–1079 (2008).
63. Y. H. Jing et al., Retardation of fetal dendritic development induced by gestational hyperglycemia is associated with brain insulin/IGF-I signals. *Int J Dev Neurosci* **37**, 15–20 (2014).
64. L. P. van der Heide, A. Kamal, A. Artola, W. H. Gispen, G. M. Ramakers, Insulin modulates hippocampal activity-dependent synaptic plasticity in a N-methyl-d-aspartate receptor and phosphatidyl-inositol-3-kinase-dependent manner. *J Neurochem* **94**, 1158–1166 (2005).
65. W. Q. Zhao, H. Chen, M. J. Quon, D. L. Alkon, Insulin and the insulin receptor in experimental models of learning and memory. *Eur J Pharmacol* **490**, 71–81 (2004).
66. G. J. Popken et al., In vivo effects of insulin-like growth factor-I (IGF-I) on prenatal and early postnatal development of the central nervous system. *Eur J Neurosci* **19**, 2056–2068 (2004).
67. A. J. D'Ercole, P. Ye, J. R. O'Kusky, Mutant mouse models of insulin-like growth factor actions in the central nervous system. *Neuropeptides* **36**, 209–220 (2002).
68. K. D. Beck, L. Powell-Braxton, H. R. Widmer, J. Valverde, F. Hefti, Igf1 gene disruption results in reduced brain size, CNS hypomyelination, and loss of hippocampal granule and striatal parvalbumin-containing neurons. *Neuron* **14**, 717–730 (1995).
69. A. J. D'Ercole et al., Brain growth retardation due to the expression of human insulin like growth factor binding protein-1 in transgenic mice: an in vivo model for the analysis of igf function in the brain. *Brain Res Dev Brain Res* **82**, 213–222 (1994).
70. W. Ni, K. Rajkumar, J. I. Nagy, L. J. Murphy, Impaired brain development and reduced astrocyte response to injury in transgenic mice expressing IGF binding protein-1. *Brain Res* **769**, 97–107 (1997).
71. J. Zhang, B. M. Moats-Staats, P. Ye, A. J. D'Ercole, Expression of insulin-like growth factor system genes during the early postnatal neurogenesis in the mouse hippocampus. *J Neurosci Res* **85**, 1618–1627 (2007).
72. J. T. Dou, M. Chen, F. Dufour, D. L. Alkon, W. Q. Zhao, Insulin receptor signaling in long-term memory consolidation following spatial learning. *Learn Mem* **12**, 646–655 (2005).
73. S. Fusco et al., Maternal insulin resistance multigenerationally impairs synaptic plasticity and memory via gametic mechanisms. *Nat Commun* **10**, 4799 (2019).
74. B. Lizarbe, A. F. Soares, S. Larsson, J. M. N. Duarte, Neurochemical modifications in the hippocampus, cortex and hypothalamus of mice exposed to long-term high-fat diet. *Front Neurosci* **12**, 985 (2018).
75. A. Sadeghi et al., The effects of maternal diabetes and insulin treatment on neurogenesis in the developing hippocampus of male rats. *J Chem Neuroanat* **91**, 27–34 (2018).
76. T. R. Bomfim et al., An anti-diabetes agent protects the mouse brain from defective insulin signaling caused by Alzheimer's disease- associated Abeta oligomers. *J Clin Invest* **122**, 1339–1353 (2012).
77. S. Gabbouj et al., Altered insulin signaling in Alzheimer's disease brain - special emphasis on PI3K-Akt pathway. *Front Neurosci* **13**, 629 (2019).
78. M. Stanley, S. L. Macauley, D. M. Holtzman, Changes in insulin and insulin signaling in Alzheimer's disease: cause or consequence? *J Exp Med* **213**, 1375–1385 (2016).
79. L. Decourtye et al., IGF-1 induces GHRH neuronal axon elongation during early postnatal life in mice. *PLoS One* **12**, e0170083 (2017).
80. M. C. Vogt et al., Neonatal insulin action impairs hypothalamic neurocircuit formation in response to maternal high-fat feeding. *Cell* **156**, 495–509 (2014).
81. A. P. Jones, D. H. Olster, B. States, Maternal insulin manipulations in rats organize body weight and noradrenergic innervation of the hypothalamus in gonadally intact male offspring. *Brain Res Dev Brain Res* **97**, 16–21 (1996).

82. J. I. Jones, D. R. Clemmons, Insulin-like growth factors and their binding proteins: Biological actions. *Endocr Rev* **16**, 3–34 (1995).
83. S. M. Steculorum, S. G. Bouret, Maternal diabetes compromises the organization of hypothalamic feeding circuits and impairs leptin sensitivity in offspring. *Endocrinology* **152**, 4171–4179 (2011).
84. A. Plagemann, I. Heidrich, F. Götz, W. Rohde, G. Dörner, Lifelong enhanced diabetes susceptibility and obesity after temporary. *Exp Clin Endocrinol* **99**, 91–95 (1992).
85. R. Schechter et al., Developmental regulation of insulin in the mammalian central nervous system. *Brain Res* **582**, 27–37 (1992).
86. R. S. Ahima, D. Prabakaran, J. S. Flier, Postnatal leptin surge and regulation of circadian rhythm of leptin by feeding. Implications for energy homeostasis and neuroendocrine function. *J Clin Invest* **101**, 1020–1027 (1998).
87. A. Baron-Van Evercooren, C. Olichon-Berthe, A. Kowalski, G. Visciano, E. Van Obberghen, Expression of IGF-I and insulin receptor genes in the rat central nervous system: A developmental, regional, and cellular analysis. *J Neurosci Res* **28**, 244–253 (1991).
88. R. Schechter et al., Developmental regulation of insulin in the mammalian central nervous system. *Brain Res* **582**, 27–37 (1992).
89. S. U. Devaskar, N. Holekamp, L. Karycki, P. Devaskar, Ontogenesis of the InsR in the rabbit brain. *Hormones* **24**, 319–327 (1986).
90. L. Dearden, S. G. Bouret, S. E. Ozanne, Nutritional and developmental programming effects of insulin. *J Neuroendocrinol* **33**, e12933 (2021).
91. K. Linder et al., Gestational diabetes impairs human fetal postprandial brain activity. *J Clin Endocrinol Metab* **100**, 4029–4036 (2015).
92. K. Linder et al., Maternal insulin sensitivity is associated with oral glucose-induced changes in fetal brain activity. *Diabetologia* **57**, 1192–1198 (2014).
93. L. Dearden, S. E. Ozanne, Early life origins of metabolic disease: Developmental programming of hypothalamic pathways controlling energy homeostasis. *Front Neuroendocrinol* **39**, 3–16 (2015).
94. F. H. Bloomfield et al., Brief undernutrition in late-gestation sheep programs the hypothalamic-pituitary-adrenal axis in adult offspring. *Endocrinology* **144**, 2933–2940 (2003).
95. A. G. Edlow, F. Guedj, D. Sverdlov, J. L. A. Pennings, D. W. Bianchi, Significant effects of maternal diet during pregnancy on the murine fetal brain transcriptome and offspring behavior. *Front Neurosci* **13**, 1335 (2019).
96. J. Costales, A. Kolevzon, The therapeutic potential of insulin-like growth factor-1 in central nervous system disorders. *Neurosci Biobehav Rev* **63**, 207–222 (2016).
97. T. F. Franke, D. R. Kaplan, L. C. Cantley, PI3K: Downstream AKTion blocks apoptosis. *Cell* **88**, 435–437 (1997).
98. G. Shambaugh, G. J. Radosevich, T. Unterman, Insulin-like growth factor-i and binding protein-1 can modulate fetal brain cell growth during maternal starvation. *Ann N Y Acad Sci* **18**, 695–703 (1993).
99. H. F. Escobar-Morreale, Polycystic ovary syndrome: definition, aetiology, diagnosis and treatment. *Nat Rev Endocrinol* **14**, 270–284 (2018).
100. R. S. Rotem et al., Associations of maternal androgen-related conditions with risk of autism spectrum disorder in progeny and mediation by cardiovascular, metabolic, and fertility factors. *Am J Epidemiol* **190**, 600–610 (2021).
101. E. Diamanti-Kandarakis, A. Dunaif, Insulin resistance and the polycystic ovary syndrome revisited: an update on mechanisms and implications. *Endocr Rev* **33**, 981–1030 (2012).
102. D. A. Ehrmann et al., Effects of race and family history of type 2 diabetes on metabolic status of women with polycystic ovary syndrome. *J Clin Endocrinol Metab* **90**, 66–71 (2005).
103. R. S. Legro, A. R. Kunselman, W. C. Dodson, A. Dunaif, Prevalence and predictors of risk for type 2 diabetes mellitus and impaired glucose tolerance in polycystic ovary syndrome: A prospective, controlled study in 254 affected women. *J Clin Endocrinol Metab* **84**, 165–169 (1999).
104. T. R. Berni, C. L. Morgan, E. R. Berni, D. A. Rees, Polycystic ovary syndrome is associated with adverse mental health and neurodevelopmental outcomes. *J Clin Endocrinol Metab* **103**, 2116–2125 (2018).
105. K. Kosidou et al., Maternal polycystic ovary syndrome and the risk of autism spectrum disorders in the offspring: A population-based nationwide study in Sweden. *Mol Psychiatry* **21**, 1441–1448 (2016).
106. K. Kosidou et al., Maternal polycystic ovary syndrome and risk for attention-deficit/hyperactivity disorder in the offspring. *Biol Psychiatry* **82**, 651–659 (2017).

107. M. M. Chi, A. L. Schlein, K. H. Moley, High insulin-like growth factor 1 (IGF-1) and insulin concentrations trigger apoptosis in the mouse blastocyst via down-regulation of the IGF-1 receptor. *Endocrinology* **141**, 4784–4792 (2000).
108. R. Homburg, R. Orvieto, I. Bar-Hava, Z. Ben-Rafael, Serum levels of insulin-like growth factor-1, IGF binding protein-1 and insulin and the response to human menopausal gonadotrophins in women with polycystic ovary syndrome. *Hum Reprod* **11**, 716–719 (1996).
109. X. K. Wu et al., Selective ovary resistance to insulin signaling in women with polycystic ovary syndrome. *Fertil Steril* **80**, 954–965 (2003).

CHAPTER THREE

Insulin Signaling Modulates Neuronal Metabolism

Qian Huang
Jialin Fu
Kelly Anne Borges and Weikang Cai

CONTENTS

INTRODUCTION

Insulin, a pancreatic hormone, is essential for regulating glucose metabolism and energy homeostasis by acting on multiple classic insulin-sensitive peripheral tissues (1–6). While the brain is traditionally considered insulin-insensitive due to the lack of acute insulin-stimulated glucose uptake, evidence from the research over the last 40 years has demonstrated that insulin signaling plays a profound role in brain metabolism and higher-order neural functions. This chapter focuses on how insulin signaling regulates neuronal metabolism and discusses the impact of impaired neuronal insulin signaling on normal neural metabolism and functions.

OVERVIEW OF NEURONAL METABOLISM AND BIOENERGETICS

The brain is a highly metabolically active organ and almost exclusively uses glucose as its fuel. Despite only accounting for ~2% of our total body weight, the brain consumes around 20% of the

DOI: 10.1201/9781003079927-3

total energy produced by glucose in the body (7). In the setting of acute or recurrent hypoglycemia, insufficient glucose supply to the brain can lead to serious consequences on cerebral functions, including impaired concentration, focal neurological deficits, seizures, and even death (8, 9). Importantly, such deterioration cannot be reversed by substituting any other alternative substrates in humans or animals, except mannose which can cross the blood-brain barrier (BBB) and subsequently be converted to fructose-6-phosphate to converge into glucose metabolism pathways.

Among all the different types of neuronal and glial cells in the brain, neurons are arguably the most energy-demanding cells. Thus, neurons consume 70%–80% of the energy produced in the brain to fuel processes such as synaptic transmission, neurotransmitter synthesis, and maintaining resting potential (10–13). Neurons almost exclusively rely on glucose as their biofuel. Like other cells in the body, neurons express all the key enzymes necessary for glucose oxidation. After entering neurons via facilitated glucose transporters, glucose is irreversibly phosphorylated into glucose-6-phosphate (G6P) by hexokinases. G6P fluxes into the classic bioenergetic pathway, glycolysis – TCA cycle – oxidative phosphorylation to generate ATP. In addition to ATP production, glucose fluxes into the pentose phosphate pathway (PPP) to generate NADPH, which is a critical component of the antioxidative system in neurons (14). While neurons commonly prefer glucose, under certain physiological and pathophysiological conditions, neurons may also utilize alternative fuel sources, such as lactate (15) and ketone bodies (16, 17) for bioenergetic processes. Many studies have shown that these alternative bioenergetic pathways may play crucial roles in multiple neural functions. Importantly, insulin plays key regulatory roles in many of these metabolic pathways through multiple mechanisms to affect neuronal metabolism and functions (Figure 3.1).

INSULIN REGULATES NEURONAL GLUCOSE METABOLISM

While insulin potently stimulates glucose uptake in peripheral tissues such as muscles and adipose tissues, it shows a very minor, if any, acute stimulatory effect on brain glucose uptake. The fluorodeoxyglucose-positron emission tomography (PET) scan study has shown that insulin can only induce a 15% increase in brain glucose uptake in humans (18). The lack of insulin-stimulated glucose uptake is partly due to the very low expression of the insulin-sensitive glucose transporter-4 (GLUT-4) in the brain (19–21). Instead, brain glucose uptake relies on glucose transporters commonly considered insulin-insensitive. Endothelial cells and astrocytes mainly express GLUT-1, while neurons predominantly express GLUT-3. Interestingly, although insulin cannot fully stimulate GLUT-3-dependent glucose uptake in neurons, it may facilitate such process. Thus, insulin induces the docking of GLUT-3-containing vesicles onto the plasma membrane, facilitating calcium-dependent membrane fusion of GLUT-3 and glucose uptake upon neuronal depolarization (22) (Figure 3.1). These results suggest that insulin is able to expand the "ready-to-go" pool of GLUT-3 and prime neurons to ramp up glucose uptake to meet the elevated energy demand due to increased neuronal activity.

In addition to GLUT-3, GLUT-8 is another highly expressed glucose transporter in neurons. Its role in glucose uptake is unclear, due to its lack of plasma membrane localization. However, GLUT-8 is transcriptionally suppressed by insulin in the brain of the mouse subjected to a hyperinsulinemic-euglycemic clamp (23). In addition, insulin triggers the translocation of GLUT-8 to the rough endoplasmic reticulum (ER) in response to insulin, indicating a potential role of GLUT-8 to transport glucose liberated from glycosylated proteins in the rough ER to the cytoplasm, thereby contributing to cellular glucose metabolism (24, 25) (Figure 3.1). The biological significance of such insulin-dependent regulations, however, awaits further investigation.

Although at a very low level, the expression of insulin-sensitive GLUT-4 has been reported in several regions of the brain such as the hippocampus, cortex, cerebellum, and mouse olfactory bulb (26, 27). Insulin regulates both the expression and membrane localization of GLUT-4 in these regions (26, 28, 29) (Figure 3.1). Subsequently, insulin increases glucose uptake and modulates physiologic functions such as spatial working memory in the hippocampus (30–34). GLUT-4 membrane translocation is especially increased at hippocampal nerve

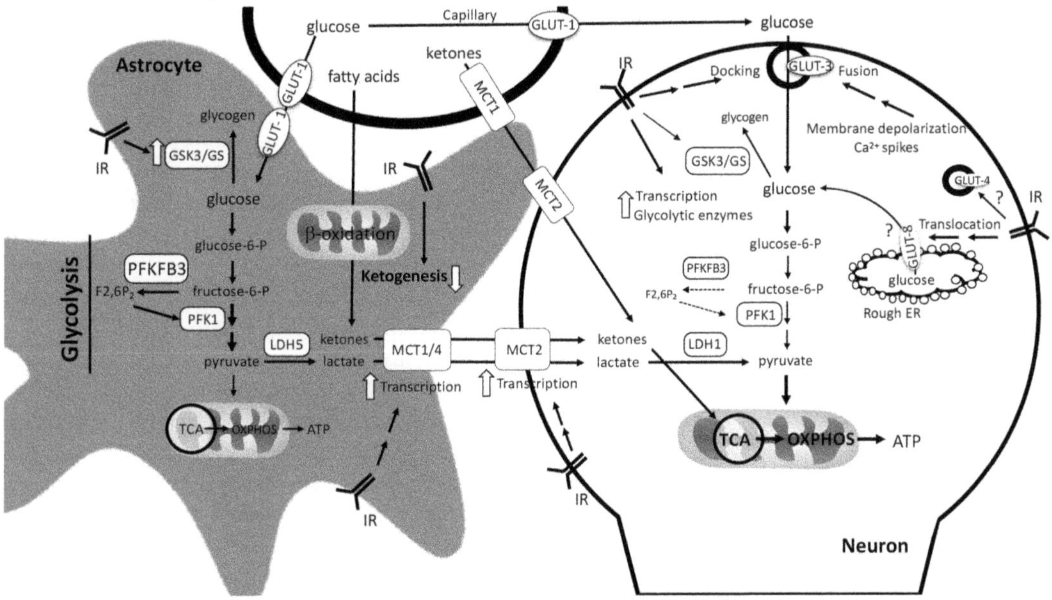

Figure 3.1 Insulin coordinates glucose metabolism in neurons and astrocytes. Insulin regulates neuronal glucose metabolism through multiple mechanisms. Insulin regulates neuron-specific GLUT-3 membrane recruitment to potentiate glucose uptake stimulated by neuronal activity. Neurons are highly oxidative. Despite that insulin increases the transcription of enzymes involved in glycolysis, the glycolytic capacity in neurons is still low and may not be sufficient to meet the energy demand during increased neuronal activity. Therefore, neurons also rely on the astrocyte-neuron lactate shuttle (ANLS) system for an additional fuel source. Astrocytes are highly glycolytic due to the high expression of PFKFB3 to activate glycolysis. During the increased neuronal activity, astrocytes increase the rate of glycolysis and the production of lactate by LDH5. The astrocyte-derived lactate then can be transported through neuron-specific MCT1/4 and astrocyte-specific MCT-2, both of which are upregulated by insulin. After entering neurons, lactate is converted back to pyruvate by astrocyte-enriched LDH1 for complete oxidation in the mitochondria. In addition to glucose metabolism, neurons also use ketone bodies as an alternative fuel source in certain physiological and pathophysiological conditions. During prolonged fasting, neurons can directly take up ketones from circulation through MCT1-MCT2. In addition, astrocytes increase the ketogenic process in response to nutrient deprivation and low insulin conditions. The resultant astrocyte-derived ketones are shuttled to neurons using the same MCT system.

terminals in response to insulin to meet the energetic demands of firing synapses during action potential firing. This process is 5′AMP-activated protein kinase (AMPK) dependent (35). In vivo study of rats shows that intracerebroventricular insulin infusion increases the translocation of GLUT-4 to the plasma membrane through phosphoinositide 3-kinase (PI3K) (32). *In vitro* cell studies have also shown that acute insulin stimulation could induce the membrane translocation of GLUT-4 in neurons (36, 37).

While insulin does not show a robust stimulatory effect on neuronal glucose uptake, it regulates the cellular metabolism of glucose through other mechanisms. For example, insulin contributes to normal mitochondrial structure and function, where glucose is further oxidized into ATP through the TCA cycle (38). A recent study from *Drosophila* has shown that insulin increases glucose content in perisomatic glia. The glucose is subsequently transferred to neurons in the olfactory memory center to fuel PPP and form long-term memory (39). Acute insulin stimulation can potently regulate gene transcription related to glucose metabolism. Thus, in primary cultured cortical neurons, insulin induces the expression of many genes involved in glycolysis (23) (Figure 3.1). Interestingly, insulin has an opposite effect on primary cultured astrocytes and strongly suppresses the expression of these glycolytic enzymes (23), indicating a complex cell-type-specific regulation of glucose metabolism in the brain. At the tissue level, insulin transcriptionally suppresses glycolysis in the hypothalamus of the mice subjected to hyperinsulinemic-euglycemic clamps (23). However, whether such transcriptional regulation has functional impacts on brain glycolysis awaits further investigation.

INSULIN SIGNALING AND ASTROCYTE-NEURON LACTATE SHUTTLING

As discussed earlier, neurons express all essential genes involved in glucose uptake and glycolysis. The glycolytic capacity of neurons, however, is relatively low compared with glial cells in the brain. This is due to the rapid neuronal degradation of 6-phosphofructo-2-kinase /fructose-2,6-bisphosphatase-3 (PFKFB3) (40, 41), which produces fructose-2,6-bisphosphate (F2,6P_2), an allosteric activator of phosphofructokinase 1 (PFK1) to potently enhance the glycolytic capacity (40, 42, 43). Therefore, neurons are unable to produce F2,6P_2, resulting in a low basal glycolytic rate and the inability to stimulate glycolysis according to substrate abundance (Figure 3.1). From this perspective, as the "execution unit" in the brain controlling neural functions, neurons lack metabolic flexibility.

To compensate for the inability to boost glycolysis, especially during the period with high neuronal activity and high energy demand, neurons may rely on more metabolically flexible glial cells for alternative substrate supply. In stark contrast with neuron's metabolic inflexibility, astrocytes express high level of PFKFB3. In response to elevated glucose influx or decreased oxidative phosphorylation, cellular G6P level in astrocytes is elevated, which triggers the production of fructose-2,6-bisphosphate by PFKFB3 to enhance the glycolytic capacity (40, 42, 43). Further, as a critical component of BBB, astrocytes send endfeet to wrap around blood capillaries in the brain. Such an anatomic structure enables astrocytes to quickly sense circulating blood glucose and other nutrients in real time and modulate energy metabolism accordingly (44). Much evidence have shown that astrocytes may directly supply energy sources to neurons through the astrocyte-neuron lactate shuttle (ANLS), which was first hypothesized by Pellerin and Magistretti in 1994 (15). Supporting this hypothesis, astrocytes primarily express lactate dehydrogenase 5 (LDH5), which favors the conversion of the glycolytic end-product pyruvate to lactate. In contrast, neurons express LDH1, which converts lactate back to pyruvate that directly enters the TCA cycle and oxidative phosphorylation to generate ATP. The shuttling of astrocyte-derived lactate to neurons is facilitated by the cell-type-specific expression of lactate transporters. Thus, the high affinity MCT-2 (lactate importer) is expressed in neurons (45–47), whereas the low affinity and high capacity MCT-1 and MCT-4 (lactate exporters) are mainly expressed in astrocytes (47–50) (Figure 3.1).

The ANLS system may be particularly beneficial under the circumstance of increased neuronal activity and high energy demand by neurons. Supporting this, astrocytes are able to increase glycolysis and lactate production in response to glutamate and K^+, the elevations of which reflect increased neuronal activities (15, 51–54). Human positron emission tomography (PET) and functional magnetic resonance imaging (fMRI) studies have shown that during neuronal stimulation and firing of action potentials, glucose utilization exceeds oxygen consumption, indicating predominant glycolytic metabolism and lactate accumulation in the brain (55–57). Further, functional studies have indicated that lactate transport plays an important role in maintaining physiological function and may also facilitate the recovery from pathological conditions in the brain. Thus, it helps to maintain ion homeostasis and normal synaptic activity, promotes adult hippocampal neurogenesis, and protects neuron from traumatic brain injury. (58–61). Conversely, mice with functional deficits of astrocyte-neuron lactate transport display impaired long-term memory formation, while showing no overt developmental defects in the brain (62). Together, these studies indicate that ANLS is important for some higher-order neural functions, such as cognition, while dispensable for neural development. This astrocytes-neuron metabolic coupling may be a unique system developed along the evolution to account for the metabolic plasticity in the central nervous system of the higher organisms.

Insulin also regulates ANLS through multiple mechanisms. Insulin is able to increase lactate production in astrocytes by increasing astrocyte glucose uptake or the presynaptic glutamate release mediated by opioid receptor-dependent disinhibition (63). Conversely, loss of insulin receptors in astrocytes results in lower glucose uptake, reduced glycolytic rate, and decreased lactate efflux (64). During a 3-hour hyperinsulinemic-euglycemic clamp, the expressions of both neuron-specific MCT-2 and astrocyte-specific MCT-1 in the hypothalamus of the mice are increased by more than 50%, indicating increased shuttling capacity of lactate from astrocytes to neurons (23) (Figure 3.1). Therefore, insulin signaling may potentiate the lactate supply to neurons during high levels of neuronal activity and high energy demand.

INSULIN SIGNALING AND KETONE BODY METABOLISM IN NEURONS

Ketone bodies are another known alternative energy substrate to neurons besides lactate (16, 17). Circulating ketone bodies can cross the BBB through MCT-1 in endothelial cells and astrocytes (48, 65, 66). Due to insulin's suppressive effect on fatty acid ketogenesis in the liver, plasma ketone body levels are usually very low under normal conditions and only contribute to less than 5% of brain metabolism (67). During prolonged fasting, however, circulating ketone body levels are increased significantly to compensate for the lack of glucose in the body. Under such condition the brain takes up significantly more ketone bodies from circulation and may rely on ketone bodies for ~60% of the brain's energy demand (68) (Figure 3.1). In addition to plasma ketones, astrocytes in the brain can serve as a production center for ketones through fatty acid beta-oxidation. Thus, fatty acids can be transported through BBB into the brain and taken up by astrocytes, where fatty acid beta-oxidation occurs (69, 70). The ketogenesis in astrocytes is regulated by several key signaling pathways, including the AMPK and insulin signaling pathways. Like elsewhere in the body, insulin suppresses many key enzymes in the ketogenic pathway. Therefore, low insulin could increase the production of ketone bodies by increasing β-hydroxy β-methylglutaryl- CoA (HMG-CoA) and AMPK activities (71, 72). Neurons take up ketone bodies such as beta-hydroxybutyrate (BHB) derived from astrocytes or periphery by MCT-2. Within neurons, BHB is first converted into acetyl-CoA, which enters the TCA cycle in the mitochondria for oxidative metabolism to produce ATP (73, 74). Together, insulin signaling indirectly regulates neuronal ketone body utilization.

Ketone body metabolism is not only able to supplement glucose metabolism but may also protect neurons under many pathological conditions such as neurodegenerative diseases and brain injuries. For instance, patients with Alzheimer's disease (AD) showed increased ketone body metabolism in the brain, indicating adapted bioenergetics in response to energy crisis-induced metabolic disturbances (75–79). Indeed, it has been hypothesized that the ketogenic diet may be beneficial and protective in AD patients, providing a potential therapeutic strategy to treat Alzheimer's disease, which is being tested in several clinical trials (80–82). Interestingly, AD patients showed impaired insulin/IGF-1 signaling in the brain, suggesting a possible molecular link between insulin signaling and ketone body metabolism in AD (83). Future studies are needed to explore such molecular mechanisms to further understand the role of brain insulin signaling in neurodegenerative disease.

INSULIN AND GLYCOGEN METABOLISM IN NEURONS

Glucose can be stored as an energy reserve in the form of glycogen when in excess. This is also true in the brain (Figure 3.1). Glycogen in the brain was first observed and reported by Koizumi and Shiraishi using biochemical assay and electron microscopy in the 1970s (84, 85). However, the glycogen appears to be present exclusively in astrocytes (86–89). Although healthy neurons express the glycogen synthase required for glycogen synthesis, glycogen is hardly detected in neurons in experimental settings (90, 91). This is likely due to the constant high rate of utilization of glycogen in neurons (92). Any resident glycogen stock within the neurons may be exhausted during the sample tissue collection and processing. In addition, the protein targeting glycogen (PTG), a potent activator of the glycogen synthesis, is rapidly degraded in neurons, resulting in low basal glycogen synthesis (90). Although it is technically challenging to detect and assess glycogen storage in neurons, disrupted glycogen synthesis/degradation results in neural dysfunction and behavioral deficit in pathological conditions like hypoxia (93) or inherited genetic diseases with abnormal glycogen storage, such as Pompe disease and Lafora disease (90, 93–96), indicating the importance of regulated neuronal glycogen storage for normal neuronal development and function.

Insulin is known to regulate glycogen metabolism via phosphorylation events on GSK3- glycogen synthase (GS) signaling cascade (97). However, whether insulin may directly affect glycogen metabolism in neurons through the same signaling pathways is not clear, due to the technical difficulties to assess neuronal glycogen storage in vivo. Nevertheless, insulin can regulate neuronal metabolism indirectly by regulating astrocytic glycogen breakdown. Thus, in response to the increase in energy demand by neurons, insulin inhibits norepinephrine uptake by neurons, resulting in the activation of the β-adrenoreceptors on astrocytes and stimulation of glucose extrusion from glycogen in astrocytes (98, 99).

INSULIN AND LIPID METABOLISM IN NEURONS

One of the major anabolic effects of insulin in the body is to potently promote de novo lipogenesis (100–102). Such anabolic effect of insulin is particularly important in the brain, due to the restrictive entry of the circulating cholesterol and long-chain fatty acids through BBB. Thus, almost all the cholesterol in the brain is synthesized in situ and is essential for normal brain development and function. Decreased cholesterol biosynthesis in the brain caused by chemical or genetic intervention has been shown to impair normal brain development, synaptic transmission, and memory in mice (103–105). Mechanistically, insulin promotes neuronal cholesterol and long-chain fatty acid biosynthesis through direct and indirect pathways (Figure 3.2). Thus, insulin directly enhances the transcription of genes involved in cholesterol and long-chain fatty acyl-CoA biosynthesis in neurons (105, 106). Alternatively, insulin stimulates the transcription of the same class of genes in astrocytes to upregulate cholesterol and long-chain fatty acid production (103, 105–107), which can be transferred to neurons in the form of ApoE-containing lipoproteins (108, 109). Therefore, when glucose supply is abundant postprandially, insulin signals both neurons and

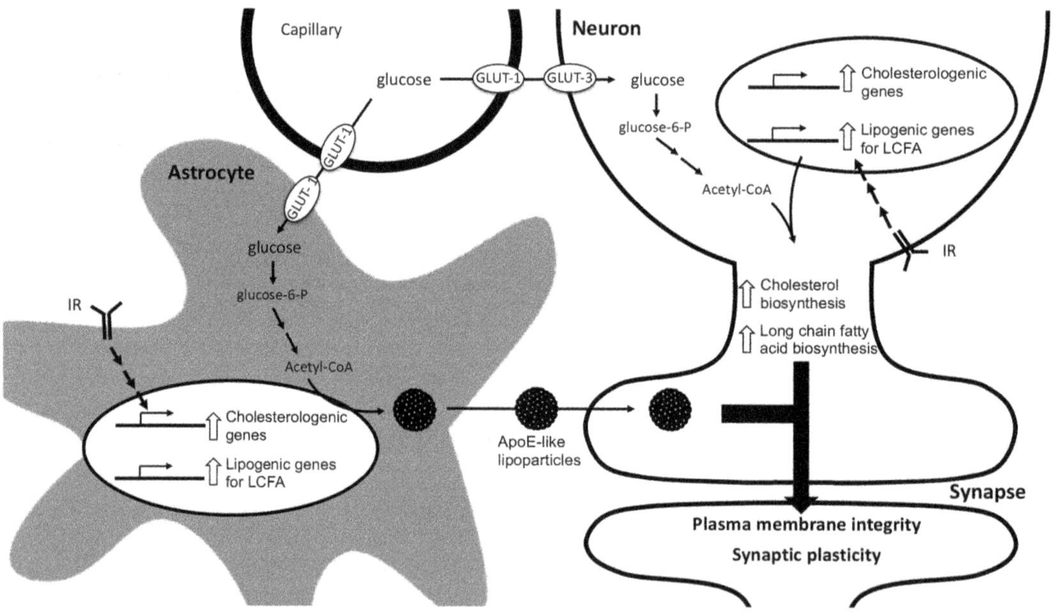

Figure 3.2 Insulin promotes de novo cholesterol and long-chain fatty acid biosynthesis. When nutrients are abundant postprandially, insulin signals both neurons and astrocytes to boost their de novo biosynthesis of cholesterol and long-chain fatty acids. Insulin up-regulates the majority of the enzymes involved in cholesterol and long-chain fatty acyl-CoA biosynthesis. The astrocyte-derived cholesterol and long-chain fatty acids can be packed into ApoE-containing lipoparticles and transported to neurons. Such an anabolic effect of insulin on de novo lipid biosynthesis may be particularly beneficial for neuronal plasma membrane integrity and synaptic plasticity.

astrocytes to reroute carbon source from glucose to the biosynthesis of cholesterol and long-chain fatty acids. Such anabolic action of insulin would facilitate the replenishment of damaged neuronal plasma membranes and remodeling of synapses, thus beneficial for healthy and normal neural functions.

INSULIN SIGNALING CONTRIBUTES TO MITOCHONDRIAL FUNCTION AND HOMEOSTASIS IN NEURONS

Neurons are predominantly oxidative and enriched for mitochondria (110, 111). Neurons require continuous energy supply in the form of ATP, the majority of which is generated by oxidative phosphorylation in the mitochondria. Thus, mitochondria are central to the normal cellular metabolism and function of neurons. Mitochondria are highly sensitive to numerous environmental stimuli and can adapt to different metabolic situations. According to nutritional status and substrate availability, neuronal mitochondria can use glucose, lactate/pyruvate, ketone body, and even fatty acid to generate ATP. In addition to the metabolic control, like in any other cells, mitochondria in neurons sense and respond to various cellular signals, including ion homeostasis, inflammation, and oxidative stress. To fulfill these roles, mitochondrial functions in neurons are tightly and dynamically controlled by many different molecular mechanisms, among which insulin signaling plays critical roles. In line with this hypothesis, animal models of diabetes/insulin resistance commonly show decreased mitochondrial respiration rate, increase mitochondrial-related oxidative stress in neurons and other glial cell types (112–114). Physical and pharmacological interventions to improve insulin sensitivity in these mouse models improve mitochondrial function in the brain (115). In addition, the whole brain insulin receptor knockout (NIRKO) mice with no apparent systemic insulin resistance display increased mitochondria fragmentation, decreased OXPHOS protein expression, and impaired mitochondrial respiration rate in the striatal neurons (112), strongly indicating that insulin signaling is a direct functional regulator of mitochondrial homeostasis.

Insulin Signaling Regulates Mitochondrial Bioenergetic Capacity

The bioenergetic capacity of the mitochondria is dynamically regulated in neurons, according to different metabolic conditions and hormonal stimulation. As a central hormone regulating cellular metabolism, insulin treatment increases the oxidative respiration and ATP production in primary neurons (115). Numerous studies from different cell and animal models have suggested that insulin may affect mitochondrial bioenergetic capacity at multiple levels (Figure 3.3).

Insulin Signaling Regulates Mitochondrial Mass by Supporting Mitochondrial Biogenesis

Peroxisome proliferator-activated receptor gamma co-activator 1-α (PGC1α) is a transcriptional co-activator and a master regulator for mitochondrial biogenesis through induction of nuclear respiratory factor 1 and 2 (NRF1/2) and mitochondrial transcriptional factor A (TFAM) (116, 117). The promoter of the PGC1α gene contains transcriptional binding sites including MEF2, FoxO1, and cAMP response element (CRE), indicating a complex transcriptional regulation by many different intracellular and extracellular stimuli. Indeed, PGC1α is transcriptionally regulated by exercise, cold, glucagon, and cytokines in different cell types throughout the body (reviewed in (116)). Diabetes and insulin resistance commonly lead to reduced levels of PGC1α/β in muscles, indicating the interaction between insulin signaling and PGC1 activity (118). Although not many studies have directly addressed the effect of insulin signaling on PGC1α activity and mitochondrial biogenesis in neurons, the mechanisms by which insulin regulates PGC1α transcription and activation in liver and muscle, however, are complex. On one hand, insulin inhibits the transcription of PGC1α via PI3K-Akt-dependent FoxO1 inactivation (119, 120). On the other hand, insulin also promotes the translation of mitochondrial proteins through the insulin-mediated mTOR signaling pathway (121, 122). In addition, insulin signaling may contribute to the regulation of PGC1α activity through posttranslational modification. Thus, Akt stabilizes PGC-1α by inhibiting GSK3β in primary neurons (123). Further, impaired insulin signaling and hyperglycemia result in increased acetylation and inactivation of PGC1α in the liver, due to the low NAD^+/NADH ratio and subsequent inactivation of Sirt1 (120). Collectively, the net effect of insulin appears to increase mitochondrial biogenesis, while insulin deficiency and insulin resistance lead to decreased

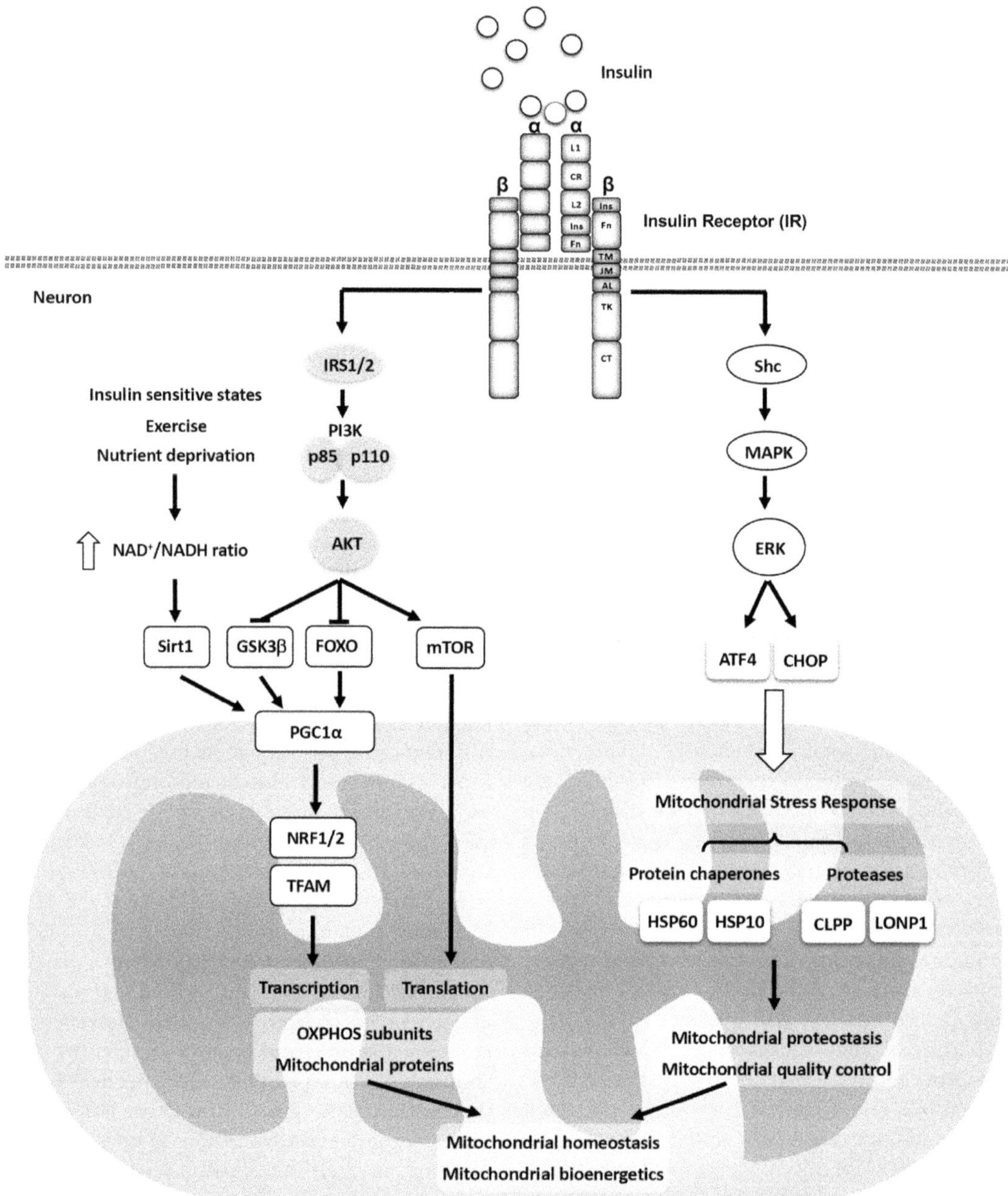

Figure 3.3 Insulin controls mitochondrial homeostasis and function in neurons. Insulin promotes mitochondrial biogenesis and bioenergetics mainly through PI3K-Akt branch downstream of the insulin receptor. Thus, insulin activates Akt to inhibit FoxO1-dependent PGC1α transcription. Meanwhile, activated Akt also inhibits GSK3β to stabilize PGC1α. In addition, the bioactivity of PGC1α is also mediated by Sirt1-dependent deacetylation. Thus, according to insulin sensitivity, nutritional status, and exercise, increased cellular NAD^+/NADH ratio activates Sirt1, which triggers deacetylation and activation of PGC1α. The activated PGC1α activates the transcription of NRF1/2 and TFAM, both of which control an array of genes involved in mitochondrial biogenesis. In addition, insulin signaling activates mTOR to enhance the protein translation of mitochondrial proteins, including OXPHOS subunits. On the other hand, insulin is also a master regulator for mitochondrial quality control. Insulin mediates transcriptional regulation of mitochondrial stress response (MSR) genes mainly through Shc/MAPK cascade. Well-known insulin-targeted MSR genes include mitochondrial chaperones like HSP60 and HSP10 and proteases like CLPP and LONP1. Insulin-mediated up-regulation of MSR proteins improves mitochondrial proteostasis and mitochondrial quality. Collectively, insulin signaling in neurons improves mitochondrial homeostasis and bioenergetics.

mitochondrial biogenesis (Figure 3.3). Whether insulin regulates PGC1α and mitochondrial biogenesis through similar mechanisms in neurons awaits further investigation.

In addition to mitochondrial biogenesis, insulin signaling also promotes oxidative phosphorylation by increasing the expression of OXPHOS protein components (124) and enhancing the mitochondrial membrane polarization (125). Thus, insulin regulates mitochondrial inner membrane potential through PI3K-Akt and cAMP response element binding protein (CREB) pathways in adult rat sensory neurons, which subsequently increases the mitochondrial ATP production (125).

Further, the morphological dynamics of the mitochondrial network are a key determinant of the bioenergetic capacity of mitochondria. Hyperfused tubular mitochondria are more efficient at producing ATP, whereas fragmented mitochondria caused by enhanced fission are associated with decreased inner membrane potential and decreased respiration capacity. The fusion and fission events are dynamically regulated by fusion proteins mitofusin 1/2 (MFN1/2), and fission proteins such as optic atrophy 1 (OPA1) and dynamin-related protein 1 (DRP1). Studies have shown that nutritional status as well as insulin sensitivity contributes to the morphological dynamics of mitochondria in the hypothalamic AgRP and POMC neurons (126, 127). Whether or how insulin signaling directly regulates mitochondrial fusion and fission awaits further investigations.

Insulin Signaling Modulates Mitochondrial Stress Response Network

Mitochondrial stress response (MSR) is an essential quality control mechanism for mitochondria. The MSR network includes chaperones responsible for the re-folding of the misfolded mitochondrial proteins, as well as proteases to degrade damaged proteins. During episodes of cellular stressors, such as nutritional deprivation and oxidative stress, MSR components can detect misfolded mitochondrial proteins for repair and degrade those beyond the threshold for repair. Thus, MSR surveils the mitochondrial proteins and repairs those with functional abnormality in time to ensure mitochondrial health and homeostasis. In disease states, such as insulin resistance, MSR system is commonly impaired, indicating decreased capability to reverse mitochondrial dysfunction.

Insulin signaling up-regulates the expression of MSR genes in an ERK-dependent manner (Figure 3.3). These regulated genes include master transcriptional factors (ATF4, CHOP), along with their target proteins including mitochondrial chaperones (HSP60 and HSP10) and proteases (ClPP and LONP1). In this way, insulin potentiates mitochondrial stress response (MSR) capacity to fight for mitochondrial damage caused by various cellular stressors (128). Supporting this, MSR protein expression is decreased in the hypothalamic neurons of diet-induced obese mice, which contributes to excessive mitochondrial damage, leading to irreversible mitochondrial fission and mitophagy (129, 130). Such functional deterioration in neurons under metabolic stressors, however, can be reversed by insulin treatment. More importantly, intranasal insulin treatment improved the metabolic phenotypes of diet-induced obesity, including decreased food intake and decreased weight gain (129, 130). Together, these data strongly suggest that insulin regulates MSR network in neurons to counteract cellular stress-induced mitochondrial protein misfolding and mitochondrial dysfunction.

Potential Link between Impaired Insulin/IGF-1 Signaling and Mitochondrial Dysfunction in Neurodegenerative Diseases

The coupling of impaired brain insulin signaling and mitochondrial dysfunction is not only evident in people with diabetes and systemic insulin resistance, but also observed in people with neurodegenerative diseases. For instance, AD brains display metabolic disturbance, such as decreased glucose utilization, accompanied by features of mitochondrial dysfunction. Thus, the expression of electron transport chain (ETC) subunits (OXPHOS) is decreased in the posterior cingulate and hippocampal neurons of patients with AD (131, 132). Accompanied with these, brains of AD patients also display impaired signaling response to insulin (83), demonstrating central insulin resistance in people with AD.

Mouse models of AD generally recapitulate the association between neuronal insulin resistance and mitochondrial dysfunction (reviewed in (133)). Thus, mouse models carrying human familial AD risk genes display increased rates of mitochondrial fission, resulting in increased fragmented synaptosomal mitochondria, and thereby impaired bioenergetics of synaptosomal mitochondria (134).

Further, hippocampal primary neurons from 3xTg-AD mouse brains are insulin resistant, displaying significantly reduced mitochondrial respiration (135). The functional interaction between insulin signaling and mitochondrial function in neurons appears to be not limited to AD, as the increased mitochondrial reactive oxygen species (ROS) and mitochondrial dysfunction in the striatal neurons from Htt mutant mice can be rescued by insulin and IGF-1 treatment (136). Similar interaction has also been found in people with Parkinson's disease (137).

Together, these data from humans and mice postulate the hypothesis that improvement of brain insulin/IGF-1 sensitivity may be beneficial to people with neurodegenerative diseases by reversing mitochondrial dysfunction. While such a hypothesis is supported by promising preclinical research in animal models (115, 138), the results from clinical trials using either intranasally administered insulin or insulin sensitizers have been mixed (139–143). Therefore, more studies are warranted to further identify the treatment window of the disease progression or to understand the specific covariance explaining the discrepancy.

INSULIN SIGNALING CONTRIBUTES TO INTRACELLULAR CA^{2+} HOMEOSTASIS IN NEURONS

Ca^{2+} is one of the most important secondary messengers in neurons. Intracellular free Ca^{2+} controls neurotransmitter secretion, synaptic plasticity, axodendritic growth and morphology, gene expression, and mitochondrial bioenergetics (144–146). While the concentration of free Ca^{2+} is very low in the cytosol at the resting state, multiple cellular stimuli can rapidly increase cytosolic Ca^{2+} level either through calcium channels on the plasma membrane or through internal stores such as the ER and mitochondria (147).

The ER is the largest reservoir and central to the intracellular Ca^{2+} homeostasis that maintains the normal metabolism and function of neurons (148–150). With the expression of high density of Ca^{2+} channels and transporters, the ER in neurons can rapidly uptake or release Ca^{2+} in response to various intracellular signaling pathways. One of the major signaling pathways controlling Ca^{2+} release from ER is G-protein-coupled receptor (GPCR)/Gq/phospholipase Cβ (PLC beta) cascade. Thus, activation of Gq-PLCβ cascade results in the hydrolysis of phosphatidylinositol 4,5-bisphosphate (PIP2), and the production of diacyl glycerol (DAG) and inositol-1,4,5-triphosphate (IP3). IP3 opens IP3 receptor (IP3R) and ryanodine receptor (RyR) on the ER, releasing Ca^{2+} stored in the lumen of the ER into the cytoplasm. Subsequently, the cytosolic Ca^{2+} concentration (i.e. $[Ca^{2+}]i$) spike elicits an array of downstream signaling events for synaptic remodeling, axon guidance, dendritic spine growth, synaptic vesicle release, and gene expression (144–146).

Insulin signaling also contributes to intracellular calcium homeostasis. Studies have shown that insulin activates AKT to phosphorylate and inhibit IP3R (151, 152). This negative regulation of IP3R allows insulin to attenuate intracellular calcium spikes triggered by neuropeptides such as ghrelin and others (153, 154). In addition to regulating Ca^{2+}-releasing channels in the ER, insulin promotes the activity of the sarco-endoplasmic reticulum Ca^{2+}-ATPase (SERCA), which is responsible for Ca^{2+} uptake from the cytosol to ER lumen. In line with this, in the dorsal root ganglia neurons from diabetic rats, Ca^{2+} uptake into ER is reduced. Both SERCA pump function and SERCA2a protein expression are suppressed in diabetic cultures (155). Overexpression of SERCA in *ob/ob* mice improves ER protein folding capacity and glycemic control (156, 157).

Collectively, the net effect of insulin appears to increase Ca^{2+} storage in the ER lumen and decrease free $[Ca^{2+}]i$, while insulin deficiency and insulin resistance lead to reduced ER Ca^{2+}. This type of insulin action may be particularly beneficial for normal ER function related to proteostasis in neurons. A high level of Ca^{2+} in ER is essential for protein processing and folding. ER Ca^{2+} depletion due to overstimulation could impair the functions of intraluminal proteins including buffering proteins (such as calsequestrin and calreticulin), chaperons [such as endoplasmin (GRP94), calnexin, ERp57, immunoglobulin binding protein (Bip), and PDI] (158), resulting in the accumulation of unfolded proteins and activation of the unfolded protein response (UPR). Insulin's effect to inhibit IP3R-dependent calcium release and boost SERCA-dependent calcium uptake can help maintain Ca^{2+} storage in the ER and prevent prolonged UPR that may ultimately trigger neuronal death. In line with this hypothesis, animal models with diabetes/aging/insulin resistance show diminished ER Ca^{2+} store, increased $[Ca^{2+}]i$, and impaired $[Ca^{2+}]i$ response to environmental stimuli (159, 160). Therapeutic interventions with a low dose of insulin are able to restore resting

$[Ca^{2+}]$i and Ca^{2+} release from ER in the sensory neurons of STZ-treated diabetic rats (161). In addition, insulin treatments reverse diabetic neuropathic changes in diabetic rats as well as humans by restoring $[Ca^{2+}]$i homeostasis and mitochondrial function (125, 162–168). All of this evidence strongly suggest that insulin signaling plays a critical role in $[Ca^{2+}]$i homeostasis in neurons.

Further, insulin action on ER and calcium homeostasis may affect mitochondrial function. Ca^{2+} is required for the activation of the TCA dehydrogenases such as pyruvate dehydrogenase, α-ketoglutarate dehydrogenase, and isocitrate dehydrogenase (169). Therefore, mitochondrial Ca^{2+} flux is critical to aerobic metabolism (170). Mitochondria take up Ca^{2+} through mitochondrial Ca^{2+} uniporter (MCU) or mitochondrial associated membrane (MAM) which is directly associated with ER (171, 172). Under the conditions of insulin resistance, decreased ER Ca^{2+} storage and increased cytosolic free Ca^{2+} may lead to increased Ca^{2+} uptake by mitochondria. Excessive Ca^{2+} accumulation in the mitochondrial results in increased ROS production, mitochondrial dysfunction, and mitophagy (173–175). In line with this, in both animal models with diabetes induced by streptozotocin (STZ) and diabetic patients, elevated intramitochondrial Ca^{2+} and ROS are commonly presented. Conversely, therapeutic treatments with insulin and neurotrophic factors normalize mitochondrial intracellular Ca^{2+} levels (176).

Therefore, insulin signaling plays an important role in the regulation of Ca^{2+} homeostasis in ER and mitochondria. Insulin resistance leads to disturbance of Ca^{2+} homeostasis (177), which is detrimental for normal neuronal function and survival.

CONCLUDING REMARKS

Insulin is the central hormone in the regulation of glucose homeostasis and energy balance in the body. While the brain is traditionally considered an insulin-insensitive organ, it is now more and more clear that insulin acts on almost all cell types in the brain to modulate brain metabolism and neural functions. Although insulin has moderate effects on glucose uptake in the brain, it plays critical role in many aspects of cellular metabolism in neurons, including glucose oxidation, glycogen synthesis, and lipid synthesis. In addition, insulin signaling is particularly important for the homeostasis of both mitochondria and ER, and their functions in neurons. Thus, insulin shows beneficial effects on ER-related calcium homeostasis, and mitochondrial-dependent cellular processes, including increasing respiration rates and ATP production, promoting protein homeostasis, and decreasing oxidative stress and mitochondrial-initiated cellular apoptosis. Conversely, in the context of diabetes and neurodegenerative diseases such as AD, impaired insulin signaling in neurons leads to mitochondrial dysfunction, calcium signaling imbalance, and metabolic disturbance, all of which contribute to the further exacerbation of cellular stresses harmful to neurons, leading to the deterioration of normal neural functions. Therefore, it is of utmost importance to continue to uncover the molecular linkage between neuronal insulin signaling and neuronal metabolism, aiming to identify potential therapeutic targets to alleviate or treat neurological symptoms caused by diabetes and neurodegeneration.

ACKNOWLEDGEMENTS

This work was supported by grants from the US National Institutes of Health (K01DK120740, R01MH125903 to W.C).

REFERENCES

1. A. R. Saltiel, C. R. Kahn, Insulin signalling and the regulation of glucose and lipid metabolism. *Nature* **414**, 799–806 (2001).
2. C. R. Kahn, G. Wang, K. Y. Lee, Altered adipose tissue and adipocyte function in the pathogenesis of metabolic syndrome. *J Clin Invest* **129**, 3990–4000 (2019).
3. Q. Huang, C. R. Kahn, E. Altindis, Viral hormones: Expanding dimensions in endocrinology. *Endocrinology* **160**, 2165–2179 (2019).
4. S. B. Biddinger, C. R. Kahn, From mice to men: insights into the insulin resistance syndromes. *Annu Rev Physiol* **68**, 123–158 (2006).
5. S. Dedrick et al., The Role of Gut Microbiota and Environmental Factors in Type 1 Diabetes Pathogenesis. *Front Endocrinol (Lausanne)* **11**, 78 (2020).
6. M. F. White, C. R. Kahn, Insulin action at a molecular level -100 years of progress. *Mol Metab* **52**, 101304 (2021).
7. F. Erbsloh, A. Bernsmeier, H. Hillesheim, [The glucose consumption of the brain & its dependence on the liver]. *Arch Psychiatr Nervenkr Z Gesamte Neurol Psychiatr* **196**, 611–626 (1958).

8. R. J. McCrimmon, Consequences of recurrent hypoglycaemia on brain function in diabetes. *Diabetologia* **64**, 971–977 (2021).
9. R. J. McCrimmon, B. M. Frier, Hypoglycaemia, the most feared complication of insulin therapy. *Diabete Metab* **20**, 503–512 (1994).
10. S. Camandola, M. P. Mattson, Brain metabolism in health, aging, and neurodegeneration. *EMBO J* **36**, 1474–1492 (2017).
11. J. J. Harris, R. Jolivet, D. Attwell, Synaptic energy use and supply. *Neuron* **75**, 762–777 (2012).
12. F. Hyder, D. L. Rothman, M. R. Bennett, Cortical energy demands of signaling and nonsignaling components in brain are conserved across mammalian species and activity levels. *Proc Natl Acad Sci U S A* **110**, 3549–3554 (2013).
13. P. J. Magistretti, I. Allaman, A cellular perspective on brain energy metabolism and functional imaging. *Neuron* **86**, 883–901 (2015).
14. J. P. Bolanos, A. Almeida, The pentose-phosphate pathway in neuronal survival against nitrosative stress. *IUBMB Life* **62**, 14–18 (2010).
15. L. Pellerin, P. J. Magistretti, Glutamate uptake into astrocytes stimulates aerobic glycolysis: a mechanism coupling neuronal activity to glucose utilization. *Proc Natl Acad Sci U S A* **91**, 10625–10629 (1994).
16. M. K. Dalsgaard, L. Nybo, Y. Cai, N. H. Secher, Cerebral metabolism is influenced by muscle ischaemia during exercise in humans. *Exp Physiol* **88**, 297–302 (2003).
17. G. van Hall et al., Blood lactate is an important energy source for the human brain. *J Cereb Blood Flow Metab* **29**, 1121–1129 (2009).
18. E. M. Bingham et al., The role of insulin in human brain glucose metabolism: an 18fluorodeoxyglucose positron emission tomography study. *Diabetes* **51**, 3384–3390 (2002).
19. K. A. Heidenrich, P. R. Gilmore, W. T. Garvey, Glucose transport in primary cultured neurons. *J Neurosci Res* **22**, 397–407 (1989).
20. R. T. Watson, J. E. Pessin, Intracellular organization of insulin signaling and GLUT4 translocation. *Recent Prog Horm Res* **56**, 175–193 (2001).
21. B. S. McEwen, L. P. Reagan, Glucose transporter expression in the central nervous system: relationship to synaptic function. *Eur J Pharmacol* **490**, 13–24 (2004).
22. E. Uemura, H. W. Greenlee, Insulin regulates neuronal glucose uptake by promoting translocation of glucose transporter GLUT3. *Exp Neurol* **198**, 48–53 (2006).
23. W. Cai et al., Peripheral insulin regulates a broad network of gene expression in Hypothalamus, Hippocampus, and Nucleus Accumbens. *Diabetes* **70**, 1857–1873 (2021).
24. G. G. Piroli et al., Peripheral glucose administration stimulates the translocation of GLUT8 glucose transporter to the endoplasmic reticulum in the rat hippocampus. *J Comp Neurol* **452**, 103–114 (2002).
25. L. P. Reagan et al., GLUT8 glucose transporter is localized to excitatory and inhibitory neurons in the rat hippocampus. *Brain Res* **932**, 129–134 (2002).
26. S. J. Vannucci et al., GLUT4 glucose transporter expression in rodent brain: effect of diabetes. *Brain Res* **797**, 1–11 (1998).
27. M. Kobayashi, H. Nikami, M. Morimatsu, M. Saito, Expression and localization of insulin-regulatable glucose transporter (GLUT4) in rat brain. *Neurosci Lett* **213**, 103–106 (1996).
28. K. Talbot et al., Demonstrated brain insulin resistance in Alzheimer's disease patients is associated with IGF-1 resistance, IRS-1 dysregulation, and cognitive decline. *J Clin Invest* **122**, 1316–1338 (2012).
29. S. El Messari et al., Immunocytochemical localization of the insulin-responsive glucose transporter 4 (Glut4) in the rat central nervous system. *J Comp Neurol* **399**, 492–512 (1998).
30. J. Pearson-Leary, V. Jahagirdar, J. Sage, E. C. McNay, Insulin modulates hippocampally-mediated spatial working memory via glucose transporter-4. *Behav Brain Res* **338**, 32–39 (2018).
31. J. Pearson-Leary, E. C. McNay, Novel roles for the insulin-regulated glucose transporter-4 in hippocampally dependent memory. *J Neurosci* **36**, 11851–11864 (2016).
32. C. A. Grillo, G. G. Piroli, R. M. Hendry, L. P. Reagan, Insulin-stimulated translocation of GLUT4 to the plasma membrane in rat hippocampus is PI3-kinase dependent. *Brain Res* **1296**, 35–45 (2009).
33. R. N. Fernando, A. L. Albiston, S. Y. Chai, The insulin-regulated aminopeptidase IRAP is colocalised with GLUT4 in the mouse hippocampus--potential role in modulation of glucose uptake in neurones? *Eur J Neurosci* **28**, 588–598 (2008).
34. G. G. Piroli et al., Corticosterone impairs insulin-stimulated translocation of GLUT4 in the rat hippocampus. *Neuroendocrinology* **85**, 71–80 (2007).
35. G. Ashrafi, Z. Wu, R. J. Farrell, T. A. Ryan, GLUT4 Mobilization supports energetic demands

of active synapses. *Neuron* **93**, 606–615 e603 (2017).
36. P. Varshney, C. S. Dey, P21-activated kinase 2 (PAK2) regulates glucose uptake and insulin sensitivity in neuronal cells. *Mol Cell Endocrinol* **429**, 50–61 (2016).
37. Y. Benomar et al., Insulin and leptin induce Glut4 plasma membrane translocation and glucose uptake in a human neuronal cell line by a phosphatidylinositol 3-kinase- dependent mechanism. *Endocrinology* **147**, 2550–2556 (2006).
38. Z. Cheng, Y. Tseng, M. F. White, Insulin signaling meets mitochondria in metabolism. *Trends Endocrinol Metab* **21**, 589–598 (2010).
39. E. de Tredern et al., Glial glucose fuels the neuronal pentose phosphate pathway for long-term memory. *Cell Rep* **36**, 109620 (2021).
40. A. Almeida, S. Moncada, J. P. Bolanos, Nitric oxide switches on glycolysis through the AMP protein kinase and 6-phosphofructo-2-kinase pathway. *Nat Cell Biol* **6**, 45–51 (2004).
41. A. Herrero-Mendez et al., The bioenergetic and antioxidant status of neurons is controlled by continuous degradation of a key glycolytic enzyme by APC/C-Cdh1. *Nat Cell Biol* **11**, 747–752 (2009).
42. C. X. Bittner et al., High resolution measurement of the glycolytic rate. *Front Neuroenergetics* **2**, (2010).
43. W. A. Zellmer, Editorial: Report of the study commission on pharmacy, I: general impressions. *Am J Hosp Pharm* **33**, 331 (1976).
44. T. M. Mathiisen, K. P. Lehre, N. C. Danbolt, O. P. Ottersen, The perivascular astroglial sheath provides a complete covering of the brain microvessels: an electron microscopic 3D reconstruction. *Glia* **58**, 1094–1103 (2010).
45. L. Mazuel et al., A neuronal MCT2 knockdown in the rat somatosensory cortex reduces both the NMR lactate signal and the BOLD response during whisker stimulation. *PLoS One* **12**, e0174990 (2017).
46. O. Chiry et al., Distribution of the monocarboxylate transporter MCT2 in human cerebral cortex: an immunohistochemical study. *Brain Res* **1226**, 61–69 (2008).
47. L. Bergersen et al., A novel postsynaptic density protein: the monocarboxylate transporter MCT2 is co-localized with delta-glutamate receptors in postsynaptic densities of parallel fiber-Purkinje cell synapses. *Exp Brain Res* **136**, 523–534 (2001).
48. K. Pierre, L. Pellerin, Monocarboxylate transporters in the central nervous system: distribution, regulation and function. *J Neurochem* **94**, 1–14 (2005).
49. K. S. Dimmer, B. Friedrich, F. Lang, J. W. Deitmer, S. Broer, The low-affinity monocarboxylate transporter MCT4 is adapted to the export of lactate in highly glycolytic cells. *Biochem J* **350** (Pt 1), 219–227 (2000).
50. S. Broer et al., Comparison of lactate transport in astroglial cells and monocarboxylate transporter 1 (MCT 1) expressing Xenopus laevis oocytes. Expression of two different monocarboxylate transporters in astroglial cells and neurons. *J Biol Chem* **272**, 30096–30102 (1997).
51. B. R. Larsen et al., Contributions of the Na(+)/K(+)-ATPase, NKCC1, and Kir4.1 to hippocampal K(+) clearance and volume responses. *Glia* **62**, 608–622 (2014).
52. N. Macaulay, T. Zeuthen, Glial K(+) clearance and cell swelling: Key roles for cotransporters and pumps. *Neurochem Res* **37**, 2299–2309 (2012).
53. C. X. Bittner et al., Fast and reversible stimulation of astrocytic glycolysis by K+ and a delayed and persistent effect of glutamate. *J Neurosci* **31**, 4709–4713 (2011).
54. A. Loaiza, O. H. Porras, L. F. Barros, Glutamate triggers rapid glucose transport stimulation in astrocytes as evidenced by real-time confocal microscopy. *J Neurosci* **23**, 7337–7342 (2003).
55. A. B. Newberg et al., Concurrent CBF and CMRGlc changes during human brain activation by combined fMRI-PET scanning. *Neuroimage* **28**, 500–506 (2005).
56. P. T. Fox, M. E. Raichle, M. A. Mintun, C. Dence, Nonoxidative glucose consumption during focal physiologic neural activity. *Science* **241**, 462–464 (1988).
57. T. L. Davis, K. K. Kwong, R. M. Weisskoff, B. R. Rosen, Calibrated functional MRI: Mapping the dynamics of oxidative metabolism. *Proc Natl Acad Sci U S A* **95**, 1834–1839 (1998).
58. Y. Lev-Vachnish et al., L-lactate promotes adult hippocampal neurogenesis. *Front Neurosci* **13**, 403 (2019).
59. J. Zhou et al., Lactate potentiates angiogenesis and neurogenesis in experimental intracerebral hemorrhage. *Exp Mol Med* **50**, 1–12 (2018).
60. E. A. Angamo, J. Rosner, A. Liotta, R. Kovacs, U. Heinemann, A neuronal lactate uptake inhibitor slows recovery of extracellular ion concentration changes in the hippocampal CA3 region by

affecting energy metabolism. *J Neurophysiol* **116**, 2420–2430 (2016).
61. Z. Alvarez et al., Neurogenesis and vascularization of the damaged brain using a lactate-releasing biomimetic scaffold. *Biomaterials* **35**, 4769–4781 (2014).
62. A. Suzuki et al., Astrocyte-neuron lactate transport is required for long-term memory formation. *Cell* **144**, 810–823 (2011).
63. T. L. Fetterly et al., Insulin bidirectionally alters NAc glutamatergic transmission: Interactions between insulin receptor activation, endogenous opioids, and glutamate release. *J Neurosci* **41**, 2360–2372 (2021).
64. C. Garcia-Caceres et al., Astrocytic insulin signaling couples brain glucose uptake with nutrient availability. *Cell* **166**, 867–880 (2016).
65. S. J. Vannucci, I. A. Simpson, Developmental switch in brain nutrient transporter expression in the rat. *Am J Physiol Endocrinol Metab* **285**, E1127–1134 (2003).
66. R. L. Leino, D. Z. Gerhart, L. R. Drewes, Monocarboxylate transporter (MCT1) abundance in brains of suckling and adult rats: A quantitative electron microscopic immunogold study. *Brain Res Dev Brain Res* **113**, 47–54 (1999).
67. S. C. Cunnane et al., Can ketones help rescue brain fuel supply in later life? Implications for cognitive health during aging and the treatment of Alzheimer's disease. *Front Mol Neurosci* **9**, 53 (2016).
68. O. E. Owen et al., Brain metabolism during fasting. *J Clin Invest* **46**, 1589–1595 (1967).
69. D. Ebert, R. G. Haller, M. E. Walton, Energy contribution of octanoate to intact rat brain metabolism measured by 13C nuclear magnetic resonance spectroscopy. *J Neurosci* **23**, 5928–5935 (2003).
70. S. Takahashi, Metabolic compartmentalization between astroglia and neurons in physiological and pathophysiological conditions of the neurovascular unit. *Neuropathology* **40**, 121–137 (2020).
71. C. Namkoong et al., Enhanced hypothalamic AMP-activated protein kinase activity contributes to hyperphagia in diabetic rats. *Diabetes* **54**, 63–68 (2005).
72. Y. Minokoshi et al., AMP-kinase regulates food intake by responding to hormonal and nutrient signals in the hypothalamus. *Nature* **428**, 569–574 (2004).
73. S. J. Koppel, R. H. Swerdlow, Neuroketotherapeutics: A modern review of a century-old therapy. *Neurochem Int* **117**, 114–125 (2018).
74. P. Puchalska, P. A. Crawford, Multi-dimensional roles of ketone bodies in fuel metabolism, signaling, and therapeutics. *Cell Metab* **25**, 262–284 (2017).
75. T. Vandoorne, K. De Bock, L. Van Den Bosch, Energy metabolism in ALS: An underappreciated opportunity? *Acta Neuropathol* **135**, 489–509 (2018).
76. Y. Zilberter, M. Zilberter, The vicious circle of hypometabolism in neurodegenerative diseases: Ways and mechanisms of metabolic correction. *J Neurosci Res* **95**, 2217–2235 (2017).
77. G. Pagano, F. Niccolini, M. Politis, Current status of PET imaging in Huntington's disease. *Eur J Nucl Med Mol Imaging* **43**, 1171–1182 (2016).
78. S. Cunnane et al., Brain fuel metabolism, aging, and Alzheimer's disease. *Nutrition* **27**, 3–20 (2011).
79. R. Bakshi, R. S. Miletich, P. R. Kinkel, M. L. Emmet, W. R. Kinkel, High-resolution fluorodeoxyglucose positron emission tomography shows both global and regional cerebral hypometabolism in multiple sclerosis. *J Neuroimaging* **8**, 228–234 (1998).
80. H. Yang, W. Shan, F. Zhu, J. Wu, Q. Wang, Ketone bodies in neurological diseases: Focus on neuroprotection and underlying mechanisms. *Front Neurol* **10**, 585 (2019).
81. M. Lilamand et al., Are ketogenic diets promising for Alzheimer's disease? A translational review. *Alzheimers Res Ther* **12**, 42 (2020).
82. B. Sridharan, M. J. Lee, Ketogenic diet: A promising neuroprotective composition for managing Alzheimer's diseases and its pathological mechanisms. *Curr Mol Med*, (2021).
83. K. Talbot et al., Demonstrated brain insulin resistance in Alzheimer's disease patients is associated with IGF-1 resistance, IRS-1 dysregulation, and cognitive decline. *J Clin Invest* **122**, 1316–1338 (2012).
84. J. Koizumi, Glycogen in the central nervous system. *Prog Histochem Cytochem* **6**, 1–37 (1974).
85. J. Koizumi, H. Shiraishi, Ultrastructural appearance of glycogen in the hypothalamus of the rabbit following chlorpromazine administration. *Exp Brain Res* **10**, 276–282 (1970).
86. M. Hamai, Y. Minokoshi, T. Shimazu, L-Glutamate and insulin enhance glycogen synthesis in cultured astrocytes from the rat brain through different intracellular mechanisms. *J Neurochem* **73**, 400–407 (1999).
87. R. Dringen, B. Hamprecht, Glucose, insulin, and insulin-like growth factor I regulate the

glycogen content of astroglia-rich primary cultures. *J Neurochem* **58**, 511–517 (1992).
88. W. Kum, S. Q. Zhu, S. K. Ho, J. D. Young, C. S. Cockram, Effect of insulin on glucose and glycogen metabolism and leucine incorporation into protein in cultured mouse astrocytes. *Glia* **6**, 264–268 (1992).
89. A. M. Cataldo, R. D. Broadwell, Cytochemical identification of cerebral glycogen and glucose-6-phosphatase activity under normal and experimental conditions. II. Choroid plexus and ependymal epithelia, endothelia and pericytes. *J Neurocytol* **15**, 511–524 (1986).
90. D. Vilchez et al., Mechanism suppressing glycogen synthesis in neurons and its demise in progressive myoclonus epilepsy. *Nat Neurosci* **10**, 1407–1413 (2007).
91. P. J. Magistretti, O. Sorg, N. Yu, J. L. Martin, L. Pellerin, Neurotransmitters regulate energy metabolism in astrocytes: implications for the metabolic trafficking between neural cells. *Dev Neurosci* **15**, 306–312 (1993).
92. J. Duran, A. Gruart, J. C. Lopez-Ramos, J. M. Delgado-Garcia, J. J. Guinovart, Glycogen in astrocytes and neurons: Physiological and pathological aspects. *Adv Neurobiol* **23**, 311–329 (2019).
93. I. Saez et al., Neurons have an active glycogen metabolism that contributes to tolerance to hypoxia. *J Cereb Blood Flow Metab* **34**, 945–955 (2014).
94. M. S. Gentry, J. J. Guinovart, B. A. Minassian, P. J. Roach, J. M. Serratosa, Lafora disease offers a unique window into neuronal glycogen metabolism. *J Biol Chem* **293**, 7117–7125 (2018).
95. J. Duran, A. Gruart, M. Garcia-Rocha, J. M. Delgado-Garcia, J. J. Guinovart, Glycogen accumulation underlies neurodegeneration and autophagy impairment in Lafora disease. *Hum Mol Genet* **23**, 3147–3156 (2014).
96. R. L. Sidman et al., Temporal neuropathologic and behavioral phenotype of 6neo/6neo Pompe disease mice. *J Neuropathol Exp Neurol* **67**, 803–818 (2008).
97. H. S. Han, G. Kang, J. S. Kim, B. H. Choi, S. H. Koo, Regulation of glucose metabolism from a liver-centric perspective. *Exp Mol Med* **48**, e218 (2016).
98. A. L. McCall, Cerebral glucose metabolism in diabetes mellitus. *Eur J Pharmacol* **490**, 147–158 (2004).
99. R. J. Schulingkamp, T. C. Pagano, D. Hung, R. B. Raffa, Insulin receptors and insulin action in the brain: review and clinical implications. *Neurosci Biobehav Rev* **24**, 855–872 (2000).
100. G. I. Smith et al., Insulin resistance drives hepatic de novo lipogenesis in nonalcoholic fatty liver disease. *J Clin Invest* **130**, 1453–1460 (2020).
101. Z. Song, A. M. Xiaoli, F. Yang, Regulation and metabolic significance of De Novo lipogenesis in adipose tissues. *Nutrients* **10**, (2018).
102. M. K. Hellerstein, De novo lipogenesis in humans: metabolic and regulatory aspects. *Eur J Clin Nutr* **53** (Suppl 1), S53–S65 (1999).
103. H. A. Ferris et al., Loss of astrocyte cholesterol synthesis disrupts neuronal function and alters whole-body metabolism. *Proc Natl Acad Sci U S A* **114**, 1189–1194 (2017).
104. R. Suzuki, H. A. Ferris, M. J. Chee, E. Maratos-Flier, C. R. Kahn, Reduction of the cholesterol sensor SCAP in the brains of mice causes impaired synaptic transmission and altered cognitive function. *PLoS Biol* **11**, e1001532 (2013).
105. R. Suzuki et al., Diabetes and insulin in regulation of brain cholesterol metabolism. *Cell Metab* **12**, 567–579 (2010).
106. A. Kleinridders, H. A. Ferris, W. Cai, C. R. Kahn, Insulin action in brain regulates systemic metabolism and brain function. *Diabetes* **63**, 2232–2243 (2014).
107. K. Park et al., Exogenous insulin infusion can decrease atherosclerosis in diabetic rodents by improving lipids, inflammation, and endothelial function. *Arterioscler Thromb Vasc Biol* **38**, 92–101 (2018).
108. M. Heni et al., Ectopic fat accumulation in human astrocytes impairs insulin action. *R Soc Open Sci* **7**, 200701 (2020).
109. I. J. Martins et al., Apolipoprotein E, cholesterol metabolism, diabetes, and the convergence of risk factors for Alzheimer's disease and cardiovascular disease. *Mol Psychiatry* **11**, 721–736 (2006).
110. A. Hamberger, H. Hyden, Inverse enzymatic changes in neurons and glia during increased function and hypoxia. *J Cell Biol* **16**, 521–525 (1963).
111. H. Hyden, P. W. Lange, A kinetic study of the neuronglia relationship. *J Cell Biol* **13**, 233–237 (1962).
112. A. Kleinridders et al., Insulin resistance in brain alters dopamine turnover and causes behavioral disorders. *Proc Natl Acad Sci U S A* **112**, 3463–3468 (2015).
113. M. Schell et al., Interplay of dietary fatty acids and cholesterol impacts brain mitochondria and insulin action. *Nutrients* **12**, (2020).

114. J. Fu, M. G. Yu, Q. Li, K. Park, G. L. King, Insulin's actions on vascular tissues: Physiological effects and pathophysiological contributions to vascular complications of diabetes. *Mol Metab* **52**, 101236 (2021).
115. G. N. Ruegsegger et al., Exercise and metformin counteract altered mitochondrial function in the insulin-resistant brain. *JCI Insight* **4**, (2019).
116. P. J. Fernandez-Marcos, J. Auwerx, Regulation of PGC-1alpha, a nodal regulator of mitochondrial biogenesis. *Am J Clin Nutr* **93**, 884S–890 (2011).
117. G. N. Ruegsegger, A. L. Creo, T. M. Cortes, S. Dasari, K. S. Nair, Altered mitochondrial function in insulin-deficient and insulin-resistant states. *J Clin Invest* **128**, 3671–3681 (2018).
118. M. E. Patti et al., Coordinated reduction of genes of oxidative metabolism in humans with insulin resistance and diabetes: Potential role of PGC1 and NRF1. *Proc Natl Acad Sci U S A* **100**, 8466–8471 (2003).
119. H. Daitoku, K. Yamagata, H. Matsuzaki, M. Hatta, A. Fukamizu, Regulation of PGC-1 promoter activity by protein kinase B and the forkhead transcription factor FKHR. *Diabetes* **52**, 642–649 (2003).
120. Z. Cheng et al., Foxo1 integrates insulin signaling with mitochondrial function in the liver. *Nat Med* **15**, 1307–1311 (2009).
121. J. T. Cunningham et al., mTOR controls mitochondrial oxidative function through a YY1-PGC-1alpha transcriptional complex. *Nature* **450**, 736–740 (2007).
122. Q. Li et al., Homozygous receptors for insulin and not IGF-1 accelerate intimal hyperplasia in insulin resistance and diabetes. *Nat Commun* **10**, 4427 (2019).
123. B. L. Olson et al., SCFCdc4 acts antagonistically to the PGC-1alpha transcriptional coactivator by targeting it for ubiquitin-mediated proteolysis. *Genes Dev* **22**, 252–264 (2008).
124. N. Zareen et al., Motor cortex and spinal cord neuromodulation promote corticospinal tract axonal outgrowth and motor recovery after cervical contusion spinal cord injury. *Exp Neurol* **297**, 179–189 (2017).
125. T. J. Huang, A. Verkhratsky, P. Fernyhough, Insulin enhances mitochondrial inner membrane potential and increases ATP levels through phosphoinositide 3-kinase in adult sensory neurons. *Mol Cell Neurosci* **28**, 42–54 (2005).
126. M. O. Dietrich, Z. W. Liu, T. L. Horvath, Mitochondrial dynamics controlled by mitofusins regulate Agrp neuronal activity and diet-induced obesity. *Cell* **155**, 188–199 (2013).
127. A. Santoro et al., DRP1 suppresses leptin and glucose sensing of POMC neurons. *Cell Metab* **25**, 647–660 (2017).
128. J. P. Castro, K. Wardelmann, T. Grune, A. Kleinridders, Mitochondrial chaperones in the brain: Safeguarding brain health and metabolism? *Front Endocrinol (Lausanne)* **9**, 196 (2018).
129. K. Wardelmann et al., Insulin action in the brain regulates mitochondrial stress responses and reduces diet-induced weight gain. *Mol Metab* **21**, 68–81 (2019).
130. K. Wardelmann et al., Central acting Hsp10 regulates mitochondrial function, fatty acid metabolism, and insulin sensitivity in the Hypothalamus. *Antioxidants (Basel)* **10**, (2021).
131. W. S. Liang et al., Alzheimer's disease is associated with reduced expression of energy metabolism genes in posterior cingulate neurons. *Proc Natl Acad Sci U S A* **105**, 4441–4446 (2008).
132. L. Zhang et al., Potential hippocampal genes and pathways involved in Alzheimer's disease: A bioinformatic analysis. *Genet Mol Res* **14**, 7218–7232 (2015).
133. W. Wang, F. Zhao, X. Ma, G. Perry, X. Zhu, Mitochondria dysfunction in the pathogenesis of Alzheimer's disease: Recent advances. *Mol Neurodegener* **15**, 30 (2020).
134. L. Wang et al., Synaptosomal mitochondrial dysfunction in 5xFAD mouse model of Alzheimer's disease. *PLoS One* **11**, e0150441 (2016).
135. J. Yao et al., Mitochondrial bioenergetic deficit precedes Alzheimer's pathology in female mouse model of Alzheimer's disease. *Proc Natl Acad Sci U S A* **106**, 14670–14675 (2009).
136. M. Ribeiro, T. R. Rosenstock, A. M. Oliveira, C. R. Oliveira, A. C. Rego, Insulin and IGF-1 improve mitochondrial function in a PI-3K/Akt-dependent manner and reduce mitochondrial generation of reactive oxygen species in Huntington's disease knock-in striatal cells. *Free Radic Biol Med* **74**, 129–144 (2014).
137. I. Aviles-Olmos, P. Limousin, A. Lees, T. Foltynie, Parkinson's disease, insulin resistance and novel agents of neuroprotection. *Brain* **136**, 374–384 (2013).
138. F. Iravanpour et al., Intranasal insulin improves mitochondrial function and attenuates motor deficits in a rat 6-OHDA model of Parkinson's disease. *CNS Neurosci Ther* **27**, 308–319 (2021).

139. S. Craft et al., Intranasal insulin therapy for Alzheimer disease and amnestic mild cognitive impairment: a pilot clinical trial. *Arch Neurol* **69**, 29–38 (2012).
140. A. T. Isik, P. Soysal, A. Yay, C. Usarel, The effects of sitagliptin, a DPP-4 inhibitor, on cognitive functions in elderly diabetic patients with or without Alzheimer's disease. *Diabetes Res Clin Pract* **123**, 192–198 (2017).
141. M. A. Reger et al., Intranasal insulin improves cognition and modulates beta-amyloid in early AD. *Neurology* **70**, 440–448 (2008).
142. A. Claxton et al., Long-acting intranasal insulin detemir improves cognition for adults with mild cognitive impairment or early-stage Alzheimer's disease dementia. *J Alzheimers Dis* **44**, 897–906 (2015).
143. S. Tzimopoulou et al., A multi-center randomized proof-of-concept clinical trial applying [(1)(8)F]FDG-PET for evaluation of metabolic therapy with rosiglitazone XR in mild to moderate Alzheimer's disease. *J Alzheimers Dis* **22**, 1241–1256 (2010).
144. G. E. Hardingham, F. J. Arnold, H. Bading, Nuclear calcium signaling controls CREB-mediated gene expression triggered by synaptic activity. *Nat Neurosci* **4**, 261–267 (2001).
145. I. Llano et al., Presynaptic calcium stores underlie large-amplitude miniature IPSCs and spontaneous calcium transients. *Nat Neurosci* **3**, 1256–1265 (2000).
146. A. Verkhratsky, E. C. Toescu, Calcium and neuronal ageing. *Trends Neurosci* **21**, 2–7 (1998).
147. T. Pozzan, R. Rizzuto, P. Volpe, J. Meldolesi, Molecular and cellular physiology of intracellular calcium stores. *Physiol Rev* **74**, 595–636 (1994).
148. A. Verkhratsky, Physiology and pathophysiology of the calcium store in the endoplasmic reticulum of neurons. *Physiol Rev* **85**, 201–279 (2005).
149. A. Verkhratsky, O. H. Petersen, The endoplasmic reticulum as an integrating signalling organelle: from neuronal signalling to neuronal death. *Eur J Pharmacol* **447**, 141–154 (2002).
150. D. Burdakov, O. H. Petersen, A. Verkhratsky, Intraluminal calcium as a primary regulator of endoplasmic reticulum function. *Cell Calcium* **38**, 303–310 (2005).
151. S. Marchi et al., Selective modulation of subtype III IP(3)R by Akt regulates ER Ca(2)(+) release and apoptosis. *Cell Death Dis* **3**, e304 (2012).
152. M. T. Khan, L. Wagner, 2nd, D. I. Yule, C. Bhanumathy, S. K. Joseph, Akt kinase phosphorylation of inositol 1,4,5-trisphosphate receptors. *J Biol Chem* **281**, 3731–3737 (2006).
153. I. Pomytkin, V. Pinelis, Insulin receptors and intracellular Ca (2+) form a double-negative regulatory feedback loop controlling insulin sensitivity. *F1000Res* **9**, 598 (2020).
154. Y. Maejima, D. Kohno, Y. Iwasaki, T. Yada, Insulin suppresses ghrelin-induced calcium signaling in neuropeptide Y neurons of the hypothalamic arcuate nucleus. *Aging (Albany NY)* **3**, 1092–1097 (2011).
155. A. Verkhratsky, P. Fernyhough, Mitochondrial malfunction and Ca2+ dyshomeostasis drive neuronal pathology in diabetes. *Cell Calcium* **44**, 112–122 (2008).
156. S. Fu et al., Aberrant lipid metabolism disrupts calcium homeostasis causing liver endoplasmic reticulum stress in obesity. *Nature* **473**, 528–531 (2011).
157. S. W. Park, Y. Zhou, J. Lee, J. Lee, U. Ozcan, Sarco(endo)plasmic reticulum Ca2+-ATPase 2b is a major regulator of endoplasmic reticulum stress and glucose homeostasis in obesity. *Proc Natl Acad Sci U S A* **107**, 19320–19325 (2010).
158. A. Gorlach, P. Klappa, T. Kietzmann, The endoplasmic reticulum: folding, calcium homeostasis, signaling, and redox control. *Antioxid Redox Signal* **8**, 1391–1418 (2006).
159. Y. Iwasaki et al., Insulin activates vagal afferent neurons including those innervating pancreas via insulin cascade and Ca(2+) influx: Its dysfunction in IRS2-KO mice with hyperphagic obesity. *PLoS One* **8**, e67198 (2013).
160. S. Maimaiti et al., Novel calcium-related targets of insulin in hippocampal neurons. *Neuroscience* **364**, 130–142 (2017).
161. T. J. Huang, N. M. Sayers, P. Fernyhough, A. Verkhratsky, Diabetes-induced alterations in calcium homeostasis in sensory neurones of streptozotocin-diabetic rats are restricted to lumbar ganglia and are prevented by neurotrophin-3. *Diabetologia* **45**, 560–570 (2002).
162. T. J. Huang et al., Insulin prevents depolarization of the mitochondrial inner membrane in sensory neurons of type 1 diabetic rats in the presence of sustained hyperglycemia. *Diabetes* **52**, 2129–2136 (2003).
163. S. Craft et al., Effects of regular and long-acting insulin on cognition and Alzheimer's disease biomarkers: A pilot clinical trial. *J Alzheimers Dis* **57**, 1325–1334 (2017).
164. K. L. Anderson et al., Impact of single or repeated dose intranasal zinc-free insulin in

young and aged F344 rats on cognition, signaling, and brain metabolism. *J Gerontol A Biol Sci Med Sci* **72**, 189–197 (2017).
165. S. Maimaiti et al., Intranasal insulin improves age-related cognitive deficits and reverses electrophysiological correlates of brain aging. *J Gerontol A Biol Sci Med Sci* **71**, 30–39 (2016).
166. D. R. Marks, K. Tucker, M. A. Cavallin, T. G. Mast, D. A. Fadool, Awake intranasal insulin delivery modifies protein complexes and alters memory, anxiety, and olfactory behaviors. *J Neurosci* **29**, 6734–6751 (2009).
167. M. A. Reger et al., Effects of intranasal insulin on cognition in memory-impaired older adults: modulation by APOE genotype. *Neurobiol Aging* **27**, 451–458 (2006).
168. C. Benedict et al., Intranasal insulin improves memory in humans. *Psychoneuroendocrinology* **29**, 1326–1334 (2004).
169. R. M. Denton, Regulation of mitochondrial dehydrogenases by calcium ions. *Biochim Biophys Acta* **1787**, 1309–1316 (2009).
170. R. Rizzuto, D. De Stefani, A. Raffaello, C. Mammucari, Mitochondria as sensors and regulators of calcium signalling. *Nat Rev Mol Cell Biol* **13**, 566–578 (2012).
171. S. Grimm, The ER-mitochondria interface: The social network of cell death. *Biochim Biophys Acta* **1823**, 327–334 (2012).
172. R. Rizzuto et al., Close contacts with the endoplasmic reticulum as determinants of mitochondrial Ca2+ responses. *Science* **280**, 1763–1766 (1998).
173. T. I. Peng, M. J. Jou, Oxidative stress caused by mitochondrial calcium overload. *Ann N Y Acad Sci* **1201**, 183–188 (2010).
174. A. J. Kowaltowski, N. C. de Souza-Pinto, R. F. Castilho, A. E. Vercesi, Mitochondria and reactive oxygen species. *Free Radic Biol Med* **47**, 333–343 (2009).
175. A. Rimessi, C. Giorgi, P. Pinton, R. Rizzuto, The versatility of mitochondrial calcium signals: from stimulation of cell metabolism to induction of cell death. *Biochim Biophys Acta* **1777**, 808–816 (2008).
176. P. Fernyhough, T. J. Huang, A. Verkhratsky, Mechanism of mitochondrial dysfunction in diabetic sensory neuropathy. *J Peripher Nerv Syst* **8**, 227–235 (2003).
177. A. P. Arruda, G. S. Hotamisligil, Calcium homeostasis and organelle function in the pathogenesis of obesity and diabetes. *Cell Metab* **22**, 381–397 (2015).

CHAPTER FOUR

Regulation of Glial Function by Insulin Peptides

Ana M. Fernandez and Laura Martinez-Rachadell
Patricia Miranda-Azpiazu
Ignacio Torres Aleman

CONTENTS

INTRODUCTION

Probably all cells in intact tissues express at least one type of ILP receptor. In the brain, all cell types analyzed express insulin and IGF-I receptors (1), while a detailed cellular localization for other ILP receptors in brain tissue is still pending. Since these two hormones are the best characterized at the brain level, whereas for the rest of the members of the family only scattered information is available, in this chapter we will focus on insulin and IGF-I. Widespread distribution of their receptors in the brain is also a characteristic of these ILPs, being present all along the brain axis, in all major regions (2). Thus, the ubiquitous, universal presence of brain ILP receptors bespeaks their relevance in brain function.

As the local synthesis of insulin and IGF-I in the adult mammalian brain is low and circumscribed to few cell types, particularly in the case of insulin (3, 4), it appears that the main source of these two brains ILPs to interact with the ample brain distribution of their receptors is from the periphery. This is a second intriguing characteristic of this family of hormones, as discussed elsewhere (2), indicating that they form part of the humoral arm of the body-brain dialog.

Somewhat surprisingly, mechanisms of passage into the brain of circulating hormones such as

DOI: 10.1201/9781003079927-4

insulin have not been thoroughly analyzed, even though its central actions were described decades ago (5). Just relatively recently, the notion that the blood-brain-barrier (BBB) contains hormone-specific transport mechanisms has been included as a normal concept in BBB physiology, thanks to the pioneer work of a few groups (6, 7). However, in the case of insulin, the precise saturable transport mechanisms involved are still under scrutiny and no consensus has been reached yet (8–10). At any rate, glial cells, and more specifically astrocytes, appear to be key cellular elements in this transport as they protrude specialized cytoplasmic extensions (end-feet) ensheathing brain vessels (11).

While neurons have been the major focus of research when addressing the role of ILPs in the nervous system (2), the different types of glial cells are increasingly becoming the object of scrutiny in this respect. These cells have been defined based on lineage and function, most of which are conserved from invertebrates to mammals. We will specifically review ILP actions in the three most important types of glial cells in vertebrates; astrocytes, oligodendrocytes, and microglia. Since there is little information in relation to insulin actions in the two latter types, most publications focus on IGF-I. In addition, while there is an increasing realization that glial cells display great heterogeneity, no information is yet available on whether ILPs differentially affect them depending on brain region or ontogeny.

IGF-I and insulin have a well-conserved role along the evolutive scale, and all organisms with a central nervous system (CNS) contain glial cells (12). Hence, it seems logical that ILPs show a wide repertoire of actions on them. Furthermore, glial cells are important components of the peripheral-brain crosstalk where ILPs actively participate - for instance, at the level of the BBB (11), and even in the interactions among ILPs (13). Thus, regulation of glial cells by ILPs should be considered an important aspect of ILPs function in the CNS. In this chapter, we will discuss recent *in vivo* observations, with underlying mechanisms explored *in vitro*, when available. Recent thorough reviews on specific aspects of ILPs actions in glial cells are available (14).

ASTROGLIA

Astrocytes were first described by Virchow (15) and are an abundant macroglial cell type in the CNS, forming intimate contacts with vessels and neurons. Astrocytes derive from the neuroepithelium, and equivalent glial cell types are present all along the phylogenic scale (16–18). Astrocytes are critical for brain health (19) exerting different homeostatic activities (20), such as the synthesis of neuromodulators, hormones, trophic factors, or modulators of synaptogenesis, forming what is collectively known as a "gliocrine" system (21). The emerging variety of actions of ILPs on astrocytes reflects the multiple roles of these heterogeneous and multifaceted cells in the brain (22). These include modulating synaptic activity, maintaining BBB integrity, regulating extracellular ion balance, participating in the neuroinflammatory response, supporting neuronal activity, and many more that are constantly updated (23). Importantly, astrocytes also participate in higher brain functions impacting behavior (24). Indeed, glial cells are sensitive to mood disturbances (25).

Physiology

Although the role of ILPs in the brain is still not fully understood, it is becoming increasingly clear that their actions on higher brain skills rely in part on astrocytes (26). For example, it was recently demonstrated that insulin signaling in astrocytes participates in mood homeostasis. Loss of IRs on astrocytes in the nucleus accumbens was sufficient to induce depressive-like behavior in mice (27). The proposed mechanism involves the impaired release of ATP from astrocytes, which can alter the activity of dopaminergic neurons and potentially other neurons. Mood disturbances were mostly observed in female mice, suggesting a sex-specific role for IR in astrocytes. Moreover, postnatal loss of the insulin receptor in astrocytes affects their morphology, circuit connectivity, and mitochondrial function (28). This data suggests that insulin can regulate multiple aspects of astrocyte function, ultimately affecting neuronal plasticity.

In turn, IGF-I signaling through astrocytes appears involved in sensory processing. Mice lacking IGF-I receptors (IGF-IR) in astrocytes show impaired whisker discrimination since cortical neuronal plasticity is compromised (29). IGF-I stimulates ATP release by astrocytes through Ca^{++} signaling, in turn activating adenosine receptors at presynaptic inhibitory terminals to modulate synaptic plasticity in neighboring neurons. Thus, IGF-I modulates sensory perception not only through neurons, as previously established (30) but also through astrocytes. Modulation of ATP release by astrocytes also underlies part of the actions of IGF-I on cognitive function(31). Hence, both insulin and IGF-I modulate ATP release by astrocytes and in this way participate in synaptic plasticity affecting mood and cognition.

These observations highlight an underlying theme in glial function; the intense astrocyte-neuronal crosstalk that constitutes a mainstay of astrocyte physiology. Already in invertebrates, astrocytes are key partners of neurons. In *drosophila*, Dilp-6, one of eight insulin peptides in this fly, is instrumental in glia-neuron interactions for reparative neurogenesis (32). In mammals, insulin acts via astrocytes to control the response of hypothalamic arcuate pro-opiomelanocortin (POMC) neurons to glucose (28) or to allow reproductive competency via gonadotropin-releasing hormone neurons (33).

Indeed, a relatively well-studied region regarding the roles of ILPs on glial cells in the hypothalamus, particularly in the case of insulin. In this region, astrocytes play a critical role as metabolic sensors (34) by modulating glucose sensing neurons that in turn control systemic metabolism, food intake, energy expenditure, and glycemia (35). However, mechanisms downstream of activation of IR and IGF-IR receptors in hypothalamic astrocytes regulating metabolism has just begun to be explored in an extensive way (36), making it likely that new observations will position astrocytes as key players in central regulation by insulin and IGF-I of peripheral metabolism.

In this vein, a wider role of astrocytes in brain glucose homeostasis is revealed by ILPs. IGF-I cooperates with insulin through their respective receptors in astrocytes to control brain glucose uptake(37, 38). The underlying mechanisms involve a network of protein-protein interactions resulting in increased Glucose transporter 1 (GluT1) at the astrocyte cell membrane in response to brain activation (37). The requirement of both astrocytic IR and IGF-IR for brain glucose uptake may explain why the brain was considered for many years insensible to the regulatory actions of insulin on glucose homeostasis. Of note, IR in astrocytes modulates fuel preference by modulating mitochondrial function (28), whereas IGF-IR maintains redox homeostasis in these cells (39).

As part of the BBB, astrocytes are ideally positioned between the cerebral vasculature and neurons to mediate nutrient uptake from the systemic circulation (40). The BBB separates the periphery from the brain, allowing the transport of blood-borne molecules in a tightly regulated fashion and protecting the brain from neurotoxins. Moreover, the BBB is involved in neurovascular coupling, whereby active brain regions increase oxygen and nutrient uptake (41) and circulating trophic factors such as IGF-I (42). Two recent studies analyzed the role played by IR and IGF-IR in astrocytes in transport through the BBB of their respective ligands. Since genetic ablation of IR or IGF-IR in astrocytes greatly reduced the entrance of circulating insulin or IGF-I into the brain, respectively, it seems that astrocytes form part of the intracellular pathway of the passage of these hormones into the brain (11). In addition, down-regulation of IR (Fernandez et al, submitted), or IGF-IR (43), also compromises neuro-vascular coupling.

Pathology

Astrocytes are emerging as key regulators of CNS responses to insult, with important implications for diverse disorders, leading to the concept of "astrocytopathies" (44, 45). For major chronic or acute neuro-pathologies, the involvement of ILPs in reactive adaptations by astrocytes is well recognized. Probably the most studied aspect of the role of astrocytes in brain pathology is neuro-inflammation, a process that recruits brain-specific cellular and molecular processes in concert with the immune system. IGF-I is well-posed to act through astrocytes in this context. Indeed, it is a potent modulator of the peripheral immune response (46) and of astrocyte responses to neuroinflammatory mediators (47). IGF-I has been shown to modulate a calcineurin/NFκB pathway in astrocytes that participate in neuroinflammatory processes underlying Alzheimer's disease (48). Pharmacological manipulation of an astrocyte-specific pro-inflammatory pathway that interferes with the association of the phosphatase calcineurin with the transcription factor FOXO3, protects neurons against

Aβ-induced cell death (49). Thus, cell-specific anti-inflammatory drugs may open a new avenue to treat neuro-inflammation associated with AD and other neurodegenerative diseases (49). Others documented that the anti-inflammatory effects of IGF-1 in association with astrocyte activation and the G protein-coupled estrogen receptor improved neurotoxin-induced motor deficits (50). In other pathologies, increased expression of IGF-I by astrocytes appears important for reparative processes, such as after traumatic brain injury (51) or ischemia (52). Conversely, reduced IGF-IR activity in astrocytes diminishes the capacity of these cells to protect neurons against oxidative damage (39), and potentially leads to excitotoxity as glutamate uptake by astrocytes is also reduced (53).

An additional impetus has been the analysis of the role of astrocytes in rare diseases. One example is Friedreich's ataxia (FA), an inherited neurodegenerative disorder that courses with ataxic gait and loss of proprioception. FA is caused by a large trinucleotide repeat expansion in the frataxin gene. Frataxin deficiency in FA can lead to oxidative damage and subsequent neurodegeneration. As gait and coordination-related symptoms are hallmarks of the disorder, cerebellar neurons and glia may be particularly disrupted as a result of frataxin deficiency. Indeed, recent studies in mice and humans found an active role of astrocytes in the disease (54, 55), opening new therapeutic avenues for this intractable disease.

MICROGLIA

Microglial cells, first described by del Pio-Hortega (56), are of myeloid origin and are considered resident macrophages in the brain, migrating into this organ during mid-development, until the BBB becomes functional (57). Hence, their role in brain homeostasis starts already at these stages of development (58), participating even in higher brain skills in the adult through modulation of synaptic function (59). Until now, their best characterized role is scavenging of cell/metabolic debris through constant surveillance of brain tissue (60) forming part of the cellular defense of the brain against homeostatic imbalance. However, as with astrocytes, their role in brain function is constantly updated (61). Conversely, their role in brain pathology is well recognized (62), and extensive data relate them to neurodegenerative diseases (63).

Physiology

Unfortunately, little is known about the role of ILPs in microglia development and function in the intact brain. Seminal observations with IGF-I or insulin knock outs (KO), did not report microglial compromise (64, 65), although at that time microglial cells were not considered brain targets of ILPs. However, recent findings indicate that IGF-I KO mice present decreased microglial cells in the ear (66), making it necessary to revisit the role of ILPs in microglial development, at least in specific brain regions. In turn, neuronal survival during development depends on IGF-I expressing microglia (67). Furthermore, a subset of neonatal microglia that participates in myelination and neurogenesis during development, express high levels of IGF-I (68). Thus, IGF-I seems to be mediating actions of microglia on developing neurons and oligodendroglia.

In the mature brain, studies in invertebrates indicate that ILPs modulate the phagocytic activity of glial cells (69), its best recognized physiological role (70). Whether this role is maintained in vertebrates is not yet documented. However, considering the conserved action of ILPs along a phylogeny, it can be expected. Thus, it is tempting to speculate that the high metabolic flexibility of vertebrate microglia, which involves mTOR (71) and is critical for its phagocytic activity, might in part rely on ILPs, known modulators of mTOR. Moreover, phagocytic activity of microglia decays with aging (72) and neurodegenerative diseases (73), two conditions where ILPs loss efficacy in the brain (2). There is also recent evidence that insulin primes microglia reactivity in the adult brain (74).

Pathology

Many observations indicate that ILPs, in particular IGF-I, act as anti-inflammatory effectors by modulating microglial activity (75), including enhancing their proliferation (76). An intense crosstalk between microglia and astroglia and their consequences on oligodendroglia and neurons in responses to injury (77) make microglial cells of paramount importance in brain pathology (78). ILPs are involved in this intricate crosstalk; for instance, the formation of a scar after spinal cord injury, that favors reparative processes and neuronal survival, depends on the microglial secretion of neuroprotective factors such as IGF-I (79). However, the production of IGF-I by microglia is not always beneficial, as it has been documented to promote brain tumor development (80). Similarly, insulin seems to activate pro-inflammatory pathways in these cells (74). The latter highlights the context-dependent actions of ILPs.

OLIGODENDROGLIA

Oligodendrocytes were also first described by del Rio-Hortega (81), and are the myelinating cells of the CNS, while Schwann cells are their counter-part in the peripheral nervous system. Both are from neuroepithelial lineage. Oligodendrocyte precursors (usually refer as NG-2 glia) remain in high numbers in the adult brain and constitute a renewal pool for these glial cells, generated all along ontogeny. Accordingly, this source of precursor cells is important for endogenous re-myelination (82) and is modulated by ILPs (83, 84). Myelinated axons are a key component of the mammalian brain, providing fast conductance capacities to neurons and forming the white matter of the brain. Other roles of oligodendroglia in the adult CNS that are increasingly emerging (85), may also be regulated by ILPs, but more work is needed.

Physiology

The role of ILPs in these glial cells was investigated years ago, but interest in this topic has drastically declined, with very little new information in late years. For this reason, we can only discuss relatively old data. A key developmental role of IGF-I in oligodendrocytes was an early finding. This growth factor, together with several others, is an oligodendrocyte differentiating signal (86). Moreover, mice overexpressing IGF-I in the brain show increased myelin content (87), whereas IGF-I KOs show specific depletion of oligodendrocytes and severe loss of myelin (64), as IGF-I is known to promote expression of myelin proteins (88). The maintenance of oligodendrocyte cell numbers in the adult also depends on proper IGF-I activity (89). As for possible actions of insulin in oligodendrocyte physiology, very little information is available to the best of our knowledge, with IR KO mice showing intact myelination (90) while double IGF-IR/IR KOs show disrupted myelination with reduced expression of fatty acid-related genes (91). In addition, studies modulating the activity of proteins involved in downstream signaling by ILPs (and other extracellular messengers) such as IRSs indicate that IRS-2 in the brain is involved in myelination (90).

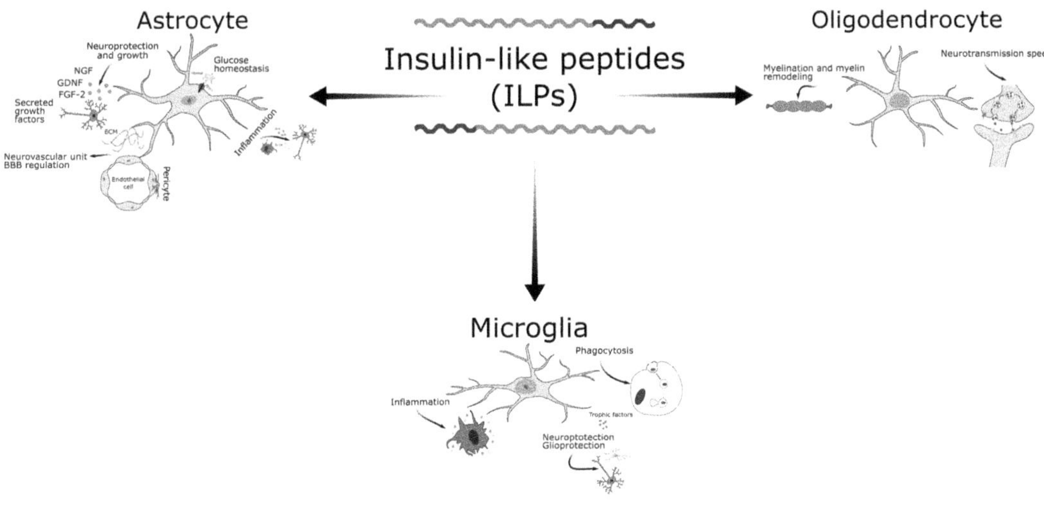

Figure 4.1 Summary of the major well-documented actions of ILPs on astrocytes, microglia, and oligodendrocytes. Astrocytes are by far the best studied glial type, with effects of ILPs described on the main known activities of these cells. In microglia, most studies are related to the protective role of these cells. In oligodendrocytes, available information is oriented to developmental actions impinging eventually on neuronal function.

TABLE 4.1
Summary of Currently Known Actions of ILPs on Glial Cells

Type of Cell	Type of Action
Astroglia	Blood-brain barrier function Glucose uptake Neuroprotection Gliotransmission Inflammatory responses Gap junction communication Glutamate handling Mitochondrial function Mood homeostasis Cognition Central control of glucose homeostasis Reproduction
Microglia	Cytoprotection Inflammation Response to injury Phagocytosis Development (?)
Oligodendroglia	Differentiation and survival Myelination and re-myelination Neurotransmitter transmission

Pathology

Demyelinating diseases are of high clinical impact (92). Unfortunately, none of them still have a cure. As a regenerative factor, IGF-I was shown time ago to participate in re-myelinating responses to various types of central injuries (93, 94), although these observations were not confirmed by others (95). Evidence indicated that astrocyte-derived IGF-I participates in re-myelinating responses to injury (96). Just recently, IGF-I, together with insulin, has been shown to be required in homeostatic reactivity to peripheral neuropathies, acting through Schwann cells (91). Despite its protective actions in pre-clinical demyelinating models, a pilot study in multiple sclerosis patients did not show changes in the disease course (97). Thus, as with most potential clinical uses of ILPs in the diseased brain, the therapeutic utility of IGF-I in demyelinating diseases has not been further pursued.

CONCLUDING REMARKS

Regulatory actions of ILPs on glial cells are understudied (see Figure 4.1). Astrocytes have been the mainstream of research in this regard, with accumulating evidence clearly pointing to a central role of these glial cells in brain ILPs actions. This merely reflects the increasing importance that is given to astrocytes in neurobiological research. For other types of glial cells, available information is limited, uneven, and most of it requires further confirmatory studies. At any rate, glial cells should be considered the main target of ILPs in the brain and merit extensive further analyses. It is tempting to predict that a major drive of future research on ILPs in the brain will include glial cells.

ACKNOWLEDGMENTS

Work at the lab is supported by grants from MINECO, Ciberned, CAM, and Ikerbasque Foundation. We thank previous members of the lab for their work.

REFERENCES

1. L. Martinez-Rachadell et al., Cell-specific expression of insulin/insulin-like growth factor-I receptor hybrids in the mouse brain. *Growth Horm IGF Res* **45**, 25–30 (2019).
2. A. M. Fernandez, I. Torres-Aleman, The many faces of insulin-like peptide signalling in the brain. *Nat Rev Neurosci* **13**, 225–239 (2012).

3. S. U. Devaskar et al., Insulin gene expression and insulin synthesis in mammalian neuronal cells. *J Biol Chem* **269**, 8445–8454 (1994).
4. T. Kuwabara et al., Insulin biosynthesis in neuronal progenitors derived from adult hippocampus and the olfactory bulb. *EMBO Mol Med* **3**, 742–754 (2011).
5. G. D. Baura et al., Saturable transport of insulin from plasma into the central nervous system of dogs in vivo. A mechanism for regulated insulin delivery to the brain. *J Clin Invest* **92**, 1824–1830 (1993).
6. W. A. Banks, J. B. Jaspan, W. Huang, A. J. Kastin, Transport of insulin across the blood-brain barrier: saturability at euglycemic doses of insulin. *Peptides* **18**, 1423–1429 (1997).
7. W. A. Banks, J. B. Owen, M. A. Erickson, Insulin in the brain: there and back again. *Pharmacol Ther* **136**, 82–93 (2012).
8. E. M. Rhea, C. Rask-Madsen, W. A. Banks, Insulin transport across the blood–brain barrier can occur independently of the insulin receptor. *J Physiol* **596**, 4753–4765 (2018).
9. S. M. Gray, K. W. Aylor, E. J. Barrett, Unravelling the regulation of insulin transport across the brain endothelial cell. *Diabetologia* **60**, 1512–1521 (2017).
10. M. Hersom et al., The insulin receptor is expressed and functional in cultured blood-brain barrier endothelial cells, but does not mediate insulin entry from blood-to-brain. *Am J Physiol Endocrinol Metab* **315**, 531–542 (2018).
11. A. M. Fernandez et al., The insulin receptor in astrocytes is involved in the entrance of circulating insulin into the brain. *bioRxiv*, 720813 (2019).
12. M. R. Freeman, D. H. Rowitch, Evolving concepts of gliogenesis: a look way back and ahead to the next 25 years. *Neuron* **80**, 613–623 (2013).
13. A. M. Fernandez, E. Hernandez, D. Guerrero-Gomez, A. Miranda-Vizuete, I. Torres Aleman, A network of insulin peptides regulate glucose uptake by astrocytes: potential new druggable targets for brain hypometabolism. *Neuropharmacology* **136**, 216–222 (2018).
14. I. Gonzalez-Garcia, T. Gruber, C. Garcia-Caceres, Insulin action on astrocytes: from energy homeostasis to behaviour. *J Neuroendocrinol* e12953 (2021).
15. V. Parpura, A. Verkhratsky, Neuroglia at the crossroads of homoeostasis, metabolism and signalling: evolution of the concept. *ASN Neuro* **4**, 201–205 (2012).
16. A. Singhvi, S. Shaham, Glia-neuron interactions in Caenorhabditis elegans. *Annu Rev Neurosci* **42**, 149–168 (2019).
17. K. Yildirim, J. Petri, R. Kottmeier, C. Klämbt, Drosophila glia: few cell types and many conserved functions. *Glia* **67**, 5–26 (2019).
18. A. Verkhratsky, M. S. Ho, V. Parpura, Evolution of neuroglia. *Adv Exp Med Biol* **1175**, 15–44 (2019).
19. M. S. McCann, K. A. Maguire-Zeiss, Environmental toxicants in the brain: a review of astrocytic metabolic dysfunction. *Environ Toxicol Pharmacol* **84**, 103608 (2021).
20. A. Verkhratsky, V. Parpura, N. Vardjan, R. Zorec, Physiology of astroglia. *Adv Exp Med Biol* **1175**, 45–91 (2019).
21. A. Verkhratsky, M. Matteoli, V. Parpura, J. P. Mothet, R. Zorec, Astrocytes as secretory cells of the central nervous system: idiosyncrasies of vesicular secretion. *EMBO J* **35**, 239–257 (2016).
22. B. S. Khakh, M. V. Sofroniew, Diversity of astrocyte functions and phenotypes in neural circuits. *Nat Neurosci* **18**, 942–952 (2015).
23. F. Zhao et al., Tak1 in the astrocytes of mediobasal hypothalamus regulates anxiety-like behavior in mice. *Glia* **69**, 609–618 (2021).
24. J. Nagai et al., Behaviorally consequential astrocytic regulation of neural circuits. *Neuron* **109**, 576–596 (2021).
25. A. Du Preez et al., Chronic stress followed by social isolation promotes depressive-like behaviour, alters microglial and astrocyte biology and reduces hippocampal neurogenesis in male mice. *Brain Behav Immun* **91**, 24–47 (2021).
26. C. A. Beddows, G. T. Dodd, Insulin on the brain: the role of central insulin signalling in energy and glucose homeostasis. *J Neuroendocrinol* e12947 (2021).
27. W. Cai et al., Insulin regulates astrocyte gliotransmission and modulates behavior. *J Clin Invest* **128**, 2914–2926 (2018).
28. C. Garcia-Caceres et al., Astrocytic insulin signaling couples brain glucose uptake with nutrient availability. *Cell* **166**, 867–880 (2016).
29. J. A. Noriega-Prieto et al., IGF-I governs cortical inhibitory synaptic plasticity by astrocyte activation. *bioRxiv*, 2020.2002.2011.942532 (2020).
30. A. R. Mardinly et al., Sensory experience regulates cortical inhibition by inducing IGF1 in VIP neurons. *Nature* **531**, 371–375 (2016).
31. S. Logan et al., Insulin-like growth factor receptor signaling regulates working memory, mitochondrial metabolism, and amyloid-beta uptake in astrocytes. *Mol Metab* **9**, 141–155 (2018).

32. N. J. Harrison et al., Regenerative neurogenic response from glia requires insulin-driven neuron-glia communication. *Elife* **10**, (2021).
33. I. H. Manaserh et al., Ablating astrocyte insulin receptors leads to delayed puberty and hypogonadism in mice. *PLoS Biol* **17**, e3000189 (2019).
34. N. Marina et al., Brain metabolic sensing and metabolic signaling at the level of an astrocyte. *Glia* **66**, 1185–1199 (2018).
35. J. A. Chowen, L. M. Frago, M. S. Fernandez-Alfonso, Physiological and pathophysiological roles of hypothalamic astrocytes in metabolism. *J Neuroendocrinol* **31**, e12671 (2019).
36. S. Camandola, Astrocytes, emerging stars of energy homeostasis. *Cell Stress* **2**, 246–252 (2018).
37. E. Hernandez-Garzon et al., The insulin-like growth factor I receptor regulates glucose transport by astrocytes. *Glia* **64**, 1962–1971 (2016).
38. A. M. Fernandez et al., Insulin regulates astrocytic glucose handling through cooperation with IGF-I. *Diabetes* **66**, 64–74 (2017).
39. L. Genis et al., Astrocytes require insulin-like growth factor I to protect neurons against oxidative injury. *F1000Res* **3**, 28 (2014).
40. N. J. Abbott, A. A. K. Patabendige, D. E. M. Dolman, S. R. Yusof, D. J. Begley, Structure and function of the blood-brain barrier. *Neurobiol Dis* **37**, 13–25 (2010).
41. C. Iadecola, The neurovascular unit coming of age: a journey through neurovascular coupling in health and disease. *Neuron* **96**, 17–42 (2017).
42. T. Nishijima et al., Neuronal activity drives localized blood-brain-barrier transport of serum insulin-like growth factor-I into the CNS. *Neuron* **67**, 834–846 (2010).
43. S. Tarantini et al., IGF1R signaling regulates astrocyte-mediated neurovascular coupling in mice: implications for brain aging. *GeroScience*, (2021).
44. D. M. de Waard, M. Bugiani, Astrocyte-oligodendrocyte-microglia crosstalk in astrocytopathies. *Front Cell Neurosci* **14**, 608073 (2020).
45. M. V. Sofroniew, Astrocyte barriers to neurotoxic inflammation. *Nat Rev Neurosci* **16**, 249–263 (2015).
46. D. Bilbao, L. Luciani, B. Johannesson, A. Piszczek, N. Rosenthal, Insulin-like growth factor-1 stimulates regulatory T cells and suppresses autoimmune disease. *EMBO Mol Med* **6**, 1423–1435 (2014).
47. A. M. Fernandez et al., Regulation of the phosphatase calcineurin by insulin-like growth factor I unveils a key role of astrocytes in Alzheimer's pathology. *Mol Psychiatry* **17**, 705–718 (2012).
48. A. M. Fernandez, S. Fernandez, P. Carrero, M. Garcia-Garcia, I. Torres-Aleman, Calcineurin in reactive astrocytes plays a key role in the interplay between proinflammatory and anti-inflammatory signals. *J Neurosci* **27**, 8745–8756 (2007).
49. A. M. Fernandez et al., Blockade of the interaction of calcineurin with FOXO in astrocytes protects against amyloid-beta-induced neuronal death. *J Alzheimers Dis* **52**, 1471–1478 (2016).
50. L. J. Yuan, M. Zhang, S. Chen, W. F. Chen, Anti-inflammatory effect of IGF-1 is mediated by IGF-1R cross talk with GPER in MPTP/MPP(+)-induced astrocyte activation. *Mol Cell Endocrinol* **519**, 111053 (2021).
51. S. K. Madathil et al., Astrocyte-specific overexpression of insulin-like growth factor-1 protects hippocampal neurons and reduces behavioral deficits following traumatic brain injury in mice. *PLoS One* **8**, e67204 (2013).
52. A. K. Okoreeh, S. Bake, F. Sohrabji, Astrocyte-specific insulin-like growth factor-1 gene transfer in aging female rats improves stroke outcomes. *Glia* **65**, 1043–1058 (2017).
53. D. Prabhu, S. M. Khan, K. Blackburn, J. P. Marshall, N. M. Ashpole, Loss of insulin-like growth factor-1 signaling in astrocytes disrupts glutamate handling. *J Neurochem* **151**, 689–702 (2019).
54. F. Loria, J. Diaz-Nido, Frataxin knockdown in human astrocytes triggers cell death and the release of factors that cause neuronal toxicity. *Neurobiol Dis* **76**, 1–12 (2015).
55. C. Franco et al., A role for astrocytes in cerebellar deficits in frataxin deficiency: protection by insulin-like growth factor I. *Mol Cell Neurosci* **80**, 100–110 (2017).
56. A. Sierra et al., The "Big-Bang" for modern glial biology: translation and comments on Pio del Rio-Hortega 1919 series of papers on microglia. *Glia* **64**, 1801–1840 (2016).
57. C. J. Bohlen, B. A. Friedman, B. Dejanovic, M. Sheng, Microglia in brain development, homeostasis, and neurodegeneration. *Annu Rev Genet* **53**, 263–288 (2019).
58. O. Matcovitch-Natan et al., Microglia development follows a stepwise program to regulate brain homeostasis. *Science* **353**, aad8670 (2016).
59. C. N. Parkhurst et al., Microglia promote learning-dependent synapse formation through

brain-derived neurotrophic factor. *Cell* **155**, 1596–1609 (2013).
60. A. Nimmerjahn, F. Kirchhoff, F. Helmchen, Resting microglial cells are highly dynamic surveillants of brain parenchyma in vivo. *Science* **308**, 1314–1318 (2005).
61. G. Saw et al., Epigenetic regulation of microglial phosphatidylinositol 3-kinase pathway involved in long-term potentiation and synaptic plasticity in rats. *Glia* **68**, 656–669 (2020).
62. O. Butovsky, H. L. Weiner, Microglial signatures and their role in health and disease. *Nat Rev Neurosci* **19**, 622–635 (2018).
63. J. A. Chowen, L. M. Garcia-Segura, Microglia, neurodegeneration and loss of neuroendocrine control. *Prog Neurobiol* **184**, 101720 (2020).
64. K. D. Beck, L. Powell-Braxton, H. R. Widmer, J. Valverde, F. Hefti, Igf1 gene disruption results in reduced brain size, CNS hypomyelination, and loss of hippocampal granule and striatal parvalbumin-containing neurons. *Neuron* **14**, 717–730 (1995).
65. J. C. Bruning et al., Role of brain insulin receptor in control of body weight and reproduction. *Science* **289**, 2122–2125 (2000).
66. V. Fuentes-Santamaría, J. C. Alvarado, L. Rodríguez-de la Rosa, J. M. Juiz, I. Varela-Nieto, Neuroglial involvement in abnormal glutamate transport in the cochlear nuclei of the igf1 (-/-) mouse. *Front Cell Neurosci* **13**, 67 (2019).
67. M. Ueno et al., Layer V cortical neurons require microglial support for survival during postnatal development. *Nat Neurosci* **16**, 543–551 (2013).
68. A. Wlodarczyk et al., A novel microglial subset plays a key role in myelinogenesis in developing brain. *EMBO J* **36**, 3292–3308 (2017).
69. D. T. Musashe, M. D. Purice, S. D. Speese, J. Doherty, M. A. Logan, Insulin-like signaling promotes glial phagocytic clearance of degenerating axons through regulation of draper. *Cell Rep* **16**, 1838–1850 (2016).
70. S. Arandjelovic, K. S. Ravichandran, Phagocytosis of apoptotic cells in homeostasis. *Nat Immunol* **16**, 907–917 (2015).
71. L. P. Bernier et al., Microglial metabolic flexibility supports immune surveillance of the brain parenchyma. *Nat Commun* **11**, 1559 (2020).
72. J. V. Pluvinage et al., CD22 blockade restores homeostatic microglial phagocytosis in ageing brains. *Nature* **568**, 187–192 (2019).
73. S. Hong et al., Complement and microglia mediate early synapse loss in Alzheimer mouse models. *Science* **352**, 712–716 (2016).
74. C. B. Haas, A. K. de Carvalho, A. P. Muller, B. J. L. Eggen, L. V. Portela, Insulin activates microglia and increases COX-2/IL-1beta expression in young but not in aged hippocampus. *Brain Res* **1741**, 146884 (2020).
75. A. Serhan, J. L. Aerts, E. Boddeke, R. Kooijman, Neuroprotection by insulin-like growth factor-1 in rats with ischemic stroke is associated with microglial changes and a reduction in neuroinflammation. *Neuroscience* **426**, 101–114 (2020).
76. S. L. O'Donnell, T. J. Frederick, J. K. Krady, S. J. Vannucci, T. L. Wood, IGF-I and microglia/macrophage proliferation in the ischemic mouse brain. *Glia* **39**, 85–97 (2002).
77. S. A. Liddelow et al., Neurotoxic reactive astrocytes are induced by activated microglia. *Nature* **541**, 481 (2017).
78. S. A. Wolf, H. W. Boddeke, H. Kettenmann, Microglia in physiology and disease. *Annu Rev Physiol* **79**, 619–643 (2017).
79. V. Bellver-Landete et al., Microglia are an essential component of the neuroprotective scar that forms after spinal cord injury. *Nat Commun* **10**, 518 (2019).
80. M. Yao et al., Astrocytic trans-differentiation completes a multicellular paracrine feedback loop required for medulloblastoma tumor growth. *Cell* **180**, 502–520.e519 (2020).
81. F. Perez-Cerda, M. V. Sanchez-Gomez, C. Matute, Pio del Rio Hortega and the discovery of the oligodendrocytes. *Front Neuroanat* **9**, 92 (2015).
82. S. D. Skaper, Oligodendrocyte precursor cells as a therapeutic target for demyelinating diseases. *Prog Brain Res* **245**, 119–144 (2019).
83. V. Skihar et al., Promoting oligodendrogenesis and myelin repair using the multiple sclerosis medication glatiramer acetate. *Proc Natl Acad Sci U S A* **106**, 17992–17997 (2009).
84. B. A. Barres, R. Schmid, M. Sendnter, M. C. Raff, Multiple extracellular signals are required for long-term oligodendrocyte survival. *Development (Cambridge, England)* **118**, 283–295 (1993).
85. L. Xiao et al., Rapid production of new oligodendrocytes is required in the earliest stages of motor-skill learning. *Nat Neurosci* **19**, 1210–1217 (2016).
86. F. A. McMorris, T. M. Smith, S. DeSalvo, R. W. Furlanetto, Insulin-like growth factor I/somatomedin C: a potent inducer of oligodendrocyte development. *Proc Natl Acad Sci U S A* **83**, 822–826 (1986).

87. M. J. Carson, R. R. Behringer, R. L. Brinster, F. A. McMorris, Insulin-like growth factor I increases brain growth and central nervous system myelination in transgenic mice. *Neuron* **10**, 729–740 (1993).
88. W. H. Lagarde et al., A non-transformed oligodendrocyte precursor cell line, OL-1, facilitates studies of insulin-like growth factor-I signaling during oligodendrocyte development. *Int J Dev Neurosci* **25**, 95–105 (2007).
89. K. Hua, M. E. Forbes, R. J. Lichtenwalner, W. E. Sonntag, D. R. Riddle, Adult-onset deficiency in growth hormone and insulin-like growth factor-I alters oligodendrocyte turnover in the corpus callosum. *Glia* **57**, 1062–1071 (2009).
90. S. Freude et al., IRS-2 branch of IGF-1 receptor signaling is essential for appropriate timing of myelination. *J Neurochem* **107**, 907–917 (2008).
91. A. R. Hackett, A. Strickland, J. Milbrandt, Disrupting insulin signaling in Schwann cells impairs myelination and induces a sensory neuropathy. *Glia* **68**, 963–978 (2020).
92. R. H. Miller, S. Mi, Dissecting demyelination. *Nat Neurosci* **10**, 1351–1354 (2007).
93. J. L. Mason, S. Xuan, I. Dragatsis, A. Efstratiadis, J. E. Goldman, Insulin-like growth factor (IGF) signaling through type 1 IGF receptor plays an important role in remyelination. *J. Neurosci* **23**, 7710–7718 (2003).
94. A. E. Lovett-Racke, P. Bittner, A. H. Cross, J. A. Carlino, M. K. Racke, Regulation of experimental autoimmune encephalomyelitis with insulin-like growth factor (IGF-1) and IGF-1/IGF-binding protein-3 complex (IGF-1/IGFBP3). *J Clin Invest* **101**, 1797–1804 (1998).
95. B. Cannella, D. Pitt, E. Capello, C. S. Raine, Insulin-like growth factor-1 fails to enhance central nervous system myelin repair during autoimmune demyelination. *Am J Pathol* **157**, 933–943 (2000).
96. S. Komoly, L. D. Hudson, H. D. Webster, C. A. Bondy, Insulin-like growth factor I gene expression is induced in astrocytes during experimental demyelination. *Proc Natl Acad Sci U S A* **89**, 1894–1898 (1992).
97. J. A. Frank et al., A pilot study of recombinant insulin-like growth factor-1 in seven multiple sderosis patients. *Mult Scler* **8**, 24–29 (2002).

CHAPTER FIVE

Neuroendocrine Interactions in the Control of Glucose- and Energy Homeostasis

Kaj Kamstra and Alexander Tups

CONTENTS

REGULATION OF ENERGY HOMEOSTASIS BY INSULIN

Insulin and its essential role in energy homeostasis are introduced elsewhere. Insulin, produced by the pancreatic β-cells, was believed to act primarily in the periphery although Claude Bernard even 70 years before the discovery of insulin in the middle of the 19th century suggested a role of the brain in the regulation of glucose homeostasis. This dogma that insulin acts exclusively in the periphery was challenged for the first time by Steve Woods and colleagues who showed that central administration of insulin in baboons led to a reduction in food intake and body weight (1). Insulin crosses the blood-brain barrier via a saturable transport to bind to insulin receptors widely expressed throughout the brain (2, 3). Neuron-specific insulin receptor deletion leads to dysregulated energy homeostasis, including increased food intake, body fat, plasma leptin and insulin levels as well as mild insulin resistance (4). Subsequent studies showing that intracerebroventricular insulin suppresses hepatic glucose production and that inhibition of insulin receptors in the arcuate nucleus (ARC), a main metabolic control centre in the hypothalamus, leads to hyperphagia and insulin resistance, confirmed that the brain is the main target organ for insulin in the control of whole-body glucose and energy metabolism (4–7). Taken together these studies suggest that insulin in the brain acts in a catabolic fashion which is fundamentally different from the action of the hormone in the periphery where it promotes the storage of energy.

The insulin receptor is expressed throughout the brain with particular abundance in neuropeptide Y/agouti-related peptide (NPY/AgRP)- as well as proopiomelanocortin/cocaine and amphetamine-regulated transcript (POMC/CART)-expressing neurons of the hypothalamic ARC (2, 8, 9). Intracerebroventricular injections of NPY and AgRP lead to increased food intake, decreased

DOI: 10.1201/9781003079927-5

energy expenditure and weight gain (10–13), identifying both neuropeptides as anabolic signals. In contrast, catabolic effects were demonstrated for both CART and α-melanocyte-stimulating hormone, which is cleaved from its precursor POMC (14, 15).

Upon binding of the extracellular α-subunits of its receptor, insulin induces the dimerization and autophosphorylation of the intracellular β-subunits, followed by tyrosine phosphorylation of insulin receptor substrate proteins, phosphoinositide 3-kinase (PI3K) activation and AKT phosphorylation. All of the major components of insulin signalling found in the periphery are also present in the brain. One signalling node that may be of particular interest for the central regulation of whole-body energy and glucose homeostasis is the enzyme glycogen synthase kinase-3β (GSK-3β). In the periphery, GSK3β is a crucial enzyme in glycogen synthesis and it is activated by glucose 6 phosphate (16). Stimulation of insulin signalling results in inhibition of GSK-3β (17–19). Pharmacological inhibition of GSK-3β in the brain has been demonstrated to rescue PI3K-AKT signalling and improve glucose homeostasis in obese and glucose-intolerant rodent models, while adeno-associated virus-mediated overexpression in the hypothalamus impaired glucose tolerance (20, 21). These findings suggest that central GSK-3β is a potent modulator of insulin sensitivity.

While the insulin signalling machinery is present and recruited by insulin in the hypothalamus the brain does not play a major role in glucose disposal. Hypothalamic insulin action is crucial for the regulation of hepatic glucose production. Insulin was shown to activate ATP-sensitive potassium (K_{ATP}) channels in certain hypothalamic areas, with specific evidence originating from NPY/AgRP neurons (22, 23). K_{ATP} channel activation, which is mediated by PI3K, regulates peripheral glucose homeostasis by inhibition of hepatic gluconeogenesis via hepatic vagal nerve activation (7) and intracerebroventricular injections of NPY induced hepatic insulin resistance via sympathetic innervation of the liver (24). However, it has recently been shown that insulin in the ARC exerts sympathoexcitatory effects through the suppression of tonic paraventricular nucleus NPY inhibition (25). These results reveal that the intricate control of central insulin action and its effects on glucoregulation remain a field of ongoing research interest (Figure 5.1).

REGULATION OF ENERGY HOMEOSTASIS BY LEPTIN

In 1994, positional cloning of the elusive *obese* gene identified the factor that caused the extreme obesity observed in leptin-deficient $Lep^{ob/ob}$ mice (26). This gene encodes a 16kDa protein of 167 amino acids that was named leptin, after the Greek word for 'lean' (λεπτός). The absence of leptin results in increased food intake, decreased energy expenditure, hyperglycaemia and severe obesity. Following this discovery, leptin was believed to have great potential as an 'anti-obesity drug' to counteract the growing obesity epidemic of our modern society. However, this initial euphoria soon abated once leptin was shown to be secreted by white adipose tissue and to circulate in proportion to body fat mass (27). However, novel research may lead to a renaissance of leptin therapy for metabolic disease as outlined later in this chapter. Like rodents, humans with diet-induced obesity (DIO) are characterized by high circulating leptin concentrations, matching their elevated adiposity. However, these elevated leptin levels are unable to normalize body weight. Similarly, whereas leptin treatment in rare cases of congenital leptin deficiency rapidly and completely normalizes the morbid obesity found in these subjects, exogenous leptin treatment fails to reduce food intake and induce weight loss in obese humans and rodents. Together, these findings have led to the concept of leptin resistance. Although several genetic mutations of the leptin system have been described that render people more susceptible to DIO, these cases are rare and cannot explain the obesity and type 2 diabetes pandemic (28, 29). Nor is there any evidence pointing towards a role for circulating antagonists in DIO rodents or humans. Thus, leptin resistance presumably results from one or more disturbances between the leptin receptor and its downstream signalling cascade, at the intracellular or intercellular level. This hypothesis is supported by the notion that high-dose exogenous leptin treatment is not only unable to rescue the metabolic syndrome found in DIO rodents, but also fails to properly induce hypothalamic phosphorylation of STAT3 in rodents (30). Interestingly, the pSTAT3 response to leptin treatment in DIO rodents is impaired most significantly in the ARC (31), suggesting that this region is the major site affected by leptin resistance. On the other hand, a

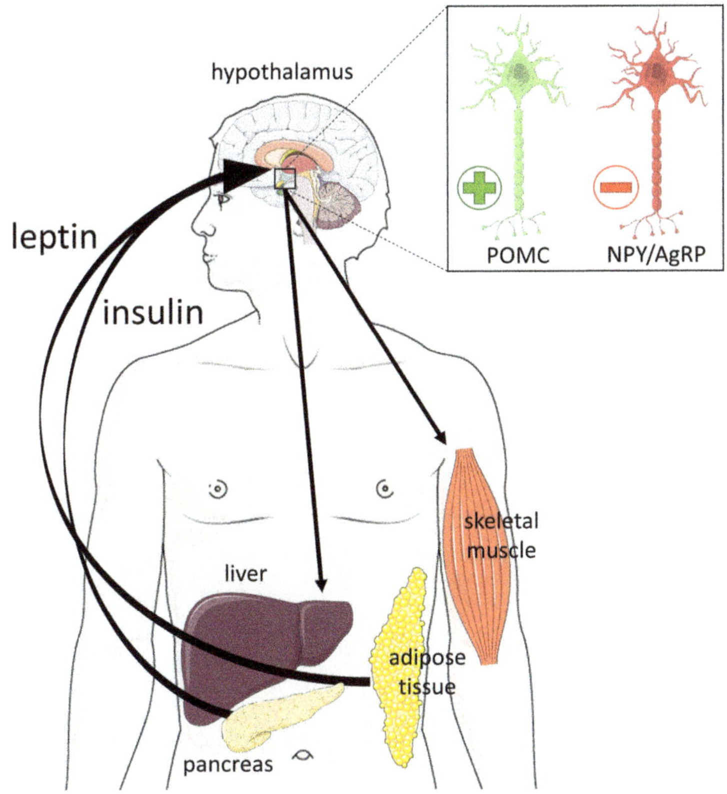

Figure 5.1 Insulin and leptin are the afferent signals in a negative feedback loop that maintains homeostatic control of energy stores. Insulin is secreted by the pancreatic β-cells, and leptin is secreted by adipose tissue. Both hormones travel through the bloodstream to act in the brain, primarily in the arcuate nucleus of the hypothalamus. Here, insulin and leptin activate anorexigenic POMC neurons, while stimulating orexigenic NPY/AgRP neurons. Using these signals, the central nervous system promotes the storage or mobilization of energy in organs including skeletal muscle and the liver.

recent report showed that the ventromedial hypothalamic nucleus (VMH) is the only nucleus that shows a correlation between STAT3 phosphorylation and inhibition of food intake (32).

The long form of the leptin receptor (LepRb) is expressed in abundance in the ARC, ventromedial hypothalamus, dorsomedial hypothalamus and lateral hypothalamic area (33, 34). Importantly, the ARC is located near a congregation of veins with fenestrated capillaries at the median eminence. The median eminence is one of the seven circumventricular organs of the brain, which are characterized by rapid neurohumoral exchange (35). This way the blood–brain barrier is bypassed, thus permitting neuronal processes access to blood and circulating hormones, including leptin and insulin. Whether the ARC or parts of the ARC reside outside of the blood–brain barrier is a contentious issue and heavily debated (36–38). Two distinct neuronal populations in the ARC co-express LepRb. These are NPY/AgRP-expressing neurons, which are inhibited by leptin (39–41), as well as POMC/CART-expressing neurons, which are activated by leptin (42, 43). At the molecular level, leptin mediates its catabolic effects by binding to LepRb, activating Janus kinase 2 (JAK2), ultimately resulting in recruitment and phosphorylation of signal transducer and activator of transcription 3 (STAT3) (44, 45). Phosphorylated STAT3 (pSTAT3) homodimerises and translocates into the nucleus, where it acts as a transcription factor and activates target gene expression (46, 47). Among those are effector molecules that cause reduced food intake and increased energy expenditure as well as the suppressor of cytokine signalling 3 (SOCS3). Through negative feedback regulation, SOCS3

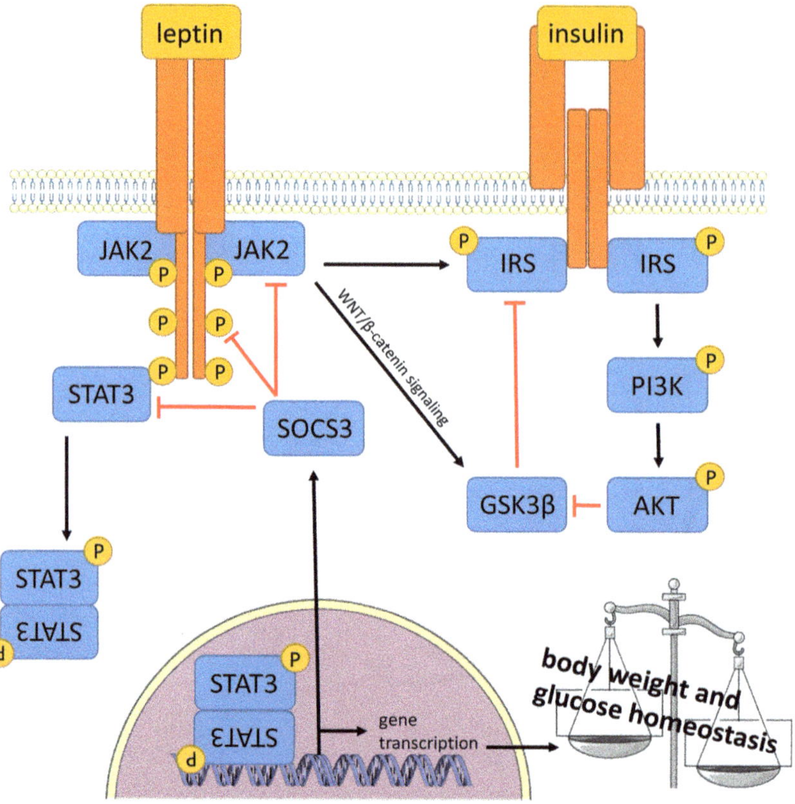

Figure 5.2 Upon binding of insulin to IR its intrinsic tyrosine kinase domain phosphorylates IRS1. Activated IRS1 phosphorylates PI3K leading to downstream activation of AKT. Activated AKT exhibits pleiotropic effects, most importantly controlling FOXO and GSK3β, thereby regulating key processes such as cell survival, proliferation, and metabolic adaptions. Binding of leptin to LepRb leads to an activation of JAK2, resulting in phosphorylation of the three intracellular receptor tyrosine residues. The phosphorylation of Tyr1138 causes recruitment of STAT3. STAT3 is subsequently phosphorylated by JAK2 at the phosphorylation site Tyr705, which leads to homodimerization and nuclear translocation of STAT3. In the nucleus, STAT3 acts as a transcription factor and induces target gene expression, including SOCS3. SOCS3 initiates a negative feedback loop inhibiting JAK2 activation, thereby decreasing leptin action. In addition to its fundamental effects on body weight via the JAK2/STAT3 pathway, leptin also acts on GSK3β, linking leptin signalling to various metabolic processes. Abbreviations: AKT, protein kinase B; GSK3β, glycogen synthase kinase 3β; IR, insulin receptor; IRS, insulin receptor substrate 1; JAK2, Janus kinase 2; PI3K, phosphoinositide 3kinase; SOCS3, suppressor of cytokine signalling 3; STAT3, signal transducer and activator of transcription 3.

binds to LepRb and inhibits JAK2 activity, identifying SOCS3 as an effective inhibitor of leptin signal transduction via JAK2/STAT3 signalling (48, 49) (Figure 5.2).

Although JAK2-STAT3-SOCS3 signalling is the most well-described and thoroughly investigated signalling cascade downstream of LEPRb, other pathways that are known to mediate insulin signal transduction are also involved in leptin signal transduction. One such pathway is the IRS-PI3K-AKT signalling pathway described above. Leptin recruitment of this pathway was first suggested by a study showing that IRS2-null mice displayed elevated leptin levels, hyperphagia and decreased energy expenditure (50). Subsequently, it was demonstrated that systemic administration of leptin in rats activates PI3K in the hypothalamus and that intracerebroventricular infusion of inhibitors of this enzyme prevented the anorexigenic effects of leptin (51). Leptin-induced PI3K activation is likely mediated through SH2B1, which recruits IRS proteins to JAK2. Activated PI3K catalyzes the phosphorylation of phosphatidylinositol-4,5-bisphosphate to phosphatidylinositol-3,4,5-tris-phosphate (PIP3), whereas tumour suppressor phosphatase and tensin homolog (Pten) initiate the opposite reaction

(52). PIP3 then binds to the Pleckstrin homology domain of Akt (also known as protein kinase B, PKB), allowing for the phosphorylation of Akt by its activating kinases, phosphoinositide-dependent kinase 1 (PDPK1, at threonine 308, and the mammalian target of rapamycin complex 2 (mTORC2, at serine 473) (53). Leptin signalling through this pathway is especially important for the regulation of glucose homeostasis, both through direct effects on glucose fluxes (54) and by regulating insulin sensitivity (55).

In wild-type mice, leptin injections increase pSTAT3 levels, presenting a marker for activated intracellular leptin signal transduction. In DIO mice, this response to exogenous leptin is disrupted, particularly in the ARC (56). Consequently, this suggests that during obesity the hypothalamus receives misinformation about the peripheral energy status and integrates these signals into an inapt neural response to control energy homeostasis. Furthermore, increased expression of SOCS3 has been shown to cause impairments of leptin signalling in rodent models of leptin resistance and obesity (49, 57).

LEPTIN HOMEOSTASIS AND ITS RELEVANCE FOR INSULIN ACTION

Unlike insulin resistance, which can be at least partially overcome by further increasing the levels of insulin (58), enhancing the circulating levels of leptin is completely ineffective. Attempts to restore the physiological role of leptin have failed, and the concept of leptin resistance remains poorly understood (59). Recent studies have provided strong evidence that increased circulating levels of leptin might be a driving force for insulin resistance. For example, it has been demonstrated that blocking central leptin signalling in DIO mice restores the ability of the liver to suppress glucose production (60). Consistently, a partial reduction of plasma leptin levels restores hypothalamic leptin sensitivity in obese mice and effectively reduces weight gain and enhances insulin sensitivity (61). Heterozygous whole-body leptin knockout mice and an adipocyte-specific congenital leptin knockout mouse reveal partial leptin deficiency and interestingly protection from DIO. Furthermore, while on the chow diet, glucose tolerance and insulin sensitivity remained unchanged, a marked improvement occurred after mice were fed HFD (62). Reducing leptin levels in DIO mice with a long-acting leptin antagonist (PESLAN) significantly improved glucose tolerance and decreased astro- and microgliosis in the hypothalamus of mice fed a high-fat diet while inducing hyperleptinemia led to the opposite effect suggesting that high levels of leptin promote hypothalamic inflammation (63). In $Lep^{ob/ob}$ mice that are extremely leptin sensitive to exogenous leptin, we found that inducing central hyperleptinemia is not sufficient to develop leptin resistance since these mice continued to lose bodyweight and consume less food. However, exposure to HFD immediately arrested the bodyweight reduction induced by hyperleptinemia. Furthermore, HFD reduced the ability of central leptin to improve glucose homeostasis (64). This insufficiency of hyperleptinemia to impair body weight regulation is in line with other studies (65, 66). For instance, in a recent study using transgenic mice where leptin expression was induced by doxycycline to negate developmental impacts, elevating leptin levels during adulthood did not show any effects on body weight gain when mice were fed chow diet (67). In conjunction with HFD feeding, however, increased circulating levels of leptin appear to potentiate the disruption of energy homeostasis leading to increased body weight gain. Some human studies support the concept that hyperleptinemia impairs energy metabolism. For instance, hyperleptinemia is strongly associated with the occurrence of low-grade systemic inflammation and metabolic dysfunction in obese subjects (68). When obese individuals were separated into different groups depending on their baseline leptin levels followed by injection with metreleptin to achieve weight loss, those individuals who had the lowest baseline leptin levels showed the most profound weight loss, suggesting increased leptin sensitivity (69). In contrast, individuals who exhibited higher baseline plasma leptin levels are more prone to regaining body weight lost (70). Individuals who undertook a weight-loss regimen showed reduced levels of leptin but the reduction in leptin did not counteract weight loss; indeed, it was correlated with further weight loss (71).

These observations imply that individual leptin levels may correlate with individualized leptin sensitivity and suggest the presence of a threshold for physiological responses to leptin. Once a specific threshold level is reached, the catabolic effects of leptin reverse to effects that lead to a disruption of energy homeostasis, leading to continuous body

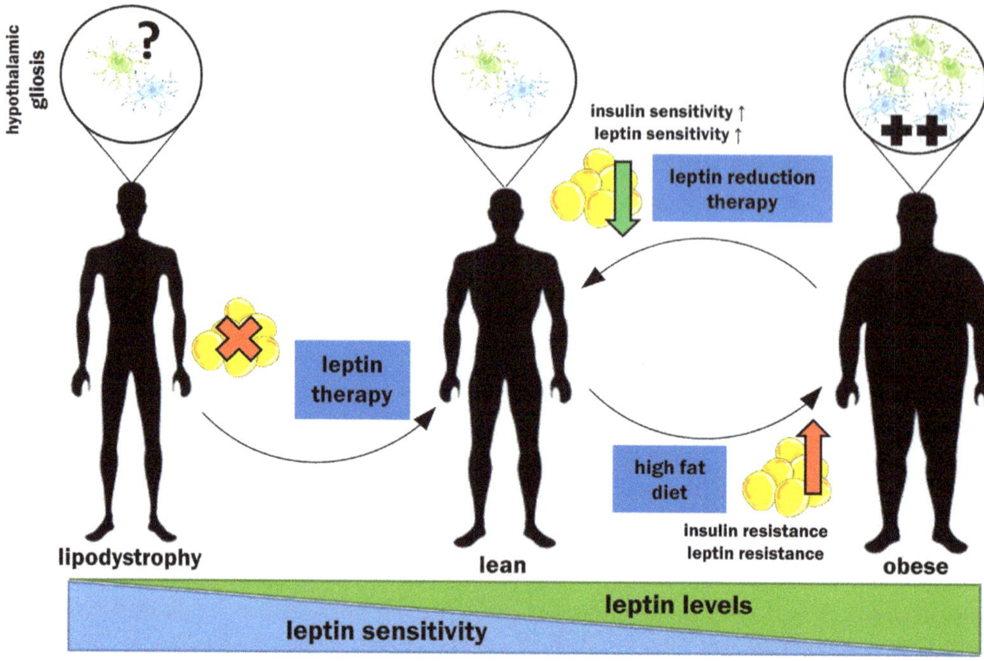

Figure 5.3 The maintenance of metabolic health depends on the balance between circulating leptin levels and leptin sensitivity. Leptin therapy is effective in treating lipodystrophy, but not in treating DIO. Partial reduction of circulating leptin restores sensitivity of the body to the remaining circulating leptin, resulting in weight loss and improvements in diabetes. In accordance, increasing leptin levels are associated with aggravated hypothalamic micro- and astrogliosis, whereas blocking excessive leptin signalling during obesity improves glucose tolerance and normalizes the number of hypothalamic microglia and astrocytes.

weight gain. Together, these studies show that improvements in insulin sensitivity can occur as a result of interventions targeted at directly lowering peripheral leptin levels. The metabolic improvements in this context depend on the reduction of leptin levels in circulation. This offers a novel basis for the explanation of the mechanism underlying the metabolic improvements observed for numerous interventions (Figure 5.3).

THE GLUCOREGULATORY ROLE OF LEPTIN

Leptin is known primarily for its adipostatic role, but it is also essential for the regulation of glucose homeostasis. The loss of leptin action leads to hyperglycaemia, hyperinsulinemia and glucose intolerance (72). The leptin receptor-deficient $Lep^{db/db}$ mouse is named after its diabetic phenotype (73). Leptin therapy in leptin-deficient rodents and humans ameliorates all aspects of dysregulated glucose homeostasis. Initially, these effects were thought to be secondary effects following a reduction in body weight. However, it has been firmly established that leptin regulates glucose homeostasis independent of its adipostatic effects by acting on various central and peripheral cells.

Various observations, mostly from studies with $Lep^{ob/ob}$ mice, suggest that leptin affects metabolism independent of its adipostatic actions. For example, low dose leptin treatment of 1mg/kg/day did not improve body weight in $Lep^{ob/ob}$ mice, but was still capable of normalizing insulin and blood glucose levels (74). Likewise, acute disruption of leptin signalling *in vivo* raises blood glucose and plasma insulin levels before altering body weight. Moreover, $Lep^{ob/ob}$ and $Lep^{db/db}$ mice become hyperinsulinemic before they become obese (75), which suggests that glucose homeostasis is already impaired before bodyweight alterations emerge. Treating $Lep^{ob/ob}$ mice with leptin proved to improve blood glucose levels, in comparison with $Lep^{ob/ob}$ mice that were pair fed the exact amount of food (41). In addition, humans and rodents that suffer from lipodystrophy, a condition characterized by degeneration and degradation of the adipose tissues that renders the body unable to store sufficient fat and thus have very low leptin levels, exhibit hyperglycaemia, hyperinsulinemia and insulin resistance. All of these symptoms are corrected by leptin therapy (76, 77). Thus, leptin can influence glucose homeostasis independent of its adipostatic effects.

Data from our laboratory show that intracerebroventricular leptin treatment that leads to negligible peripheral leptin levels is sufficient to restore euglycemia and insulin sensitivity in leptin-deficient mice (78), consistent with data from insulin-deficient (79) and DIO mice (80). Thus, central signalling alone is sufficient for the glucoregulatory actions of leptin. Significant progress has been made over the past years to elucidate which brain areas are essential for the leptin-mediated control of glucose homeostasis. In diabetic rats, leptin injection into the VMH restored blood glucose levels and hepatic glucose production to normal levels, but did not affect glucose uptake in brown adipose tissue or muscle (81). Selective restoration of leptin receptor function in arcuate POMC neurons completely reversed hyperglycaemia and improved insulin sensitivity. However, loss of leptin receptors specifically in the POMC neurons of wild type mice did not block the glucose lowering actions of leptin. Selective leptin receptor re-expression in AgRP neurons of $Lep^{db/db}$ mice restored blood glucose levels to normal levels (82). Importantly, AgRP neuron-specific loss of leptin receptor blocks the glucoregulatory actions of leptin. Similarly, in $Lep^{ob/ob}$-MC4R knockout mice and diabetic MC4R-knockout rats, leptin is unable to improve glucose homeostasis (83). Thus, AgRP neurons, but not POMC neurons, and their downstream MC4 receptors are both sufficient and required for the glucoregulatory actions of leptin. Furthermore, AgRP neurons are GABAergic and co-secrete NPY, but AgRP-selective knockout of either GABA or NPY does not prevent the glucoregulatory effect of leptin. Consistently, leptin is still able to lower blood glucose levels in LEPR GABA knockout mice that lack insulin (84). Together, these studies indicate AgRP to be the main mediator of glucoregulation by leptin.

The fact that insulin's glucoregulatory action is deteriorated in the absence of functional leptin signalling, as observed in $Lep^{ob/ob}$ and $Lepr^{db/db}$, as well as $Lepr^{s/s}$ mice, which express a mutant leptin receptor with selective disruption of STAT3 signal transduction. All three mutant strains exhibit hyperglycaemia and hyperinsulinemia, implying that intact leptin action is required for insulin to function. Whether this also applies *vice versa* and insulin is required for leptin to function has long been a matter of debate. To answer this question, Neumann et al. conducted a study in mice in which insulin signalling was maximally blocked. Mice with streptozotocin (STZ)-induced insulin deficiency were additionally treated with insulin receptor antagonists, as well as mice with global insulin knockout were treated with leptin. In both models, Neumann et al. reported that leptin was still efficacious at lowering glucose levels, even though global insulin knockout mice displayed volatile changes in glucose levels (85). These data demonstrate that leptin is able to restore glucose homeostasis in the total absence of insulin, but minimal insulin is still required to stabilize glucose levels indefinitely. The finding that leptin can effectively restore euglycemia in STZ-induced rodent models of insulin-deficient T1DM (86) sparked interest in leptin as a therapeutical replacement for insulin in the treatment of T1DM in the hope to avoid the recurring hypoglycaemic episodes caused by insulin therapy. Leptin was found to be capable of reversing the catabolic consequences of a total lack of insulin (87). In a clinical trial testing the effect of metreleptin on glycaemic control in human T1DM patients, it was demonstrated that leptin led to a subtle decrease in body weight (88), potentially due to the reduction in leptin levels of approximately 50% compared with those observed in T1DM patients (89, 90). Although the authors furthermore reported a significant reduction in the daily insulin dose required to maintain glycaemic control, they concluded that leptin did not improve any parameters for glycaemic control (88). We believe, however, that comparable to results from rodent models of T1DM (91), these findings suggest that leptin promotes insulin sensitivity in human T1DM patients. The molecular mechanisms underlying the interaction of central insulin and leptin are not entirely understood. However, the WNT/β-catenin pathway appears as a potential integration site for these metabolic signals, as discussed in the next section.

WNT/β-CATENIN SIGNALLING AS AN INTEGRATOR OF CENTRAL INSULIN AND LEPTIN ACTION ON ENERGY HOMEOSTASIS

The WNT pathway was first described in 1976 when it was reported that Drosophila Melanogaster has a wingless phenotype when this pathway is mutated (92). In 1982, the same signalling cascade was found to promote tumour formation in mice, and therefore given the term 'integration-1 (int1)' (93). 'Wingless' and 'int1' were later combined, and thus, the WNT pathway was coined. The WNT signalling pathway is evolutionarily highly conserved and is classically known for its role in embryogenesis and tumorigenesis (94). Its ligands (WNTs) are involved in three different pathways: the WNT/β-catenin pathway (also known as the canonical WNT pathway), the planar cell polarity pathway, and the WNT/Ca2+ pathway. The canonical WNT pathway is activated when a WNT ligand binds to the frizzled (Fzd) receptor, which subsequently forms a complex with the co-receptor lipoprotein related protein (LRP) 5/6. This causes dishevelled (Dvl) to phosphorylate LRP, which then inactivates GSK3β. Next, GSK3β inactivation decreases phosphorylation of the transcriptional co-activator β-catenin. Stabilized β-catenin then enters the nucleus where it associates with transcription factors of the lymphoid enhancer factor (LEF)/T cell factor (TCF) family, to ultimately regulate the transcription of downstream target genes such as cyclin D1 and axin 2 (95).

One of the first indications that WNT signalling plays an essential role in the adult brain is that this signalling cascade regulates not only embryonic, but also postembryonic hypothalamic progenitor differentiation in both zebrafish and mice, thereby emphasizing the evolutionary conservation of WNT action in the brain (96). Using *in situ* hybridization for the individual TCF members and for GFP in a zebrafish reporter line that expresses canonical WNT activity, it was found that all TCF members are expressed in the hypothalamus, especially in the posterior part where postmitotic precursor cells are found (97). Interestingly, the expression of TCF7l2 is confined to a small area right above the future hypophysis. Consistent with this finding is a study that reports negative regulation of pituitary development by TCF7l2 in mouse embryos (98). Furthermore, zebrafish larvae were found to express LEF1 and TCF7 in the caudal zone of the periventricular hypothalamus. No specific expression of TCF7l1a/b or TCF7l2 was found. A similar expression pattern was found in adult fish, where TOP:dGFP fluorescence, a canonical WNT reporter construct, persisted in cells adjacent to the diencephalic ventricle. Importantly, TOP:dGFP was only found in the hypothalamus and optic tectum of adult fish.

Canonical WNT signalling has been widely implicated in the expansion of progenitor poos or neural stem cells, suggesting that this pathway is a required factor for neural stem cell maintenance. However, at least ten distinct regions have been identified as proliferation zones within the adult zebrafish brain (99). This suggests that the WNT-responsive cells found in the hypothalamus of adult zebrafish serve another purpose. The change in function of WNT signalling during development and in adult organisms is far from fully understood. In this regard, it is important to note that the key loci for WNT signalling in the adult hypothalamus represent a stem cell population that consists of tanycytes lining the third ventricle and arcuate nucleus neurons (100, 101). Tanycytes are a subtype of ependymal cells with long processes that reach the fenestrated capillary network, while their tight junctions prevent the diffusion of blood-borne molecules into the adjacent cerebrospinal fluid. Studies from divergent fields have demonstrated the significance of tanycytes and ependymal cells in the neuroendocrine control of feeding and energy balance (102). However, a study by Wang et al. shows that WNT pathway activation decreases the number of hypothalamic tanycytes (96). This excludes WNT signalling from being a necessary requirement for tanycyte formation and maintenance, thereby complicating things further.

In the last decade, strong evidence linked the WNT/β-catenin pathway to the pathogenesis of type 2 diabetes and other metabolic disorders. A 2006 genome-wide association study identified a link between polymorphisms in the TCF7l2 gene, a transcription factor of the WNT pathway, and the risk of type 2 diabetes (103). Out of all common genetic mutations implicated in T2DM, TCF7l2 is most significantly associated with disease susceptibility. Another GWAS-identified risk gene for T2DM is Cyclin-dependent Kinase 5 Regulatory Subunit-Associated Protein 1-like 1 (CDKAL1) (104). Interestingly, CDKAL1 increases protein levels of β-catenin and its active unphosphorylated form in the nucleus, thereby promoting Wnt target gene expression, suggesting that CDKAL1 activates canonical WNT signalling (105). Finally, a recent report showed that R-spondin-3 (RSPO3), which has previously been implicated by a GWAS as a risk gene for increased BMI and insulin resistance (106), regulates energy homeostasis in association with WNT signalling (107). Together, these studies strongly suggest a role for canonical WNT signalling in the regulation of energy homeostasis.

The emerging role of WNT/β-catenin signalling in the neuroendocrine control of glucose homeostasis was first highlighted by the *in vitro* finding that glucose activates the Wnt pathway (108). Expression of DKK1, the main antagonist of the Wnt/β-catenin signalling pathway that acts by isolating the LRP6 co-receptor, is increased in the ARC of $Lep^{ob/ob}$ mice, and ICV DKK-1 treatment in wild type mice leads to severely impaired glucose homeostasis (21). In accordance, humans with type 2 diabetes have markedly higher circulating DKK-1 levels compared with healthy controls (109). Pharmacological inhibition of DKK reduces basal blood glucose levels and improves glucose tolerance in mice, and DKK-2 knockout mice have increased hepatic glycogen accumulation and decreased hepatic glucose output, coupled with increased WNT activity in the intestines (110). Together, these data implicate an important role for WNT receptor antagonism in the pathogenesis of type 2 diabetes. Besides DKK-mediated receptor antagonization, the main endogenous WNT pathway inhibitor is GSK3β. Importantly, GSK3β is phosphorylated and inhibited by AKT in L6 myotubes (19), skeletal muscle (111) and in HeLa cells (112), thus making it a target of insulin signalling. Indeed, GSK3β inhibition is required for insulin stimulation of glycogenesis (20), hence the name glycogen synthase kinase. Systemic GSK3β inhibition ameliorates glucose intolerance in $Lep^{ob/ob}$ mice (113), suggesting a tonic inhibitory effect of GSK3β on glucose homeostasis. While some studies point towards a role for GSK3β in skeletal muscle as a mediator of glucose metabolism (114), data from our laboratory show that GSK3β action specifically in the hypothalamus is essential for glucose homeostasis. $Lep^{ob/ob}$ mice were found to have upregulated hypothalamic GSK3β expression, and glucose intolerance in these mice was acutely ameliorated by ICV injection of a GSK3β inhibitor. Furthermore, neuron-specific overexpression of GSK3β in the hypothalamus exacerbated the effects of a high-fat diet in wild-type mice compared with mice on a standard diet, measured as increased hyperphagia, obesity and glucose intolerance (21). In this light, it appears that WNT pathway inhibition leads to detrimental effects on energy homeostasis.

INCRETINS AND ENERGY HOMEOSTASIS

The incretins, consisting of glucose-dependent insulinotropic polypeptide (GIP) and glucagon-like peptide 1 (GLP-1), are hormones that are released from the enteroendocrine K and L cells respectively, when digested food reaches the intestine (115, 116). GIP and GLP-1 in turn stimulate glucose-dependent insulin secretion from the pancreatic β-cells. Incretins are estimated to account for up to 65% of total insulin secretion after a meal (117, 118). The molecular mechanisms of action of GIP and GLP-1 have been reviewed elsewhere (119).

Besides their well-known actions in the endocrine pancreas, both GIP and GLP-1 have several extrapancreatic effects. GIP receptors are found on rat adipocytes and in 3T3-L1 cells for example (120). GIP has been deemed unsuitable for treating type 2 diabetes, due to its lack of insulinotropic action during hyperglycaemia (121), but some studies suggest that GIP might protect against obesity. GIP receptor expression is negatively correlated with adiposity, and transgenic mice overexpressing GIP were found to have improved glucose tolerance, enhanced insulin sensitivity, and protection against DIO (122). Furthermore, it has been shown that body weight and glucose control are improved in DIO mice with central deletion of the GIP receptor and that central loss of this receptor is paired with a decrease in food intake without alterations in energy expenditure (123). GLP-1 receptors are found in the hypothalamus and the brainstem (124, 125), and it has been shown that GLP-1 decreases food intake in a dose-dependent manner by acting on these brain regions (126). Importantly, recent studies now show that GLP-1 can not only act on the brain, it is produced in the brain as well. GLP-1 is secreted by neurons in the nucleus tractus solitarius of the brainstem (127), in a process stimulated by leptin and gastric distension, but not by peripheral GLP-1 (128, 129). Acute activation of GLP-1-expressing neurons in the brainstem was found to reduce basal glucose production, increase glucose tolerance, and improve hepatic insulin sensitivity (130). It is now clear that central and peripheral GLP-1 systems both suppress food consumption while operating independently from each other (131). The ability of GLP-1 receptor agonists to induce weight loss has been well established by clinical trials. The long-acting GLP-1 receptor agonists Liraglutide (132) and Semiglutide (133) are currently considered the most promising drugs for treating obesity (134). Despite the lack of insulinotropic action of GIP, the possibility of dual agonism between GIP and GLP-1 as a treatment for metabolic syndrome has been suggested by studies showing that a dual GIP and GLP-1 receptor agonist, LY3298176, leads to significantly higher efficacy with regard to glucose control and weight loss compared with GLP-1-specific agonism (123, 135). The crosstalk between GLP-1, GIP, insulin and leptin might be a promising avenue for weight loss therapy.

WNT/β-catenin signalling appears to play a role in incretin-mediated insulin secretion. Reducing the level of β-catenin in pancreatic β-cells was found to attenuate glucose-dependent insulin secretion. On the other hand, pharmacological manipulation to increase β-catenin levels in β-cells resulted in increased insulin secretion. The same study showed that β-catenin is required for the glucose- and incretin-induced secretion of insulin (136). Genetic studies have shown that WNT/β-catenin signalling increases the expression of both incretins through evolutionarily conserved elements located within their proximal promoters (137, 138). Furthermore, WNT/β-catenin signalling appears to modulate incretin signalling in the pancreas and vice versa. Expression levels of the WNT/β-catenin transcription factor TCF7l2 correlate with pancreatic incretin receptor levels (139). On the other hand, both GIP and GLP-1 can enhance WNT/β-catenin signalling by inducing cAMP accumulation, leading to stabilization of β-catenin (140, 141). Thus, incretin and WNT/β-catenin signalling appear to be interconnected on multiple levels.

CONCLUDING REMARKS

Insulin and leptin are two central hormones for the regulation of metabolic homeostasis. Secreted peripherally, they act on the brain, playing key roles in signalling and orchestrating energy homeostasis. Although their respective physiologic roles are distinct, insulin and leptin display a high degree of crosstalk. An important role of leptin in glucoregulation has been revealed. Typically regarded as being controlled by insulin, the control of glucose homeostasis critically depends on functional leptin action. The signalling cross-talk between insulin and leptin is likely conveyed by the WNT/β- catenin pathway. Another exciting development in the field is the emergence of leptin reduction therapy. Improvements in insulin sensitivity can occur as a result of interventions targeted at directly reducing circulating levels of leptin. Finally, a link between insulin, leptin and the incretin system is beginning to emerge.

ACKNOWLEDGEMENTS

We thank the Department of Physiology, University of Otago and the Maurice Wilkins Centre for Molecular Biodiscovery for financial support

REFERENCES

1. Woods SC, Lotter EC, McKay LD, Porte D, Jr. Chronic intracerebroventricular infusion of insulin reduces food intake and body weight of baboons. *Nature*. 1979;282(5738):503–5.
2. Baura GD, Foster DM, Porte D, Jr., Kahn SE, Bergman RN, Cobelli C, et al. Saturable transport of insulin from plasma into the central nervous system of dogs in vivo. A mechanism for regulated insulin delivery to the brain. *J Clin Invest*. 1993;92(4):1824–30.
3. Banks WA, Jaspan JB, Huang W, Kastin AJ. Transport of insulin across the blood-brain barrier: saturability at euglycemic doses of insulin. *Peptides*. 1997;18(9):1423–9.
4. Bruning JC, Gautam D, Burks DJ, Gillette J, Schubert M, Orban PC, et al. Role of brain insulin receptor in control of body weight and reproduction. *Science*. 2000;289(5487):2122–5.
5. Obici S, Feng Z, Karkanias G, Baskin DG, Rossetti L. Decreasing hypothalamic insulin receptors causes hyperphagia and insulin resistance in rats. *Nature Neurosci*. 2002;5(6):566–72.
6. Obici S, Zhang BB, Karkanias G, Rossetti L. Hypothalamic insulin signaling is required for inhibition of glucose production. *Nat Med*. 2002;8(12):1376–82.
7. Pocai A, Lam TK, Gutierrez-Juarez R, Obici S, Schwartz GJ, Bryan J, et al. Hypothalamic K(ATP) channels control hepatic glucose production. *Nature*. 2005;434(7036):1026–31.
8. Marks JL, Porte D, Jr., Stahl WL, Baskin DG. Localization of insulin receptor mRNA in rat brain by in situ hybridization. *Endocrinology*. 1990;127(6):3234–6.
9. Schwartz MW, Woods SC, Porte D, Jr., Seeley RJ, Baskin DG. Central nervous system control of food intake. *Nature*. 2000;404(6778):661–71.
10. Stanley BG, Kyrkouli SE, Lampert S, Leibowitz SF. Neuropeptide Y chronically injected into the hypothalamus: a powerful neurochemical inducer of hyperphagia and obesity. *Peptides*. 1986;7(6):1189–92.
11. Billington CJ, Briggs JE, Grace M, Levine AS. Effects of intracerebroventricular injection of

neuropeptide Y on energy metabolism. *Am J Physiol.* 1991;260(2 Pt 2):R321–7.

12. Hagan MM, Rushing PA, Benoit SC, Woods SC, Seeley RJ. Opioid receptor involvement in the effect of AgRP- (83–132) on food intake and food selection. *Am J Physiol Regul, Integr Comparat Physiol.* 2001;280(3):R814–21.
13. Rossi M, Kim MS, Morgan DG, Small CJ, Edwards CM, Sunter D, et al. A C-terminal fragment of Agouti-related protein increases feeding and antagonizes the effect of alpha-melanocyte stimulating hormone in vivo. *Endocrinology.* 1998;139(10):4428–31.
14. Fan W, Boston BA, Kesterson RA, Hruby VJ, Cone RD. Role of melanocortinergic neurons in feeding and the agouti obesity syndrome. *Nature.* 1997;385(6612):165–8.
15. Kristensen P, Judge ME, Thim L, Ribel U, Christjansen KN, Wulff BS, et al. Hypothalamic CART is a new anorectic peptide regulated by leptin. *Nature.* 1998;393(6680):72–6.
16. Bouskila M, Hunter RW, Ibrahim AF, Delattre L, Peggie M, van Diepen JA, et al. Allosteric regulation of glycogen synthase controls glycogen synthesis in muscle. *Cell Metab.* 2010;12(5):456–66.
17. Sutherland C, Leighton IA, Cohen P. Inactivation of glycogen synthase kinase-3 beta by phosphorylation: new kinase connections in insulin and growth-factor signalling. *Biochem J.* 1993;296 (Pt 1):15–9.
18. Stambolic V, Woodgett JR. Mitogen inactivation of glycogen synthase kinase-3 beta in intact cells via serine 9 phosphorylation. *Biochem J.* 1994;303 (Pt 3):701–4.
19. Cross DA, Alessi DR, Cohen P, Andjelkovich M, Hemmings BA. Inhibition of glycogen synthase kinase-3 by insulin mediated by protein kinase B. *Nature.* 1995;378(6559):785–9.
20. Cline GW, Johnson K, Regittnig W, Perret P, Tozzo E, Xiao L, et al. Effects of a novel glycogen synthase kinase-3 inhibitor on insulin-stimulated glucose metabolism in Zucker diabetic fatty (fa/fa) rats. *Diabetes.* 2002;51(10):2903–10.
21. Benzler J, Ganjam GK, Kruger M, Pinkenburg O, Kutschke M, Stohr S, et al. Hypothalamic glycogen synthase kinase 3beta has a central role in the regulation of food intake and glucose metabolism. *Biochem J.* 2012;447(1):175–84.
22. Könner AC, Janoschek R, Plum L, Jordan SD, Rother E, Ma X, et al. Insulin action in AgRP-expressing neurons is required for suppression of hepatic glucose production. *Cell Metab.* 2007;5(6):438–49.
23. Spanswick D, Smith MA, Mirshamsi S, Routh VH, Ashford ML. Insulin activates ATP-sensitive K+ channels in hypothalamic neurons of lean, but not obese rats. *Nature Neurosci.* 2000;3(8):757–8.
24. van den Hoek AM, van Heijningen C, Schröder-van der Elst JP, Ouwens DM, Havekes LM, Romijn JA, et al. Intracerebroventricular administration of neuropeptide Y induces hepatic insulin resistance via sympathetic innervation. *Diabetes.* 2008;57(9):2304–10.
25. Cassaglia PA, Shi Z, Brooks VL. Insulin increases sympathetic nerve activity in part by suppression of tonic inhibitory neuropeptide Y inputs into the paraventricular nucleus in female rats. *Am J Physiol Regul, Integr Comparat Physiol.* 2016;311(1):R97–r103.
26. Zhang Y, Proenca R, Maffei M, Barone M, Leopold L, Friedman JM. Positional cloning of the mouse obese gene and its human homologue. *Nature.* 1994;372(6505):425–32.
27. Considine RV, Sinha MK, Heiman ML, Kriauciunas A, Stephens TW, Nyce MR, et al. Serum immunoreactive-leptin concentrations in normal-weight and obese humans. *New Engl J Med.* 1996;334(5):292–5.
28. Dávalos-Salas M, Montgomery MK, Reehorst CM, Nightingale R, Ng I, Anderton H, et al. Deletion of intestinal Hdac3 remodels the lipidome of enterocytes and protects mice from diet-induced obesity. *Nat Commun.* 2019;10(1):5291.
29. Wang S, Wang C, Turdi S, Richmond KL, Zhang Y, Ren J. ALDH2 protects against high fat diet-induced obesity cardiomyopathy and defective autophagy: role of CaM kinase II, histone H3K9 methyltransferase SUV39H, Sirt1, and PGC-1α deacetylation. *Int J Obesity (2005).* 2018;42(5):1073–87.
30. El-Haschimi K, Pierroz DD, Hileman SM, Bjørbaek C, Flier JS. Two defects contribute to hypothalamic leptin resistance in mice with diet-induced obesity. *J Clin Invest.* 2000;105(12):1827–32.
31. Münzberg H, Flier JS, Bjørbaek C. Region-specific leptin resistance within the hypothalamus of diet-induced obese mice. *Endocrinology.* 2004;145(11):4880–9.
32. Harris RBS. Phosphorylation of STAT3 in hypothalamic nuclei is stimulated by lower doses of leptin than are needed to inhibit food intake. *Am J Physiol Endocrinol Metab.* 2021;321(1):E190–e201.

33. Mercer JG, Hoggard N, Williams LM, Lawrence CB, Hannah LT, Trayhurn P. Localization of leptin receptor mRNA and the long form splice variant (Ob-Rb) in mouse hypothalamus and adjacent brain regions by in situ hybridization. *FEBS Lett.* 1996;387(2–3):113–6.
34. Tartaglia LA, Dembski M, Weng X, Deng N, Culpepper J, Devos R, et al. Identification and expression cloning of a leptin receptor, OB-R. *Cell.* 1995;83(7):1263–71.
35. Ganong WF. Circumventricular organs: definition and role in the regulation of endocrine and autonomic function. *Clin Exp Pharmacol Physiol.* 2000;27(5–6):422–7.
36. Morita-Takemura S, Wanaka A. Blood-to-brain communication in the hypothalamus for energy intake regulation. *Neurochem Int.* 2019;128:135–42.
37. Mirzadeh Z, Alonge KM, Cabrales E, Herranz-Pérez V, Scarlett JM, Brown JM, et al. Perineuronal net formation during the critical period for neuronal maturation in the hypothalamic arcuate nucleus. *Nat Metab.* 2019;1(2):212–21.
38. Campbell JN, Macosko EZ, Fenselau H, Pers TH, Lyubetskaya A, Tenen D, et al. A molecular census of arcuate hypothalamus and median eminence cell types. *Nat Neurosci.* 2017;20(3):484–96.
39. Mercer JG, Hoggard N, Williams LM, Lawrence CB, Hannah LT, Morgan PJ, et al. Coexpression of leptin receptor and preproneuropeptide Y mRNA in arcuate nucleus of mouse hypothalamus. *J Neuroendocrinol.* 1996;8(10):733–5.
40. Elmquist JK, Coppari R, Balthasar N, Ichinose M, Lowell BB. Identifying hypothalamic pathways controlling food intake, body weight, and glucose homeostasis. *J Comp Neurol.* 2005;493(1):63–71.
41. Schwartz MW, Baskin DG, Bukowski TR, Kuijper JL, Foster D, Lasser G, et al. Specificity of leptin action on elevated blood glucose levels and hypothalamic neuropeptide Y gene expression in ob/ob mice. *Diabetes.* 1996;45(4):531–5.
42. Cheung CC, Clifton DK, Steiner RA. Proopiomelanocortin neurons are direct targets for leptin in the hypothalamus. *Endocrinology.* 1997;138(10):4489–92.
43. Cowley MA, Smart JL, Rubinstein M, Cerdán MG, Diano S, Horvath TL, et al. Leptin activates anorexigenic POMC neurons through a neural network in the arcuate nucleus. *Nature.* 2001;411(6836):480–4.
44. Rizk NM, Stammsen D, Preibisch G, Eckel J. Leptin and tumor necrosis factor-alpha induce the tyrosine phosphorylation of signal transducer and activator of transcription proteins in the hypothalamus of normal rats in vivo. *Endocrinology.* 2001;142(7):3027–32.
45. Hübschle T, Thom E, Watson A, Roth J, Klaus S, Meyerhof W. Leptin-induced nuclear translocation of STAT3 immunoreactivity in hypothalamic nuclei involved in body weight regulation. *J Neurosci.* 2001;21(7):2413–24.
46. Banks AS, Davis SM, Bates SH, Myers MG, Jr. Activation of downstream signals by the long form of the leptin receptor. *J Biol Chem.* 2000;275(19):14563–72.
47. Bjorbaek C, Kahn BB. Leptin signaling in the central nervous system and the periphery. *Recent Prog Horm Res.* 2004;59:305–31.
48. Bjorbaek C, Elmquist JK, Frantz JD, Shoelson SE, Flier JS. Identification of SOCS-3 as a potential mediator of central leptin resistance. *Molecular cell.* 1998;1(4):619–25.
49. Bjorbaek C, El-Haschimi K, Frantz JD, Flier JS. The role of SOCS-3 in leptin signaling and leptin resistance. *J Biol Chem.* 1999;274(42):30059–65.
50. Withers DJ, Gutierrez JS, Towery H, Burks DJ, Ren JM, Previs S, et al. Disruption of IRS-2 causes type 2 diabetes in mice. *Nature.* 1998;391(6670):900–4.
51. Niswender KD, Morton GJ, Stearns WH, Rhodes CJ, Myers MG, Jr., Schwartz MW. Intracellular signalling. Key enzyme in leptin-induced anorexia. *Nature.* 2001;413(6858):794–5.
52. Backer JM, Myers MG, Jr., Shoelson SE, Chin DJ, Sun XJ, Miralpeix M, et al. Phosphatidylinositol 3'-kinase is activated by association with IRS-1 during insulin stimulation. *EMBO J.* 1992;11(9):3469–79.
53. Sarbassov DD, Guertin DA, Ali SM, Sabatini DM. Phosphorylation and regulation of Akt/PKB by the rictor-mTOR complex. *Science.* 2005;307(5712):1098–101.
54. Liu L, Karkanias GB, Morales JC, Hawkins M, Barzilai N, Wang J, et al. Intracerebroventricular leptin regulates hepatic but not peripheral glucose fluxes. *J Biol Chem.* 1998;273(47):31160–7.
55. Morton GJ, Gelling RW, Niswender KD, Morrison CD, Rhodes CJ, Schwartz MW. Leptin regulates insulin sensitivity via phosphatidylinositol-3-OH kinase signaling in mediobasal hypothalamic neurons. *Cell Metab.* 2005;2(6):411–20.
56. Munzberg H, Flier JS, Bjorbaek C. Region-specific leptin resistance within the

hypothalamus of diet-induced obese mice. *Endocrinology*. 2004;145(11):4880–9.
57. Tups A, Ellis C, Moar KM, Logie TJ, Adam CL, Mercer JG, et al. Photoperiodic regulation of leptin sensitivity in the Siberian hamster, Phodopus sungorus, is reflected in arcuate nucleus SOCS-3 (suppressor of cytokine signaling) gene expression. *Endocrinology*. 2004;145(3): 1185–93.
58. Brown MS, Goldstein JL. Selective versus total insulin resistance: a pathogenic paradox. *Cell Metab*. 2008;7(2):95–6.
59. Myers MG, Jr., Leibel RL, Seeley RJ, Schwartz MW. Obesity and leptin resistance: distinguishing cause from effect. *Trends Endocrinol Metab*. 2010;21(11):643–51.
60. Balland E, Chen W, Dodd GT, Conductier G, Coppari R, Tiganis T, et al. Leptin signaling in the arcuate nucleus reduces insulin's capacity to suppress hepatic glucose production in obese mice. *Cell Rep*. 2019;26(2):346–55.e3.
61. Zhao S, Zhu Y, Schultz RD, Li N, He Z, Zhang Z, et al. Partial leptin reduction as an insulin sensitization and weight loss strategy. *Cell Metab*. 2019.
62. Zhao S, Li N, Zhu Y, Straub L, Zhang Z, Wang MY, et al. Partial leptin deficiency confers resistance to diet-induced obesity in mice. *Mol Metab*. 2020;37:100995.
63. Pretz D, Le Foll C, Rizwan MZ, Lutz TA, Tups A. Hyperleptinemia as a contributing factor for the impairment of glucose intolerance in obesity. *FASEB J*. 2021;35(2):e21216.
64. Koch CE, Lowe C, Pretz D, Steger J, Williams LM, Tups A. High-fat diet induces leptin resistance in leptin-deficient mice. *J Neuroendocrinol*. 2014;26(2):58–67.
65. Ravussin Y, LeDuc CA, Watanabe K, Mueller BR, Skowronski A, Rosenbaum M, et al. Effects of chronic leptin infusion on subsequent body weight and composition in mice: Can body weight set point be reset? *Mol Metab*. 2014;3(4):432–40.
66. Ravussin Y, Edwin E, Gallop M, Xu L, Bartolomé A, Kraakman MJ, et al. Evidence for a non-leptin system that defends against weight gain in overfeeding. *Cell Metab*. 2018;28(2):289–99.e5.
67. Skowronski AA, LeDuc CA, Foo KS, Goffer Y, Burnett LC, Egli D, et al. Physiological consequences of transient hyperleptinemia during discrete developmental periods on body weight in mice. *Sci Transl Med*. 2020;12(524).
68. Leon-Cabrera S, Solís-Lozano L, Suárez-Álvarez K, González-Chávez A, Béjar YL, Robles-Díaz G, et al. Hyperleptinemia is associated with parameters of low-grade systemic inflammation and metabolic dysfunction in obese human beings. *Front Integrat Neurosci*. 2013;7:62.
69. Depaoli A, Long A, Fine GM, Stewart M, Rahilly S. Efficacy of metreleptin for weight loss in overweight and obese adults with low leptin levels. *Diabetes*. 2018;67(Supplement 1):296-LB.
70. Crujeiras AB, Goyenechea E, Abete I, Lage M, Carreira MC, Martínez JA, et al. Weight regain after a diet-induced loss is predicted by higher baseline leptin and lower ghrelin plasma levels. *J Clin Endocrinol Metab*. 2010;95(11):5037–44.
71. Neseliler S, Hu W, Larcher K, Zacchia M, Dadar M, Scala SG, et al. Neurocognitive and hormonal correlates of voluntary weight loss in humans. *Cell Metab*. 2019;29(1):39–49.e4.
72. Coleman DL. Obese and diabetes: two mutant genes causing diabetes-obesity syndromes in mice. *Diabetologia*. 1978;14(3):141–8.
73. Wua J, Wang HM, Li J, Men XL. [The research applications of db/db mouse]. *Sheng li ke xue jin zhan [Progress in physiology]*. 2013;44(1):12–8.
74. Pelleymounter MA, Cullen MJ, Baker MB, Hecht R, Winters D, Boone T, et al. Effects of the obese gene product on body weight regulation in ob/ob mice. *Science*. 1995;269(5223):540–3.
75. Genuth SM, Przybylski RJ, Rosenberg DM. Insulin resistance in genetically obese, hyperglycemic mice. *Endocrinology*. 1971;88(5):1230–8.
76. Oral EA, Simha V, Ruiz E, Andewelt A, Premkumar A, Snell P, et al. Leptin-replacement therapy for lipodystrophy. *New Engl J Med*. 2002;346(8): 570–8.
77. Shimomura I, Hammer RE, Ikemoto S, Brown MS, Goldstein JL. Leptin reverses insulin resistance and diabetes mellitus in mice with congenital lipodystrophy. *Nature*. 1999;401(6748):73–6.
78. Koch C, Augustine RA, Steger J, Ganjam GK, Benzler J, Pracht C, et al. Leptin rapidly improves glucose homeostasis in obese mice by increasing hypothalamic insulin sensitivity. *J Neurosci*. 2010;30(48):16180–7.
79. Pocai A, Morgan K, Buettner C, Gutierrez-Juarez R, Obici S, Rossetti L. Central leptin acutely reverses diet-induced hepatic insulin resistance. *Diabetes*. 2005;54(11):3182–9.
80. Hidaka S, Yoshimatsu H, Kondou S, Tsuruta Y, Oka K, Noguchi H, et al. Chronic central leptin infusion restores hyperglycemia independent of

food intake and insulin level in streptozotocin-induced diabetic rats. *FASEB J.* 2002;16(6):509–18.
81. Meek TH, Matsen ME, Dorfman MD, Guyenet SJ, Damian V, Nguyen HT, et al. Leptin action in the ventromedial hypothalamic nucleus is sufficient, but not necessary, to normalize diabetic hyperglycemia. *Endocrinology.* 2013;154(9):3067–76.
82. Goncalves GH, Li W, Garcia AV, Figueiredo MS, Bjorbaek C. Hypothalamic agouti-related peptide neurons and the central melanocortin system are crucial mediators of leptin's antidiabetic actions. *Cell Rep.* 2014;7(4):1093–103.
83. da Silva AA, Spradley FT, Granger JP, Hall JE, do Carmo JM. Brain-mediated antidiabetic, anorexic, and cardiovascular actions of leptin require melanocortin-4 receptor signaling. *J Neurophysiol.* 2015;113(7):2786–91.
84. Xu Y, Chang JT, Myers MG, Jr., Xu Y, Tong Q. Euglycemia restoration by central leptin in type 1 diabetes requires STAT3 signaling but not fast-acting neurotransmitter release. *Diabetes.* 2016;65(4):1040–9.
85. Neumann UH, Denroche HC, Mojibian M, Covey SD, Kieffer TJ. Insulin knockout mice have extended survival but volatile blood glucose levels on leptin therapy. *Endocrinology.* 2016;157(3):1007–12.
86. Chinookoswong N, Wang JL, Shi ZQ. Leptin restores euglycemia and normalizes glucose turnover in insulin-deficient diabetes in the rat. *Diabetes.* 1999;48(7):1487–92.
87. Yu X, Park BH, Wang MY, Wang ZV, Unger RH. Making insulin-deficient type 1 diabetic rodents thrive without insulin. *Proc Natl Acad Sci U S A.* 2008;105(37):14070–5.
88. Vasandani C, Clark GO, Adams-Huet B, Quittner C, Garg A. Efficacy and safety of metreleptin therapy in patients with type 1 diabetes: a pilot study. *Diabetes Care.* 2017;40(5):694–7.
89. Hanaki K, Becker DJ, Arslanian SA. Leptin before and after insulin therapy in children with new-onset type 1 diabetes. *J Clin Endocrinol Metab.* 1999;84(5):1524–6.
90. Hathout EH, Sharkey J, Racine M, Ahn D, Mace JW, Saad MF. Changes in plasma leptin during the treatment of diabetic ketoacidosis. *J Clin Endocrinol Metab.* 1999;84(12):4545–8.
91. Denroche HC, Levi J, Wideman RD, Sequeira RM, Huynh FK, Covey SD, et al. Leptin therapy reverses hyperglycemia in mice with streptozotocin-induced diabetes, independent of hepatic leptin signaling. *Diabetes.* 2011;60(5):1414–23.
92. Sharma RP, Chopra VL. Effect of the Wingless (wgl) mutation on wing and haltere development in Drosophila melanogaster. *Develop Biol.* 1976;48(2):461–5.
93. Nusse R, Varmus HE. Many tumors induced by the mouse mammary tumor virus contain a provirus integrated in the same region of the host genome. *Cell.* 1982;31(1):99–109.
94. Nusse R, Varmus HE. Wnt genes. *Cell.* 1992;69(7):1073–87.
95. Dale TC. Signal transduction by the Wnt family of ligands. *Biochem J.* 1998;329 (Pt 2):209–23.
96. Wang X, Kopinke D, Lin J, McPherson AD, Duncan RN, Otsuna H, et al. Wnt signaling regulates postembryonic hypothalamic progenitor differentiation. *Dev Cell.* 2012;23(3):624–36.
97. Wang X, Lee JE, Dorsky RI. Identification of Wnt-responsive cells in the zebrafish hypothalamus. *Zebrafish.* 2009;6(1):49–58.
98. Brinkmeier ML, Potok MA, Davis SW, Camper SA. TCF4 deficiency expands ventral diencephalon signaling and increases induction of pituitary progenitors. *Develop Biol.* 2007;311(2):396–407.
99. Grandel H, Kaslin J, Ganz J, Wenzel I, Brand M. Neural stem cells and neurogenesis in the adult zebrafish brain: origin, proliferation dynamics, migration and cell fate. *Develop Biol.* 2006;295(1):263–77.
100. Helfer G, Ross AW, Russell L, Thomson LM, Shearer KD, Goodman TH, et al. Photoperiod regulates vitamin A and Wnt/beta-catenin signaling in F344 rats. *Endocrinology.* 2012;153(2):815–24.
101. Bolborea M, Pollatzek E, Benford H, Sotelo-Hitschfeld T, Dale N. Hypothalamic tanycytes generate acute hyperphagia through activation of the arcuate neuronal network. *Proc Natl Acad Sci U S A.* 2020;117(25):14473–81.
102. Lewis JE, Ebling FJ. Tanycytes as regulators of seasonal cycles in neuroendocrine function. *Front Neurol.* 2017;8:79.
103. Grant SF, Thorleifsson G, Reynisdottir I, Benediktsson R, Manolescu A, Sainz J, et al. Variant of transcription factor 7-like 2 (TCF7L2) gene confers risk of type 2 diabetes. *Nat Genet.* 2006;38(3):320–3.
104. Steinthorsdottir V, Thorleifsson G, Reynisdottir I, Benediktsson R, Jonsdottir T, Walters GB, et al. A variant in CDKAL1 influences insulin response and risk of type 2 diabetes. *Nat Genet.* 2007;39(6):770–5.

105. Take K, Waki H, Sun W, Wada T, Yu J, Nakamura M, et al. CDK5 regulatory subunit-associated protein 1-like 1 negatively regulates adipocyte differentiation through activation of Wnt signaling pathway. *Sci Rep.* 2017;7(1):7326.
106. Shungin D, Winkler TW, Croteau-Chonka DC, Ferreira T, Locke AE, Mägi R, et al. New genetic loci link adipose and insulin biology to body fat distribution. *Nature.* 2015;518(7538):187–96.
107. Loh NY, Minchin JEN, Pinnick KE, Verma M, Todorčević M, Denton N, et al. RSPO3 impacts body fat distribution and regulates adipose cell biology in vitro. *Nat Commun.* 2020;11(1):2797.
108. Anagnostou SH, Shepherd PR. Glucose induces an autocrine activation of the Wnt/beta-catenin pathway in macrophage cell lines. *Biochem J.* 2008;416(2):211–8.
109. Lattanzio S, Santilli F, Liani R, Vazzana N, Ueland T, Di Fulvio P, et al. Circulating dickkopf-1 in diabetes mellitus: association with platelet activation and effects of improved metabolic control and low-dose aspirin. *J Am Heart Assoc.* 2014;3(4).
110. Li X, Shan J, Chang W, Kim I, Bao J, Lee HJ, et al. Chemical and genetic evidence for the involvement of Wnt antagonist Dickkopf2 in regulation of glucose metabolism. *Proc Natl Acad Sci U S A.* 2012;109(28):11402–7.
111. Stambolic V, Woodgett JR. Mitogen inactivation of glycogen synthase kinase-3 beta in intact cells via serine 9 phosphorylation. *Biochem J.* 1994;303 (Pt 3)(Pt 3):701–4.
112. Sutherland C, Leighton IA, Cohen P. Inactivation of glycogen synthase kinase-3 beta by phosphorylation: new kinase connections in insulin and growth-factor signalling. *Biochem J.* 1993;296 (Pt 1)(Pt 1):15–9.
113. Kaidanovich-Beilin O, Eldar-Finkelman H. Long-term treatment with novel glycogen synthase kinase-3 inhibitor improves glucose homeostasis in ob/ob mice: molecular characterization in liver and muscle. *J Pharmacol Exp Therapeut.* 2006;316(1):17–24.
114. Nikoulina SE, Ciaraldi TP, Mudaliar S, Carter L, Johnson K, Henry RR. Inhibition of glycogen synthase kinase 3 improves insulin action and glucose metabolism in human skeletal muscle. *Diabetes.* 2002;51(7):2190–8.
115. Dupre J, Ross SA, Watson D, Brown JC. Stimulation of insulin secretion by gastric inhibitory polypeptide in man. *J Clin Endocrinol Metab.* 1973;37(5):826–8.
116. Kreymann B, Williams G, Ghatei MA, Bloom SR. Glucagon-like peptide-1 7–36: a physiological incretin in man. *Lancet (London, England).* 1987;2(8571):1300–4.
117. Muscelli E, Mari A, Natali A, Astiarraga BD, Camastra S, Frascerra S, et al. Impact of incretin hormones on beta-cell function in subjects with normal or impaired glucose tolerance. *Am J Physiol Endocrinol Metab.* 2006;291(6):E1144–50.
118. Nauck MA, Homberger E, Siegel EG, Allen RC, Eaton RP, Ebert R, et al. Incretin effects of increasing glucose loads in man calculated from venous insulin and C-peptide responses. *Am J Physiol Endocrinol Metab.* 1986;63(2):492–8.
119. Campbell JE, Drucker DJ. Pharmacology, physiology, and mechanisms of incretin hormone action. *Cell Metab.* 2013;17(6):819–37.
120. Yip RG, Boylan MO, Kieffer TJ, Wolfe MM. Functional GIP receptors are present on adipocytes. *Endocrinology.* 1998;139(9):4004–7.
121. Chia CW, Carlson OD, Kim W, Shin YK, Charles CP, Kim HS, et al. Exogenous glucose-dependent insulinotropic polypeptide worsens post prandial hyperglycemia in type 2 diabetes. *Diabetes.* 2009;58(6):1342–9.
122. Kim SJ, Nian C, Karunakaran S, Clee SM, Isales CM, McIntosh CH. GIP-overexpressing mice demonstrate reduced diet-induced obesity and steatosis, and improved glucose homeostasis. *PloS One.* 2012;7(7):e40156.
123. Zhang Q, Delessa CT, Augustin R, Bakhti M, Colldén G, Drucker DJ, et al. The glucose-dependent insulinotropic polypeptide (GIP) regulates body weight and food intake via CNS-GIPR signaling. *Cell Metab.* 2021;33(4):833–44.e5.
124. Holst JJ. The physiology of glucagon-like peptide 1. *Physiol Rev.* 2007;87(4):1409–39.
125. Sandoval DA, D'Alessio DA. Physiology of proglucagon peptides: role of glucagon and GLP-1 in health and disease. *Physiol Rev.* 2015;95(2):513–48.
126. Verdich C, Flint A, Gutzwiller JP, Näslund E, Beglinger C, Hellström PM, et al. A meta-analysis of the effect of glucagon-like peptide-1 (7–36) amide on ad libitum energy intake in humans. *J Clin Endocrinol Metab.* 2001;86(9):4382–9.
127. Holt MK, Richards JE, Cook DR, Brierley DI, Williams DL, Reimann F, et al. Preproglucagon neurons in the nucleus of the solitary tract are the main source of brain GLP-1, mediate stress-induced hypophagia, and limit unusually large intakes of food. *Diabetes.* 2019;68(1):21–33.

128. Vrang N, Phifer CB, Corkern MM, Berthoud HR. Gastric distension induces c-Fos in medullary GLP-1/2-containing neurons. *Am J Physiol Regul, Integr Comp Physiol*. 2003;285(2):R470–8.
129. Hisadome K, Reimann F, Gribble FM, Trapp S. Leptin directly depolarizes preproglucagon neurons in the nucleus tractus solitarius: electrical properties of glucagon-like Peptide 1 neurons. *Diabetes*. 2010;59(8):1890–8.
130. Shi X, Chacko S, Li F, Li D, Burrin D, Chan L, et al. Acute activation of GLP-1-expressing neurons promotes glucose homeostasis and insulin sensitivity. *Mol Metab*. 2017;6(11):1350–9.
131. Brierley DI, Holt MK, Singh A, de Araujo A, McDougle M, Vergara M, et al. Central and peripheral GLP-1 systems independently suppress eating. *Nat Metab*. 2021;3(2):258–73.
132. Pi-Sunyer X, Astrup A, Fujioka K, Greenway F, Halpern A, Krempf M, et al. A Randomized, Controlled Trial of 3.0 mg of Liraglutide in Weight Management. *New Engl J Med*. 2015;373(1):11–22.
133. O'Neil PM, Birkenfeld AL, McGowan B, Mosenzon O, Pedersen SD, Wharton S, et al. Efficacy and safety of semaglutide compared with liraglutide and placebo for weight loss in patients with obesity: a randomised, double-blind, placebo and active controlled, dose-ranging, phase 2 trial. *Lancet (London, England)*. 2018;392(10148):637–49.
134. Lin CH, Shao L, Zhang YM, Tu YJ, Zhang Y, Tomlinson B, et al. An evaluation of liraglutide including its efficacy and safety for the treatment of obesity. *Expert Opin Pharmacother*. 2020;21(3):275–85.
135. Frias JP, Nauck MA, Van J, Kutner ME, Cui X, Benson C, et al. Efficacy and safety of LY3298176, a novel dual GIP and GLP-1 receptor agonist, in patients with type 2 diabetes: a randomised, placebo-controlled and active comparator-controlled phase 2 trial. *Lancet (London, England)*. 2018; 392(10160):2180–93.
136. Sorrenson B, Cognard E, Lee KL, Dissanayake WC, Fu Y, Han W, et al. A critical role for β-catenin in modulating levels of insulin secretion from β-cells by regulating actin cytoskeleton and insulin vesicle localization. *J Biol Chem*. 2016;291(50):25888–900.
137. Yi F, Brubaker PL, Jin T. TCF-4 mediates cell type-specific regulation of proglucagon gene expression by beta-catenin and glycogen synthase kinase-3beta. *J Biol Chem*. 2005;280(2):1457–64.
138. García-Martínez JM, Chocarro-Calvo A, Moya CM, García-Jiménez C. WNT/beta-catenin increases the production of incretins by entero-endocrine cells. *Diabetologia*. 2009;52(9):1913–24.
139. Shu L, Matveyenko AV, Kerr-Conte J, Cho JH, McIntosh CH, Maedler K. Decreased TCF7L2 protein levels in type 2 diabetes mellitus correlate with downregulation of GIP- and GLP-1 receptors and impaired beta-cell function. *Human Mol Genet*. 2015;24(10):3004.
140. Wheeler MB, Gelling RW, Hinke SA, Tu B, Pederson RA, Lynn F, et al. Characterization of the carboxyl-terminal domain of the rat glucose-dependent insulinotropic polypeptide (GIP) receptor. A role for serines 426 and 427 in regulating the rate of internalization. *J Biol Chem*. 1999;274(35):24593–601.
141. Liu Z, Habener JF. Glucagon-like peptide-1 activation of TCF7L2-dependent Wnt signaling enhances pancreatic beta cell proliferation. *J Biol Chem*. 2008;283(13):8723–35.

CHAPTER SIX

Hypothalamic Neuronal Circuits Are Modulated by Insulin and Impact Metabolism[1]

Tadeu de Oliveira Diz
Sabela Casado, Rubén Nogueiras, and Sulay Tovar

CONTENTS

[1] Tadeu de Oliveira Diz and Sabela Casado have contributed equally.

DOI: 10.1201/9781003079927-6

INTRODUCTION

More than 150 years ago, in 1855, the French physiologist Claude Bernard (1) identified that the central nervous system (CNS) plays a critical role in the regulation of peripheral blood glucose levels. The first demonstration of the glucose-lowering effect of a pancreatic extract was in 1916 followed by the insulin discovery in 1921 by Banting and Best (2). These first studies were focused on the peripheral actions of insulin in the control of glucose metabolism denying any involvement of the brain, which was considered to be an insulin-insensitive organ.

It was necessary to wait until the late 1970s in 1977 when Woods and Porte (3) demonstrated that insulin had a role in CNS showing that insulin infusion into the brain reduced food intake. A year later, in 1978, Havrankova et al. (4) found that measurements of insulin concentrations in extracts of whole-brain averaged about 25-fold greater than the concentration in plasma. These two discoveries highlighted the importance of insulin in the brain.

However, while the functions of insulin signalling in peripheral tissues have been studied widely, its actions in the CNS are far less understood. Only they started to gain attention over the past 30 years, when studies in both animals and humans provided convincing evidence that central insulin signalling is directly involved in both energy and glucose homeostasis.

THE HYPOTHALAMIC RELEVANCE OF INSULIN

The hypothalamus is a vital region of the CNS that is essential for most basic physiological functions. It is located below the thalamus and is part of the limbic system, and it contains distinct neuronal populations implicated in the regulation of many homeostatic processes, such as energy balance, behaviour, reproduction, biological rhythms, stress responses and autonomic functions (5). The hypothalamus orchestrates and coordinates the neuroendocrine regulation of energy balance through a complex pathway via afferent signals received from the periphery depending on nutrient status, and efferent pathways including the sympathetic nervous system, which promotes energy expenditure, and the parasympathetic nervous system, which promotes energy storage (6). There is now convincing evidence that insulin has a significant impact on the brain. Hypothalamic insulin signalling functions as a key neuromodulator in the regulation of energy homeostasis, insulin sensitivity and glucose homeostasis (7).

Insulin acts directly on different neuronal populations to adapt energy homeostasis in the organism. Until now, the most studied functions were the suppression of food intake and body weight reduction, with the improvement of glucose metabolism. So, insulin could act as an inhibitory, anorexigenic and leptogenic signal, interacting with neurotransmitters and regulatory peptides. The hypothalamic relevance of insulin effects on the organism is due to the fact that insulin receptors (INSRs) are widely expressed in the brain and co-localize with both AgRP (agouti-related peptide) and POMC (pro-opiomelanocortin) neurons in the hypothalamic arcuate nucleus (ARC) (8, 9). Activation of the insulin signalling pathway in the ARC decreases NPY and AgRP expression as well as increases the expression of POMC (9–12), producing an anorexigenic effect, reducing food intake and body weight (3). Normal insulin signalling in the hypothalamus is essential for the regulation of several important physiological functions (13, 14). However, excessive insulin in the brain affects the regulation of reproduction, life span, body weight and nutrient metabolism (3, 15). Thus, insulin action in the brain is physiologically beneficial when properly controlled, whereas insulin dysregulation at the hypothalamus can lead to metabolic disorders.

HOW INSULIN TARGETS THE HYPOTHALAMUS

Pancreatic β-cells secrete insulin into the circulation, and the principal regulator of insulin secretion is blood glucose. Post-meal spikes in blood glucose induce rapid insulin secretion into blood circulation that affects several organs, including the hypothalamus. The level of hypothalamic insulin depends on the amount of blood insulin transported in the brain and on insulin potentially originating from local sources.

Peripheral circulating insulin reaches the mediobasal hypothalamus (MBH), where an interplay between circulating insulin and the hypothalamus happens. The mechanism by which insulin is transferred from the bloodstream to the hypothalamus to reach the arcuate nucleus (ARC) is unclear, but several routes have been proposed and can be divided into three major groups:

1. *Receptor-mediated transcytosis, across the blood-brain barrier (BBB)*: The BBB is a highly selective semipermeable border of endothelial cells, joined by tight junctions, that regulate the passage of insulin through paracellular diffusion into the hypothalamus (16). Circulating insulin binds to insulin receptors (INSRs) expressed on the endothelial cells of the BBB. Once bound to the INSRs, insulin is internalized into vesicles and transported across the endothelial cells (17, 18). There, it is presented by receptor-mediated exocytosis to the abluminal side of the endothelial cell, where it enters the hypothalamus (18, 19). This is a saturable transport system in a concentration-dependent way (20).
2. *Passive extravasation, in the median eminence* (ME): The ME is a circumventricular organ that is situated below the MBH, ventral to the third ventricle and adjacent to the ARC. Insulin can enter the brain via passive diffusion through the highly permeable capillaries devoid of a BBB and ependymal cells of the ME (21).
3. *Tanycyte-mediated transport, from the cerebrospinal fluid* (CSF): Insulin is also present within the CSF, where it can be transported into the brain via tanycytes that line the floor of the third ventricle and project into the hypothalamus, including the ARC, ventromedial hypothalamus (VMH) and the ME (22–24).

Future studies are needed to fully outline the relative contributions of the BBB, ME, and blood-CSF barrier in the insulin transport from the bloodstream to the hypothalamus to reach the ARC (Figure 6.1).

INSULIN AND THE HYPOTHALAMIC NEURONAL CIRCUITS

The Arcuate Nucleus, Key Hypothalamic Region in the Insulin Signalling

Insulin directly acts on distinct hypothalamic neuronal populations to adapt energy homeostasis and glucose metabolism. Once the insulin reaches the hypothalamus, it contacts the **arcuate nucleus (ARC)**, due to its accessible structural location. The ARC is a population of neurons located in the

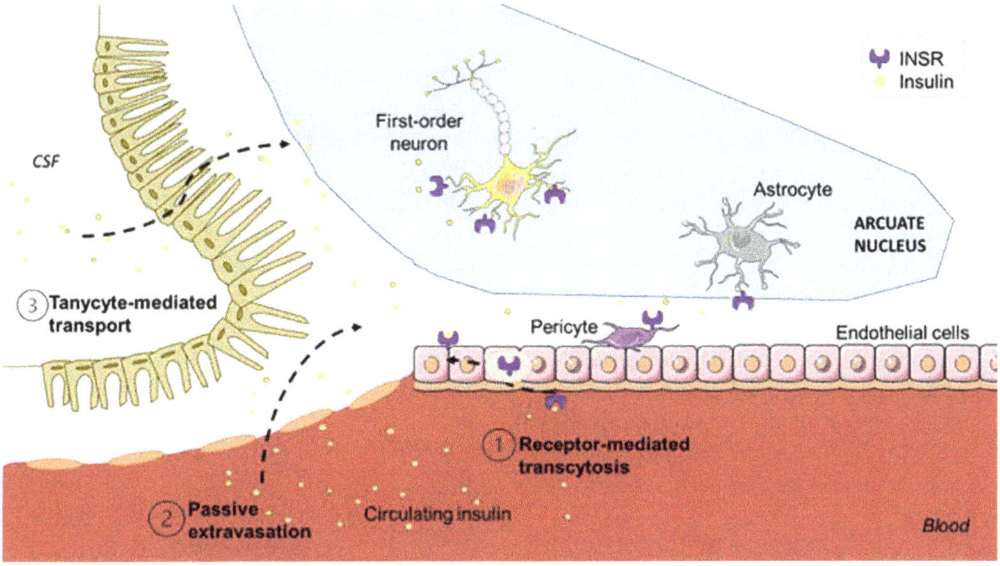

Figure 6.1 Proposed mechanisms in which insulin reaches the medio basal hypothalamus. Circulating insulin targets the hypothalamus via (1) Receptor-mediated transcytosis (2) Passive extravasation from blood to the media eminence (ME) (3) Tanycyte-mediated transport, from the cerebrospinal fluid (CSF). *Graphic abbreviations*: INSR: Insulin receptor, CSF: Cerebrospinal fluid, ARC: Hypothalamic arcuate nucleus, ME: Median eminence, 3V: Third ventricle.

medio basal hypothalamus (MBH), adjacent to the median eminence (ME) and at the base of the third ventricle. It integrates circulating hormonal and metabolic signals to control energy expenditure and intake (25).

The melanocortin system has been considered a promising model of neuronal control in the regulation of long-term energy balance. The ARC is critical in this model as it contains two functionally opposing neuronal populations: agouti-related peptide (AgRP) and pro-opiomelanocortin (POMC). They are one of the first-order cells through which insulin signals in the brain. They play critical roles in ingestion behaviour, controlling energy and glucose homeostasis, and the INSRs co-localize with them (8, 9).

- The '*AgRP neurons*' are NPY (neuropeptide Y)/AgRP (agouti-related peptide) expressing neurones. These orexigenic neurons play a critical role in energy balance increasing food intake and body weight and decreasing energy expenditure and have a role in glucose homeostasis (26, 27). The activation of these neurons generates a liberation of AgRP, NPY and gamma-aminobutyric acid (GABA) to target other different hypothalamic neuronal populations (28).
- The '*POMC neurons*' are POMC (pro-opiomelanocortin)/CART (cocaine and amphetamine-regulated transcript) neurons. They are anorexigenic neurons and promote negative energy balance, decreasing food intake and body weight, and increasing energy expenditure (29, 30). Activation of POMC neurons reduces appetite (31, 32), whereas inhibition of POMC neurons modestly increases feeding (32). The activation of POMC neurons results in POMC and CART peptides liberation. POMC is processed to generate α-melanocyte-stimulating hormone (α-MSH), which is released and agonizes melanocortin-4 receptors (MC4R) on neuron populations in other brain regions (33).

Within the arcuate nucleus, activation of the insulin receptor substrate (IRS), the phosphatidylinositol 3-kinase (PI3K) by insulin occurs in neurons expressing IRS-2, like AgRP and POMC neurons. In these neurons, insulin auto phosphorylates the insulin receptor, leading to the activation of the PI3K enzyme system (PI3K/Akt pathway, see part 3). These neurons respond to insulin signalling by altering neuronal excitability (short term) and gene expression and transcription (long term) to modulate the control of food intake, energy expenditure and glucose homeostasis (9, 33).

Within *AgRP neurons*, insulin activates the INSR signalling, activating PI3K promoting a robust increase in the phosphorylation of AKT, activating ATP-sensitive potassium (K-ATP) channels (34). This has been shown to cause hyperpolarization of the AgRP neuron, decreasing the firing rate and inhibiting AgRP and NPY transcription and decreasing the release of AgRP, NPY, GABA and other transmitters from AgRP neurons. This reduces appetite and increases energy expenditure (33, 35, 36).

In *POMC neurons*, insulin activates the INSR signalling, the activation of the insulin pathway in the POMC neurons generates an increase in the phosphorylation of AKT augmenting the activation of the K-ATP channels through the action of PI3K (34). This causes an hyperpolarization with a reduction in the firing rate of POMC neurons (36, 37) which increases the POMC mRNA expression (9) and generation an increase in POMC and CART peptides release, and other transmitters from POMC neurons. POMC is then cleaved by prohormone convertases to generate α-melanocyte-stimulating hormone (α-MSH) among other different bioactive products (β-MSH, adenocorticotropin and β-endorphin), which is released and agonizes melanocortin 4 receptors (MC4R) in other hypothalamic regions (33).

Thus, both classes of neurons account for insulin signalling in the hypothalamus: POMC is activated to release anorexigenic peptides, and AgRP is inhibited with a consequent reduction in the release of orexigenic peptides. Moreover, there is a crosstalk between AgRP and POMC neurons, when POMC neurons are activated, in this case by insulin, the release of POMC acts on melanocortin 3 receptors (MC3R) of the AgRP neurons to inactivate its activity. Altogether, the effect of insulin signalling over POMC neurons generates a decrease in food intake, body weight, and increased energy expenditure (29, 30).

However, there is some controversy in the insulin action on the POMC neurons as some studies report that insulin can both stimulate (38) or inhibit (36) POMC neurons. There is a publication showing that insulin in POMC neurons can display either excitation (11%), inhibition (51%) or no-responsivity (38%) and that in POMC neurons of diet-induced obese mice the neuronal response is quite different (3% excitation, 65% inhibition, 32% no-response) (39). Interestingly, disruption of insulin signalling from POMC neurons did not affect energy or glucose homeostasis (36), but it is known that POMC has a critical role in the regulation of metabolism as some human studies show that some POMC gene mutations display early-onset obesity (40, 41).

In animals and humans, peripheral administration, intranasal administration, as well as intracerebroventricular infusion of insulin or insulin mimetics activates the insulin signalling pathway in the ARC. Insulin action generates a hyperpolarization of AgRP and POMC neurons (10, 11, 42), thus generating a decrease in food intake, body weight and an increase in energy balance. On the contrary, pharmacological, or genetic blockage of the insulin signalling in the ARC leads to an increase in food intake and body weight and a reduction in energy balance (15, 42). However, insulin's anorectic effect appears to be sex-specific. Whereas intracerebroventricular injection of insulin acutely inhibits food intake in male rats, it fails to do so in female rats (43). Moreover, in humans, insulin intranasal administration lowers food intake in men, but there is no change in food intake in women (44). Moreover, there are some studies pointing out that INSR expression in neurons other than, or in addition to, POMC or AgRP cells is required for insulin's ability to control energy homeostasis, and they provide direct evidence that neurons other than AgRP and POMC mediate insulin's ability to suppress food intake (15, 36, 45). In addition, in genetic experiments with a deletion in INSRs using Cre-Lox recombination within AgRP and POMC neurons in development stages in mice, these neuronal populations do not seem to be vital as they show limited effects on body weight or food intake (27, 36, 46). Likewise, the fact that during development stages other neuronal populations that do not contain POMC nor AgRP may generate a compensatory mechanism for the regulation of feeding and energy homeostasis by insulin (26) indicates the relevance of this pathway in survival.

All in all, there is enough evidence and studies supporting the notion that INSR signalling in the hypothalamus mostly reduces the firing rate of POMC neurons, inducing POMC transcription, and at the same time inhibits AgRP expression. This generates a release of anorexigenic peptides that leads to a reduction in food intake, body weight and an increase in energy expenditure (29, 30). And although POMC nor AgRP seem to be crucial, the fact that a compensatory regulation exists highlights the importance of this insulin signalling route for survival (26).

OTHER HYPOTHALAMIC CIRCUITS MODULATED BY INSULIN

The projections from the AgRP and POMC neurons, extensively communicate with the "second-order" neurons, located in other hypothalamic nuclei.

Insulin action over POMC neurons in the ARC stimulates the release of POMC peptide that is processed to generate α-melanocyte-stimulating hormone (α-MSH). This α-MSH activates second-order neurons through direct interaction with the melanocortin receptors (MCRs) which are part of the central melanocortin system, essential for the regulation of long-term energy homeostasis. The two main MCRs in the hypothalamus are melanocortin 3 receptor (MC3R) and melanocortin 4 receptor (MC4R), they are G protein-coupled receptors with a high level of constitutive activity and play a critical role in mediating the anorexigenic signal through α-MSH and the orexigenic signal through AgRP, respectively (28, 47). The activation of MC4R by α-MSH signals decreases energy intake. Moreover, POMC neurons' action over distinct MC4R expressing neuronal populations causes a decrease in insulin secretion or increases insulin sensitivity independently of their effect on food intake and energy expenditure (48).

The MC4R is extensively located in the *paraventricular nucleus (PVN)*. The PVN lies adjacent to the third ventricle and within the periventricular zone. There, MC4R is targeted by a-MSH that acts as an agonist and potentiates GABA-mediated response by acting presynaptically provoking food intake reduction (49, 50). However, the direct role of insulin on neurons residing in the PVN is It is so far quite unknown. Moreover, the MCRs also have the presence in other hypothalamic nucleus such as the *dorsomedial hypothalamus (DMH)* (51) and *lateral hypothalamic area (LHA)* (52) also generating food intake reduction.

Insulin also targets steroidogenic factor 1 (SF1) neurons located in the *ventromedial hypothalamus*

(*VMH*) under conditions of hyperinsulinemia (53, 54). The VMH is a pear-shaped structure in the tuberal region of the hypothalamus and represents another effector site for insulin action in the control of energy homeostasis. The SF1 positive VMH neurons are a representative neuronal population involved in many aspects of metabolic regulation and they have emerged as a critical component of the mechanism underpinning glucose homeostasis. They also co-localize with the INSR. When insulin targets this neuron population, they activate the INSR and immediately hyperpolarize SF1 positive VMH neurons via PI3K-dependent activation of K-ATP channels (PI3K/Akt pathway, see part 3) providing glutamatergic input on POMC neurons (54, 55). As it was previously demonstrated that VMH and SF1 neurons send glutamatergic projections to arcuate POMC neurons (54, 56), the acute inhibition of SF1 neurons by insulin could result in the suppression of an anorexigenic pathway from SF1 neuron to POMC neurons. Specific deletion of the forkhead box O (FOXO), a downstream target of PI3K, in the SF1 neurons of mice promotes energy expenditure and improves glucose metabolism, this suggests that the PI3K-FOXO1 pathway is a potential signalling pathway for glycaemic homeostasis in SF1 neurons (57). Insulin can also dictate diet-mediated anabolic actions that can temporally elevate insulin levels in the brain. It was found that when rodents were fed a high-fat diet (HFD), central insulin levels were increased, which indirectly inhibited POMC neurons in the ARC. This action was through activation of the VMH SF1 neurons, which ultimately suppressed satiety and resulted in an increase in appetite and food intake (54). Mice with deletion of INSRs in VMH-specific SF1 neurons are protected from diet-induced obesity (54), which suggests that INSRs expressed by SF1 neurons promote obesity on an HFD.

The ***dorsomedial hypothalamic nucleus (DMH)*** is a brain structure adjacent to the third ventricle, which is located caudal and ventral to the paraventricular nucleus of the hypothalamus (PVN). Its lateral portion is adjacent to both the fornix and the lateral hypothalamic area. The DMH is subdivided into two distinct portions: a diffuse and a compact zone. It contains a high level of NPY terminals and α-MSH terminals originating from the ARC (58). The DMH also contains MCRs in which α-MSH acts generating an anorexigenic response (51). NPY is also expressed by neurons within the DMH. The function of NPY in this nucleus has yet to be fully determined but we know that NPY overexpression in the DMH induces obesity and overeating, whereas NPY downregulation in DMH ameliorates hyperphagia and obesity and delays high fat diet-induced obesity (58, 59) it also improves insulin sensitivity and glucose tolerance, prevented diet-induced hyperglycaemia and hyperinsulinemia (60).

The ***lateral hypothalamic area (LHA)*** is located along the midline of the anterior hypothalamus, closely below the anterior commissure and above the optic chiasm. It is a brain region directly implicated in body weight control and in the regulation of feeding control and is considered the most extensively interconnected area of the hypothalamus (61). Experiments in rats reveal that after injury to the lateral hypothalamus, the animals showed an acute decrease in body weight and food intake (62–64) with higher metabolic rates and increased activity (63). Also, it has been reported that the activation of some glutamatergic receptors in the lateral hypothalamus of rats stimulates feeding behaviour (65, 66). LHA contains two important neuronal populations: the MCH expressing neurons and the orexin expressing neurons that produce the orexigenic neuropeptides MCH and orexin A and B. The melanin-concentrating hormone (MCH) expressing neurons are an LHA neuron population that regulates energy and glucose homeostasis. In MCH neurons, insulin activates the PI3K pathway in a cell-autonomous manner independent of synaptic transmission, regulating neuronal excitability (67), this inhibits the release of the orexigenic peptide MCH, potentiating the anorexigenic effect of insulin, moreover cell-type-specific deletion of the insulin receptor (IR) abrogates this response (67). Insulin also targets the LHA via AgRP and α-MSH peptides, α-MSH acts on the orexigenic neuronal populations MCH and orexin to inhibit the release of the orexigenic peptides potentiating the anorexigenic response. The LHA also contains MCRs in which α-MSH acts generating an anorexigenic response (52) (Figure 6.2).

NON-NEURONAL INSULIN TARGETING PLACES IN THE HYPOTHALAMUS

Although insulin signalling in hypothalamic neurons is established, emerging evidence suggests that may be important functions governed by insulin activity in non-neuronal cells.

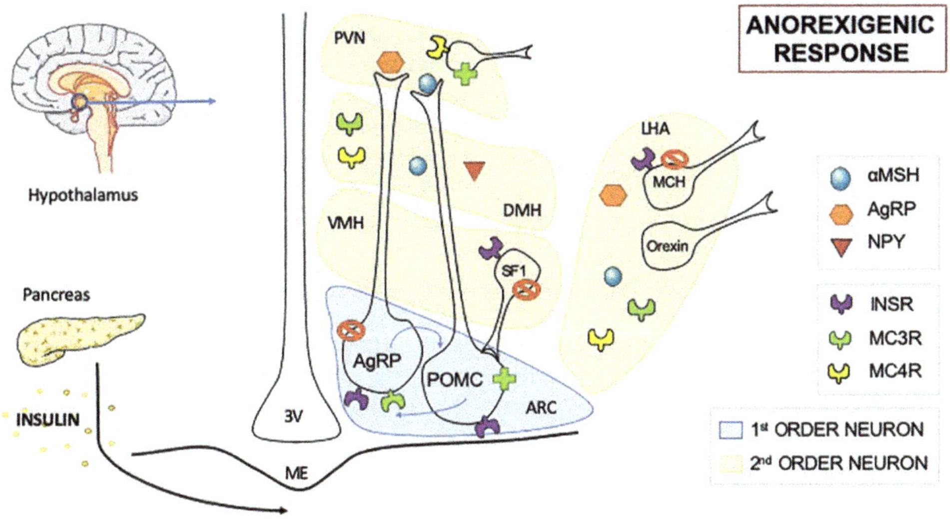

Figure 6.2 Insulin-responsive hypothalamic circuits in control of energy balance. After insulin is released from the pancreatic β cells, circulating insulin targets the brain and acts on the hypothalamus to adapt energy homeostasis and glucose metabolism. In the ARC, insulin signalling induces both AgRP and POMC hyperpolarization, leading to an increased Pomc transcription, and at the same time an inhibition of Agrp expression, generating an anorexigenic response. Moreover, these neurons project to the PVN (among other nuclei) where α-MSH (derived from the precursor protein POMC) bind to MC4Rs to induce the anorexigenic response on PVN neurons. Besides, under conditions of hyperinsulinemia, insulin hyperpolarizes SF1 neurons located in the VMH, which provide glutamatergic input to POMC neurons. In the DMH, NPY and α-MSH activities are not fully determined yet. Insulin also acts directly on MCH neurons in the LHA but the precise functional role of these MCH neurons in insulin signalling remains unclear. *Graphic abbreviations:* ME: Median eminence, 3V: Third ventricle, INSR: Insulin receptor, MC3R: Melanocortin-3 receptor, MC4R: Melanocortin-4 receptor, ARC: Hypothalamic arcuate nucleus, PVN: Paraventricular nucleus of the hypothalamus, LHA: Lateral hypothalamic area, VMH: Ventromedial hypothalamus, DMH: Dorsomedial hypothalamic nucleus, AgRP: Agouti-related protein, POMC: Pro-opiomelanocortin, NPY: Neuropeptide Y, αMSH: Alpha-melanocyte stimulating hormone, SF-1: Steroidogenic factor 1.

Astrocytes

One of these cell types that are beginning to become relevant in the field of neuroendocrinology are astrocytes. Their main function is the biochemical support of endothelial cells of the BBB, but they also provide nutrients to the nervous tissue, maintain the extracellular ion balance, regulate the cerebral blood flow, and have a role in the repair and scarring process after infection and traumatic injuries (68, 69). Within the hypothalamus, astrocytes play a key role with a specific contribution to the insulin signalling, regulating metabolic processes and behavioural response through coupling brain glucose uptake with nutrient availability to maintain energy balance and systemic glucose homeostasis (73).

Pericytes

Another cell type that may play a role in insulin signalling in hypothalamus are the pericytes, as they are a key component of the neurovascular unit, which includes endothelial cells, astrocytes, and neurons. The pericytes are multi-functional mural cells of the microcirculation that wrap around the endothelial cells that line the capillaries and venules throughout the body. In the brain, they help to maintain homeostatic and haemostatic functions, they are also implicated in the sustain and permeability of the BBB (68). Pericytes have also been found to express INSRs and reside within the hypothalamus. Intriguingly, hypothalamic neurons show increased sensitivity to insulin and enhanced Akt phosphorylation in response to pericyte-conditioned media, an effect not seen in response to other cell-conditioned media (74).

INSULIN RECEPTOR SIGNALLING IN THE HYPOTHALAMUS

The insulin receptor (INSR) is a transmembrane receptor that belongs to the receptor tyrosine kinase superfamily (75). The INSR gene undergoes differential splicing that generates two INSR isoforms: INSR-B and INSR-A (76, 77). As for the brain distribution of INSR, both isoforms are present in the brain and abundantly expressed in the hypothalamus (78). The physiological roles of INSR isoforms are incompletely understood and appear to be determined by their different binding affinities for insulin-like growth factors (IGFs), particularly for IGF-2. Insulin can also bind to insulin-like growth factor (IGF)-1 receptors (IGF-1R), which are also ubiquitously expressed within the brain, including the hypothalamus (81, 82). The wide expression of this receptor in the brain demonstrates a relevant role of insulin in the neuronal pathways, energy balance and feeding behaviour (83).

THE CANONICAL INSULIN SIGNALLING PATHWAY: PI3K-AKT PATHWAY

The phosphoinositol-3 kinase (PI3K)-AKT pathway is the main insulin signalling pathway. The link between insulin and INSR is characterized by the consequent autophosphorylation of tyrosine residues that allows the union and phosphorylation of the insulin receptor substrate (IRS) and Shc proteins (84). IRS-1 and IRS-2 are the most relevant substrates in the regulation and activation of the PI3K in the insulin signalling pathway (86). After its activation, the enzyme PI3K phosphorylates the substrate phosphatidylinositol 4,5-biphosphate (PIP2) to form phosphatidylinositol 3,4,5-triphosphate PIP3, responsible for binding to Akt/PKB kinase and allowing phosphoinositide-dependent kinase-1 (PDK1) and target of rapamycin complex 2 (mTORC2) to also phosphorylate and activate the same enzyme (87–90).

For PKB/AKT activation from PIP2 to occur, PI3K is recruited and acts directly on PIP2 to convert it to PIP3 (90). With the formation of PIP3, the enzyme PDK1 is recruited and thus promoting Ser473 phosphorylation of Akt (92, 93). With AKT activated, forkhead box-containing protein-O1 (FOXO1) can be phosphorylated and inactivated allowing its action in the AGRP and POMC neurons (94, 95). FOXO1 acts in ARC, increasing the expression of AGRP and decreasing the expression of POMC, which consequently causes food intake and body weight gain (95–97). Nevertheless, the effects can be opposed with the use of AKT-FoxO1 pathway inhibitors (94, 98).

In the brain, PI3K activation is directly related to neuronal development and plasticity, formation of synapses, neuroinflammation, neurotransmission, autophagy, memory, energy balance and glucose metabolism (108, 109). However, it is known that any abnormality in one stage of the PIK3 pathway, can cause different damages to the cellular function and in the homeostasis of the organism causing brain malformations, tumours, neurodegenerative diseases, cancer and epilepsy (110–112). In the hypothalamus, activation of the PI3K pathway by insulin plays an important role in energy homeostasis (113). Some studies report that the inhibition of insulin receptors in the arcuate nucleus promotes a change in the expression of some neuronal populations implicated in energy balance. It has been observed that local inhibition of insulin receptors in the brain of rats can increase the expression of the orexigenic neurons NPY and AgRP. Consequently, this alteration caused hyperphagia and an increase in adipose tissue, thus blocking the regulatory role of insulin receptors in the hypothalamus (114). In addition, central insulin inhibition seems to affect the control of plasma glucose levels (45). Another factor that also alters the role of insulin in the hypothalamus is the dysregulation of enzymes present in this signalling pathway. After binding insulin to its receptors, the IRS is recruited to continue the signal triggered by the hormone, phosphorylating sequential enzymes in this cascade to modulate the effects of insulin on the ARC (98). Nonetheless, this signal can be hindered by a malfunction of this enzyme. Some evidence indicates that deregulation of IRS-2 in its signalling pathway is responsible for causing an increase in food consumption, significant weight gain and an increase in insulin and leptin levels in plasma (115–117). Like the insulin receptor substrate (IRS), PI3K-Akt can also compromise energy homeostasis in case of changes in the PI3K-Akt pathway. A study on mice with reduced PI3K activity in the neurons of the ventromedial hypothalamic nucleus (VMH), reports a wide imbalance in the body's ability to control body weight, thermogenesis and energy expenditure (113). Animals with this inhibited pathway have more difficulties in their energy control, both under normal and specific conditions, such as food deprivation and diet-induced obesity. When compared to the control group, animals with PI3K inhibition gain more weight when subjected to an HFD, and lose less weight when in food deprivation (113). A possible cause of alteration in the PIK3-Akt pathway is chronic exposure to HFD. A study on chronic treatment of rats with an HFD shows that deregulation of this pathway can be detected in the brain, specifically in the hypothalamus and hippocampus. It was observed that rats with access to an HFD for 4 months, present deregulation of PI3K in the hippocampus and deregulation of Akt in the hypothalamus (118). In addition, another study reported that mice with access to a high-fat diet have hyper-activation of the PI3K pathway in VHM neurons by insulin, causing glucose intolerance and the onset of obesity (54).

The PI3K K-ATP Downstream Pathway

ATP-sensitive potassium (K-ATP) channels are transmembrane proteins, composed of inwardly rectifying K+ channel (Kir6.x) and sulfonylurea receptor (SUR) sub-units with distribution in various tissues, such as skeletal and cardiac muscle, vascular smooth muscles, pancreatic β-cells and brain tissue, which can act by depolarizing the plasma membrane, controlling the metabolism, protecting against hypoxia, regulating apoptosis as well as several other functions (119–124).

In the brain, these K-ATP channels play an important role in neuronal excitability in different areas. Moreover, it has a punctual role in homeostatic glucose control at a hypothalamic level (125, 126). Activation of the K-ATP channels in the MB hypothalamus has been shown to promote a decrease in plasma glucose levels by inhibiting gluconeogenesis in the liver. Thus, the K-ATP channels demonstrate an important role in glucose homeostasis, since their central activation results in inhibiting gluconeogenesis and its peripheral function in pancreatic cells decreases insulin secretion (126). Besides, it is important to note that the increase in blood glucose levels leads to a change in the expression of the K-ATP channels (127). In the VMH, the K-ATP channels play a role in glucose regulation, in situations of hypoglycaemia, the K-ATP channels are activated and respond to this stimulus through the secretion of glucagon, which consequently will promote the liver glucose production (34, 128). In the ARC, the activation of the insulin pathway in the POMC neurons, increases the activation of the K-ATP channels, through the action of PIP3K (129). There is a study in which this consequent increase in the activity of the K-ATP channels through PIP3K action promoted hyperphagia and a body weight increase in mice (130). Therefore, we can conclude that possible dysregulation in the K-ATP channels in the hypothalamus may be a key factor to understand the pathophysiology of diseases such as obesity and DM2.

mTOR/S6K Downstream Pathway

The mammalian target of rapamycin (mTOR) is a serine-threonine-kinase that, when activated, generates that is able to activate an important cascade pathway found in different cell types and involved in different roles, from cell growth to energy homeostasis (131). The mTOR is characterized by having two complexes, mTORC1 and mTORC2, which have different constitutions and act in different ways and in different locations (132). The activity of the mTORC1 complex in the hypothalamus depends on insulin binding with its receptor in the plasma membrane. Upon activation of IRS1, PI3K is stimulated and consequently phosphorylates PDK1 and AKT. Then, AKT is responsible for phosphorylating the TSC1-TSC2 (Tuberous sclerosis proteins 1 and 2) complex, leading to its inactivation and thus enabling the Rheb-GTP (Ras homolog enriched in brain) activity on mTORC1 (133, 134). With mTORC1 activated, the S6K1 (Ribosomal protein S6 kinase β-1) protein becomes a target for phosphorylation to generate a cellular response.

The activation of this pathway in the hypothalamus is predominantly activated by nutrients, neurotrophic factors, and neurotransmitters such as insulin (135), promoting the regulation of energy expenditure, food intake, satiety, hunger, body weight, glucose homeostasis and lipid metabolism (97, 135–137). Several studies demonstrate that the activation of mTORC1 in the hypothalamus is essential for the regulation of energy homeostasis (136). The mTORC2 complex appears to have some connection with the energy control regulated by mTORC1 (135, 137–139). The activation by insulin of the mTORC1 pathway is well consolidated. Activation of the mTORC1 complex in the hypothalamus for energy control initially depends on the binding of mediators such as amino acids, growth factors and transmitters (135, 137). Studies show that some mediators such as ghrelin (140), leptin (141), are activators of the mTORC1 pathway; however, insulin and the growth hormone play a key role in this activation (136, 137). It is possible to see an increase in the expression of the mTORC1 pathway in the arcuate nucleus when exogenous insulin is administered into the mouse brain (142). In addition, the anorectic effects and weight reduction caused by insulin can be reversed by the inhibition of mTORC through rapamycin. It is also important to highlight that, the excessive activation of this pathway on POMC neurons can lead to the development of hyperphagia and obesity (143). Moreover, it was observed in the hypothalamus of rats treated with a hypercaloric diet for a short period of time, an increase in the activation of S6K, a fact that contributes to hepatic insulin resistance, in response to excess nutrients in a short period of time (144).

AMPK Downstream Pathway

The AMP-activated protein kinase (AMPK) is an intracellular serine/threonine protein kinase complex activated when there is energy depletion. Through catabolic processes, cells will produce more ATP, thus maintaining the balance between their production and consumption. A balance is necessary for a normal cellular functioning (145). The activation of this pathway is responsible for coordinating cell growth, autophagy and metabolism (145–147). It is known that the AMPK complex is expressed in many tissues, but it has a high expression in tissues such as liver, skeletal muscle, and brain. Regarding the brain, AMPK has been described to play an important role in the hypothalamus. Studies have shown that AMPK acts on energy homeostasis, food intake and glucose sensitivity through hypothalamic neurons POMC, AgRP and NPY (148–150). During fasting periods, the activation of the AMPK pathway in NPY neurons can be observed, in order to promote food intake (149). In addition, a study carried out with knockout mice for AMPK in POMC and AgRP neurons demonstrates a direct effect on body weight, fat percentage and cumulative intake of these animals. It was observed that the deletion of AMPK in anorectic POMC neurons generates an increase in body mass as well as body fat. The AMPK pathway in the MB hypothalamus plays a key role in glucose homeostasis. In the hypothalamus, AMPK activity is inhibited by the performance insulin (147). It has been shown that AMPK inhibition consequently promotes a decrease in glucose production in the liver (151). Also, it has been observed that its inhibition in the ventromedial nucleus, promotes an increase in thermogenesis and energy dissipation in brown adipose tissue (152). Inhibition of the AMPK pathway in the brain results in a sympathetic flow, consequently norepinephrine is released from neurons and binds to β-receptors in brown adipose tissue (153).

THE MAPK/ERK PATHWAY

Mitogen-activated protein kinase (MAPK) cascades is a complex pathway, with a fundamental role in the transduction of extracellular signals for several cellular responses, such as proliferation, differentiation, and apoptosis (154, 155).

In the hypothalamus, some studies show that the MAPK/ERK pathway can be activated by insulin (160, 161) regulating NPY and AgRP neurons, promoting a decrease in mRNA expression of these neurons (160). Another interesting study demonstrated that central insulin administration in rats promotes an activation of the MAPK/ERK pathway, which consequently induces a regional sympathetic activation that increases the sympathetic flow of thermogenic BAT (162). In another study, systemic insulin administration in mice promoted a phosphorylation of the extracellular signal-regulated kinase 5 (ERK5) in the hypothalamus (157). Furthermore, with the inactivation of ERK5, a worsening of glucose homeostasis, an increase in white adipose tissue and a decrease in physical activity, energy intake and energy expenditure is observed (157) (Figure 6.3).

HYPOTHALAMIC INSULIN SIGNALLING IMPACT PERIPHERAL METABOLISM

Neuronal Insulin Signalling Regulates Peripheral Metabolism

As mentioned before, insulin secreted from pancreatic islets crosses the BBB to bind to INSR in the ARC where POMC and AgRP neurons integrate peripheral signals and adapt their electrical activity according to energy availability. Until now we have seen that hypothalamic insulin signalling regulates food intake by acting as an anorexigenic signal (33, 35, 36). However, its function in the organism is more widely as it controls nutrient partitioning in peripheral organs such as the adipose tissue or the liver via the autonomic nervous system. The manipulation of brain insulin signalling through a variety of methods has been shown to regulate and/or disrupt glucose, lipid, and amino acid metabolism in several

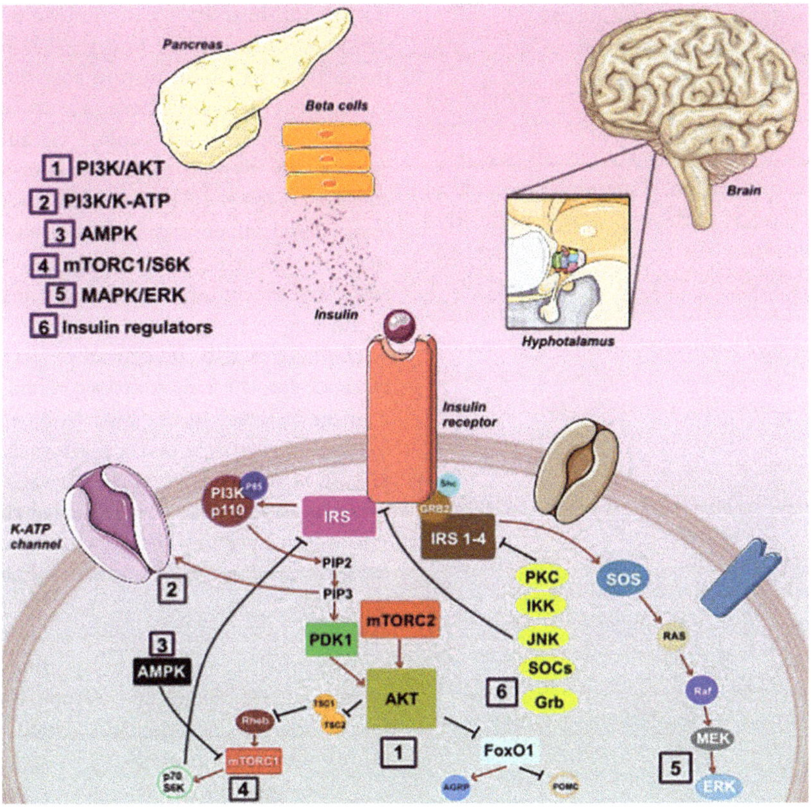

Figure 6.3 Roles of insulin-activated signalling pathways in the hypothalamus. Effects of insulin binding to its receptor in the hypothalamus promote the activation of different pathways acting over energy homeostasis. (1) IRS/PI3K/AKT; the canonical pathway of insulin signalling and the main pathway for insulin action in the hypothalamus. The inhibition of the proteins IRS, IRS2 and PI3K cause alterations in energy control, body weight, food consumption, lipid profile and glucose homeostasis. (2) PI3K/K-ATP downstream pathway. Its activation is related to glucose homeostasis. In the ARC hypothalamic activation generates a peripheral response in organs such as liver, inhibiting gluconeogenesis; in the POMC neurons, increases the activation of the K-ATP channels, consequently promoting weight gain and hyperphagia. In the VMH, this pathway emits a stimulus that acts on the pancreas, enabling the secretion of glucagon. (3) AMPK downstream pathway. Insulin inhibits the AMPK pathway, the inhibition in POMC neurons is related to weight gain and increased white adipose tissue depots; its inhibition in the AgRP neurons decreases food consumption and increases sensitivity to melanocortin agonists, the hypothalamic inhibition of AMPK promotes peripheral effects such as increased thermogenesis and reduction of glucose production. (4) mTORC/S6K downstream pathway. Central insulin administration in the ARC promotes increased expression of mTORC. Inhibition and hyper activation of mTORC consecutively promote the reversal of the anorectic effects of insulin and increased weight and food intake. (5) MAPK/ERK; a non-canonical pathway. It acts by regulating energy balance in key neurons present in the hypothalamus, with an effect on the sympathetic flow culminating the increase of thermogenesis. In addition, it is observed that the inhibition of ERK5 decreases glucose homeostasis and promotes the increase of white adipose tissue depots. (6) Modulators of the insulin signalling pathway in the hypothalamus. *Graphic abbreviations*: AKT: Protein kinase B; IRS-IRS2: Insulin receptor substrate-2; PI3K: Phosphatidylinositol 3-kinases; PIP2: Phosphatidylinositol-4, 5-bisphosphate; PIP3: Phosphatidylinositol 3,4,5-triphosphate; K-ATP: ATP-sensitive potassium channels; AMPK: AMP-activated protein kinase; mTORC: Mammalian target of rapamycin complex; S6K: Ribosomal S6 kinase; POMC: Proopiomelanocortin; AgRP: Agouti-related protein; NPY: Neuropeptide Y; MEK: Mitogen-activated protein kinase; ERK-ERK5: Extracellular signal-regulated kinases-5.

peripheral tissues, in particular by controlling metabolic pathways in the liver and adipose tissue. A lot of research in animals using different approaches has indicated that brain insulin action influences peripheral metabolism. Moreover, in humans, although it is challenging to study, intranasal insulin administration assay, which is consider a brain-specific delivery mode for insulin, has also provided some insights into how brain insulin impacts peripheral metabolism (163–165).

Brain Insulin Action in Hepatic Glucose Production and Triglyceride Export

The liver plays a major role in maintaining glucose homeostasis, as it is the main organ for glucose storage, in the form of glycogen, as well as endogenous glucose production. The term *hepatic glucose production* (HGP) refers to the appearance of de novo glucose generated by the liver through gluconeogenesis and glycogenolysis in the blood (166). Central insulin actions in the brain act over gluconeogenesis via AgRP neuronal sensitive potassium channels, which attenuates HGP via vagal efferent neurons innervating the liver (45, 167). This results in a stimulation of signal transducer and activator of transcription 3 (STAT3) activation by tyrosine phosphorylation in the liver parenchymal cells (168). Moreover, via α7-nicotinic acetylcholine receptors, promotes hepatic interleukin-6 (IL-6) secretion from hepatic Kupffer cells (169), which reduces HGP via gluconeogenesis by reducing the expression of Phosphoenolpyruvate carboxykinase 1 (PCK1) and Glucose 6-phosphatase (G6Pase). However, this regulation seems to be complex, and the detailed mechanisms involved needs to be fully determined.

Evidence from animal studies: The role of brain insulin signalling in the regulation of hepatic glucose metabolism was first suggested in studies that reported the intracerebroventricular infusion of insulin suppressed HGP in rats (45). Central injection of insulin reduces glucose production in the liver, whereas, blocking insulin signalling, diminishes the ability of insulin to suppress peripheral glucose output. This regulation may involve electric transmission via the autonomic nervous system since transection of the vagal nerve abolishes this effect. Besides, the work of Könner et al. demonstrated that insulin action in the CNS controls HGP specifically via AgRP neurons (36). Moreover, a branched-pathway model suggests that insulin receptor knock-in in POMC neurons promotes HGP and activates the energy expenditure through melanocortin pathway whereas insulin receptor knock-in in AgRP neurons decreases HGP (170).

Evidence from human studies: intranasal insulin administration promotes hypothalamic activity and improves insulin sensitivity, implicating the hypothalamus as a key site coordinating central insulin's actions on glycaemic control (171). However, while some studies demonstrate that intranasal insulin administration can inhibit HGP in humans, the slow onset of the effect (>3 hours) supports the notion that brain insulin action is insufficient to suppress HGP anywhere near to what systemic insulin is able to accomplish within minutes as assessed by hyperinsulinemic clamps in several species (172–174). These studies suggest that the roles of brain insulin and/or its downstream mediators are of only moderate potency in regulating HGP but, admittedly, this might still be important in the long-term regulation of glucose homeostasis.

The liver is also the central organ for fatty acid metabolism, the *triglyceride export* refers to the mobilization of lipids in the liver to be eliminated by oxidation within the cell or for its secretion into the plasma within triglyceride-rich very low-density lipoproteins (VLDL) to the rest of the organism. The secretion of triglycerides in the form of VLDL protects the liver from excessive lipid accumulation, steatosis and non-alcoholic fatty liver disease (NAFLD) (175). The intracerebroventricular infusion of insulin promotes triglyceride export from the liver (176). This anti-steatotic effect of brain insulin occurs independent of peripheral (and more specifically direct hepatic) insulin signalling and does not induce de novo lipogenesis in the liver, a key contributor to hepatic steatosis (177).

Brain Insulin and White Adipose Tissue Metabolism

Insulin is considered the most important regulator in WAT metabolism as an anti-lipolytic and pro lipogenic regulator (178). Hypothalamic insulin signalling acts over lipid metabolism via POMC neurons/melanocortin pathway (179) reducing the sympathetic outflow to the adipose tissue. This generates a reduction in hormone-sensitive lipase (HSL) activation and perilipin phosphorylation (180) inhibiting WAT lipolysis, stimulating de novo lipogenesis, and thus enabling WAT to store lipids (180, 181). As WAT lipolysis provides the gluconeogenic precursor glycerol and energy substrates in the form of fatty acids, which drive hepatic gluconeogenesis and therefore HGP; impaired brain insulin action might contribute to overfeeding-induced hepatic insulin resistance by unrestraining adipose tissue lipolysis (186, 187).

Evidence from animal studies: intracerebroventricular insulin infusion restrain lipolysis reducing sympathetic outflow in WAT. The suppresses lipolysis precedes the suppression of HGP. Also, the same data from genetically modified mice with conditional deletion of the insulin receptor in neurons produces an increase of lipolytic rates in fasting and suppression of free fatty acids (FFA) release after refeeding. This suppression of circulating FFAs tallies animal data that suggest that insulin action in the brain reduces lipolysis in adipose tissue (180).

Evidence from human studies: due to the intranasal insulin administration results on this topic are conflicting (182). The findings of one study with intranasal insulin supported the hypothesis that insulin in the brain suppresses systemic lipolysis (183). Other studies could not replicate this finding presumably because of different study designs (171, 184, 185). Hence, further research is needed to clarify the potential effects of brain insulin on systemic lipolysis in humans.

Brain Insulin and Brown Adipose Tissue Thermogenesis

Brown adipose tissue (BAT) is a key organ in energy homeostasis due to its role in thermogenesis (188). There is evidence for the role of brain insulin action in thermogenesis and energy expenditure. In addition to classic BAT, brown-like (or beige) adipocytes embedded in WAT depots also contain multilocular lipid droplets and express the uncoupling protein one (UCP1) and therefore have thermogenic properties (189). Beige adipocytes can switch from energy storage to expenditure upon environmental stimuli such as cold, a process referred to as WAT browning. Although the relative contribution of brown versus beige adipocytes to whole energy expenditure is not well defined, both cell types can significantly improve energy metabolism and insulin sensitivity in mammals (190).

Evidence from animal studies: intracerebroventricular insulin infusion in fed but not in fasted rats promotes BAT activation (191). Conversely, NIRKO mice, which lack the insulin receptor in all neurons, are intolerant to the cold and show a rapid drop in body temperature in response to a 4°C cold exposure challenge, which is consistent with a defect in thermogenesis (163). Moreover, intracerebroventricular co-infusion of insulin and leptin can affect WAT browning and energy expenditure, resulting in a lean phenotype (192). These effects were partially inhibited by sympathetic denervation of inguinal adipose depots or pharmacological inhibition of the PI3K pathway in the arcuate nucleus, which suggests that insulin and leptin act via a central mechanism.

Evidence from human studies: Human data on brain insulin action and thermogenesis are scarce. It was provided by Benedict et al. (188), that intranasal insulin in humans, administered before ingestion of a high caloric meal, moderately increases postprandial energy expenditure, and thus increase their thermogenesis, using indirect calorimetry (188). Further studies are required to define the role of brain insulin action in BAT function and WAT browning in humans.

HYPOTHALAMIC INSULIN RESISTANCE IMPACT PERIPHERAL METABOLISM

Insulin resistance is a condition characterized by a lack of cellular response, when the insulin receptor does not elicit a response, preventing the entry of glucose into the cell. This condition affects various insulin-sensitive tissues, such as adipose, muscle, liver, and also centrally, specifically in the hypothalamus (193). The emergence of insulin resistance is directly related to overnutrition, inflammatory processes and dysregulation of insulin signalling pathways (194). It is already known that a sedentary lifestyle and chronic excessive caloric intake can lead to insulin resistance (195, 196), and consequently, lead to the development of obesity, DM2, metabolic syndrome, visceral adiposity, cardio vascular disease, hypertension and fatty liver diseases (84, 197–199).

Proposed Mechanism by Which Brain Insulin Resistance Occurs

Brain insulin resistance seems to be due to an inflammatory process in the hypothalamus, where insulin receptors are expressed at high density. This impairs local signalling systems by inducing mitochondrial dysfunction and endoplasmic reticulum stress in the hypothalamus, and cause glucose and energy metabolism disorders. Through studies with obesity models induced by a high-fat diet, it was observed that animals submitted to this type of diet, presented an inflammatory process and decreased sensitivity to insulin in the hypothalamus (200–202). Insulin resistance in most cases is believed to be manifest at the cellular level via post-receptor defects in insulin signalling. Possible mechanisms include down-regulation, deficiencies, or genetic polymorphisms of tyrosine phosphorylation of the insulin receptor, IRS proteins or PIP-3 kinase, or may involve abnormalities of glucose transporter type 4 function (203). Despite promising findings in experimental animals with respect to a range of insulin signalling defects, their relevance to human insulin resistance is presently unclear.

Brain Insulin Resistance and Its Metabolic Sequelae

Overnutrition generates metabolic stress in the organism which induces an impaired physiological action of insulin in the brain, also called brain insulin resistance. This brain insulin resistance can be due to, a reduced transport of insulin into the brain, or to an impaired molecular signalling originating from the insulin receptor or the IGF-1 receptor. This hypothalamic insulin resistance caused by excessive nutrition occurs more rapidly than that in other insulin-sensitive tissues. The physiological consequence of impaired brain insulin resistance is an impaired ability to regulate central and peripheral metabolism.

In rodents fed with HFD, ICV insulin does not suppress food intake (54) or have an effect on hepatic glucose production (144). In humans, intranasal insulin administration in obese individuals has no effect on food intake or endogenous glucose production (182).

Obesity leads to hypothalamic insulin resistance and may cause dyslipidaemia and hyperglycaemia in the organism. Ultimately this generates unrestrained lipolysis in white adipose tissue, which drives adipose tissue dysfunction, ectopic lipid deposition and lipotoxicity. It also affects the liver, in which the increased HGP, the activation of de novo lipogenesis, and the reduction in triglyceride export generate NAFLD. This also generates a malfunction of pancreatic beta-cell function and survival due to glucolipotoxicity driven by the deleterious effects of elevated glucose and fatty acid levels (45, 176, 180, 181, 183).

Therefore, targeting the brain to prevent, improve or even reverse this insulin resistance is a promising pharmacological option with a low risk of hypoglycaemia that could reverse the lipotoxicity and metabolic disease. However, the best approach is still weight loss, as appropriate studies for certain pharmacologic approaches in humans are still required to address this issue (Figure 6.4).

Hypothalamic Insulin Resistance, a Maladaptive Response

The question of why brain insulin resistance develops is still unanswered; however, Scherer et al. (204) have proposed that brain insulin resistance may have been an adaptative response during mammalian evolution. From an evolutionary point of view, the prevention of hypoglycaemia to enable survival in the settings of metabolic stress was an advantage as the brain critically depends on glucose as an energy substrate. They postulate that brain insulin resistance can be understood as a physiological adaptation (205, 206) to periodic starvation (207), cold exposure (208) and infection (209). These conditions lead to increased lipolysis and HGP, but in this scenario, this substrate mobilization was compensated by increased substrate consumption and hence does not lead to metabolic disease.

Nowadays, we know that overnutrition rapidly promotes brain insulin resistance, and this brain insulin resistance appears before the insulin signalling in peripheral organs is impaired. Hyperalimentation along with a sedentary lifestyle, social stress, disruption of the circadian rhythm and ageing also causes metabolic stress in the organism. This activates a maladaptive hypothalamic response, leading to a dysregulation of the autonomous nervous system, generating a situation of dyslipidaemia and hyperglycaemia in the organism. In this condition of overnutrition, this fuel excess is not going to be consumed, but wrongly stored causing an ectopic lipid deposition among with glucolipotoxicity, conditions that fuel systemic insulin resistance and which are hallmarks of the metabolic syndrome and DM2.

CONCLUDING REMARKS

Insulin secreted from pancreatic islets reaches the medio basal hypothalamus through three proposed mechanisms: receptor-mediated transcytosis, passive extravasation from blood to the media

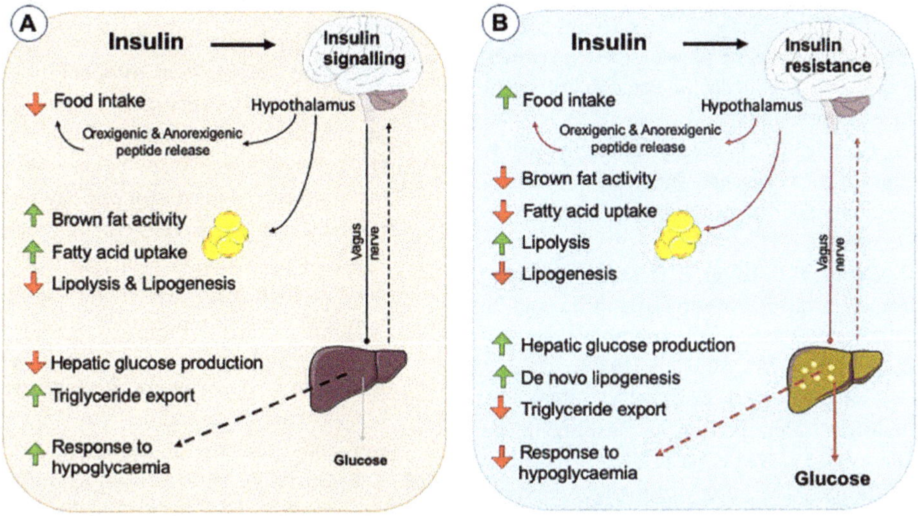

Figure 6.4 Central insulin impacts peripheral metabolism. Insulin secreted from the pancreas crosses the blood-brain barrier to bind to insulin receptors expressed throughout the brain. (a) Physiological functions of brain insulin action: Insulin might act as an anorexic signal but, more importantly, it controls metabolism in peripheral organs such as the adipose tissue or the liver via the autonomic nervous system. Brain insulin action thereby regulates lipolysis and lipogenesis, fatty acid uptake, hepatic triglyceride secretion, thermogenesis, and hepatic glucose production. (b) Brain insulin resistance and its metabolic sequelae: Overnutrition induces brain insulin resistance that plays an important role in overnutrition-induced hepatic steatosis and adipose tissue dysfunction, which in turn disrupt glucose homeostasis. In the end, this leads to dyslipidaemia, hyperglycaemia and glucolipotoxicity, which end up in pathologic processes such as DM2 and other comorbidities of obesity. *Graphic abbreviations*: NTS: nucleus of the solitary tract. DM2: type 2 diabetes mellitus.

eminence (ME), or tanycyte-mediated transport, from the cerebrospinal fluid (CSF).

Once in the brain, insulin binds to the INSR and it is targeted in the ARC. POMC and AgRP neurons integrate this insulin signalling and adapt their electrical activity and neuropeptide transcription in response. Insulin signals via POMC neurons as "first-neurons", generating signal to the "second-order" neurons in other hypothalamic regions. However, the hypothalamus is very complex, not only composed of distinct neuronal populations expressing a specific combination of neuropeptides, neurotransmitters, and receptors but also of non-neuronal cell types such as astrocytes and pericytes. They all communicate with each other overlapping their physiological roles directly and indirectly. Thus, many of the mechanistic pathways remain understudied and is still being uncovered and the functionality and importance of insulin signalling within are only beginning to emerge. The physiological result of this hypothalamic insulin signalling is an anorexigenic tone, along with complex inter-organ crosstalk, orchestrating nutrient partitioning by regulating not only appetite but also lipolysis, and lipogenesis, triglyceride secretion and uptake, thermogenesis, and HGP in peripheral tissues.

Last but not least, the dysfunction of hypothalamic insulin signalling plays a substantial role in the development of metabolic diseases. For this reason, in order to provide better therapeutics for the treatment of obesity and DM2, we must fully understand the effects of insulin not only in the periphery but also in the brain, to provide some insight into how insulin may mediate the development of obesity-related insulin resistance.

ACKNOWLEDGEMENTS

This work has been supported by grants from Ministerio de Ciencia Innovación-Agencia Estatal de Investigación (ST: PID2020-116741RB-100) Xunta de Galicia (RN:HR18-00100), Centro de Investigación Biomédica en Red (CIBER) de Fisiopatología de la Obesidad y Nutrición (CIBERobn). CIBERobn is an initiative of the Instituto de Salud Carlos III (ISCIII) of Spain which is supported by FEDER funds.

REFERENCES

1. C. Bernard, Leçons de physiologie expérimentale appliquée à la médecine faites au Collège de France (France, 1855).
2. F. G. Banting, C. H. Best, J. B. Collip, W. R. Campbell, A. A. Fletcher, Pancreatic extracts in the treatment of diabetes mellitus. *Can. Med. Assoc. J.* **12**, 141–146 (1922).
3. S. C. Woods, E. C. Lotter, L. D. McKay, D. Porte, Chronic intracerebroventricular infusion of insulin reduces food intake and body weight of baboons. *Nature*. **282**, 503–505 (1979).
4. J. Havrankova, D. Schmechel, J. Roth, M. Brownstein, Identification of insulin in rat brain. *PNAS*. **75**, 5737–5741 (1978).
5. C. M. Brooks, The history of thought concerning the hypothalamus and its functions. *Brain Res. Bull.* **20**, 657–667 (1988).
6. P. Seoane-Collazo, J. Fernø, F. Gonzalez, C. Diéguez, R. Leis, R. Nogueiras, M. López, Hypothalamic-autonomic control of energy homeostasis. *Endocrine*. **50**, 276–291 (2015).
7. B. F. Belgardt, J. C. Brüning, CNS leptin and insulin action in the control of energy homeostasis. *Ann. N. Y. Acad. Sci.* **1212**, 97–113 (2010).
8. J. L. Marks, D. J. Porte, W. L. Stahl, D. G. Baskin, Localization of insulin receptor mRNA in rat brain by in situ hybridization. *Endocrinology*. **127**, 3234–3236 (1990).
9. S. C. Benoit, E. L. Air, L. M. Coolen, R. Strauss, A. Jackman, D. J. Clegg, R. J. Seeley, S. C. Woods, The catabolic action of insulin in the brain is mediated by melanocortins. *J. Neurosci.* **22**, 9048–9052 (2002).
10. M. W. Schwartz, A. J. Sipols, J. L. Marks, G. Sanacora, J. D. White, A. Scheurink, S. E. Kahn, D. G. Baskin, S. C. Woods, D. P. Figlewicz, Inhibition of hypothalamic neuropeptide Y gene expression by insulin. *Endocrinology*. **130**, 3608–3616 (1992).
11. A. J. Sipols, D. G. Baskin, M. W. Schwartz, Effect of intracerebroventricular insulin infusion on diabetic hyperphagia and hypothalamic neuropeptide gene expression. *Diabetes*. **44**, 147–151 (1995).
12. L. Plum, B. F. Belgardt, J. C. Brüning, Central insulin action in energy and glucose homeostasis. *J. Clin. Invest.* **116**, 1761–1766 (2006).
13. G. J. Morton, D. E. Cummings, D. G. Baskin, G. S. Barsh, M. W. Schwartz, Central nervous system control of food intake and body weight. *Nature*. **443**, 289–295 (2006).
14. M. W. Schwartz, S. C. Woods, D. J. Porte, R. J. Seeley, D. G. Baskin, Central nervous system control of food intake. *Nature*. **404**, 661–671 (2000).
15. J. C. Brüning, D. Gautam, D. J. Burks, J. Gillette, M. Schubert, P. C. Orban, R. Klein, W. Krone, D. Müller-Wieland, C. R. Kahn, Role of brain insulin receptor in control of body weight and reproduction. *Science*. **289**, 2122–2125 (2000).
16. P. Ballabh, A. Braun, M. Nedergaard, The blood-brain barrier: an overview: structure, regulation, and clinical implications. *Neurobiol. Dis.* **16**, 1–13 (2004).
17. E. M. Rhea, W. A. Banks, Role of the blood-brain barrier in central nervous system insulin resistance. *Front. Neurosci.* **13**, 521 (2019).
18. K. R. Duffy, W. M. Pardridge, Blood-brain barrier transcytosis of insulin in developing rabbits. *Brain Res.* **420**, 32–38 (1987).
19. G. L. King, S. M. Johnson, Receptor-mediated transport of insulin across endothelial cells. *Science*. **227**, 1583–1586 (1985).
20. W. A. Banks, J. B. Jaspan, W. Huang, A. J. Kastin, Transport of insulin across the blood-brain barrier: saturability at euglycemic doses of insulin. *Peptides*. **18**, 1423–1429 (1997).
21. M. van Houten, B. I. Posner, B. M. Kopriwa, J. R. Brawer, Insulin binding sites localized to nerve terminals in rat median eminence and arcuate nucleus. *Science*. **207**, 1081–1083 (1980).
22. A. Mullier, S. G. Bouret, V. Prevot, B. Dehouck, Differential distribution of tight junction proteins suggests a role for tanycytes in blood-hypothalamus barrier regulation in the adult mouse brain. *J. Comp. Neurol.* **518**, 943–962 (2010).
23. V. Prevot, F. Langlet, B. Dehouck, Flipping the tanycyte switch: how circulating signals gain direct access to the metabolic brain. *Aging (Albany. NY)*. **5** (2013), 332–334.
24. E. M. Rhea, C. Rask-Madsen, W. A. Banks, Insulin transport across the blood-brain barrier can occur independently of the insulin receptor. *J. Physiol.* **596**, 4753–4765 (2018).
25. M. J. McKinley, D. A. Denton, P. J. Ryan, S. T. Yao, A. Stefanidis, B. J. Oldfield, From sensory circumventricular organs to cerebral cortex: Neural pathways controlling thirst and hunger. *J. Neuroendocrinol.* **31**, e12689 (2019).
26. S. Luquet, F. A. Perez, T. S. Hnasko, R. D. Palmiter, NPY/AgRP neurons are essential for feeding in adult mice but can be ablated in neonates. *Science*. **310**, 683–685 (2005).

27. E. Gropp, M. Shanabrough, E. Borok, A. W. Xu, R. Janoschek, T. Buch, L. Plum, N. Balthasar, B. Hampel, A. Waisman, G. S. Barsh, T. L. Horvath, J. C. Brüning, Agouti-related peptide-expressing neurons are mandatory for feeding. *Nat. Neurosci.* **8**, 1289–1291 (2005).
28. M. M. Ollmann, B. D. Wilson, Y. K. Yang, J. A. Kerns, Y. Chen, I. Gantz, G. S. Barsh, Antagonism of central melanocortin receptors in vitro and in vivo by agouti-related protein. *Science.* **278**, 135–138 (1997).
29. G. W. Millington, The role of proopiomelanocortin (POMC) neurones in feeding behaviour. *Nutr. Metab. (Lond).* **4**, 18 (2007).
30. A. J. Mercer, S. T. Hentges, C. K. Meshul, M. J. Low, Unraveling the central proopiomelanocortin neural circuits. *Front. Neurosci.* **7**, 19 (2013).
31. C. Zhan, J. Zhou, Q. Feng, J.-E. Zhang, S. Lin, J. Bao, P. Wu, M. Luo, Acute and long-term suppression of feeding behavior by POMC neurons in the brainstem and hypothalamus, respectively. *J. Neurosci.* **33**, 3624–3632 (2013).
32. M. Koch, L. Varela, J. G. Kim, J. D. Kim, F. Hernández-Nuño, S. E. Simonds, C. M. Castorena, C. R. Vianna, J. K. Elmquist, Y. M. Morozov, P. Rakic, I. Bechmann, M. A. Cowley, K. Szigeti-Buck, M. O. Dietrich, X.-B. Gao, S. Diano, T. L. Horvath, Hypothalamic POMC neurons promote cannabinoid-induced feeding. *Nature.* **519**, 45–50 (2015).
33. L. Varela, T. L. Horvath, Leptin and insulin pathways in POMC and AgRP neurons that modulate energy balance and glucose homeostasis. *EMBO Rep.* **13**, 1079–1086 (2012).
34. M. L. Evans, R. J. McCrimmon, D. E. Flanagan, T. Keshavarz, X. Fan, E. C. McNay, R. J. Jacob, R. S. Sherwin, Hypothalamic ATP-sensitive K+ channels play a key role in sensing hypoglycemia and triggering counterregulatory epinephrine and glucagon responses. *Diabetes.* **53**, 2542–2551 (2004).
35. K. Loh, H. Herzog, Y.-C. Shi, Regulation of energy homeostasis by the NPY system. *Trends Endocrinol. Metab.* **26**, 125–135 (2015).
36. A. C. Könner, R. Janoschek, L. Plum, S. D. Jordan, E. Rother, X. Ma, C. Xu, P. Enriori, B. Hampel, G. S. Barsh, C. R. Kahn, M. A. Cowley, F. M. Ashcroft, J. C. Brüning, Insulin action in AgRP-expressing neurons is required for suppression of hepatic glucose production. *Cell Metab.* **5**, 438–449 (2007).
37. K. W. Williams, L. O. Margatho, C. E. Lee, M. Choi, S. Lee, M. M. Scott, C. F. Elias, J. K. Elmquist, Segregation of acute leptin and insulin effects in distinct populations of arcuate proopiomelanocortin neurons. *J. Neurosci.* **30**, 2472–2479 (2010).
38. J. Qiu, C. Zhang, A. Borgquist, C. C. Nestor, A. W. Smith, M. A. Bosch, S. Ku, E. J. Wagner, O. K. Rønnekleiv, M. J. Kelly, Insulin excites anorexigenic proopiomelanocortin neurons via activation of canonical transient receptor potential channels. *Cell Metab.* **19**, 682–693 (2014).
39. G. T. Dodd, N. J. Michael, R. S. Lee-Young, S. P. Mangiafico, J. T. Pryor, A. C. Munder, S. E. Simonds, J. C. Brüning, Z.-Y. Y. Zhang, M. A. Cowley, S. Andrikopoulos, T. L. Horvath, D. Spanswick, T. Tiganis, Insulin regulates POMC neuronal plasticity to control glucose metabolism. *Elife.* **7**, 1–30 (2018).
40. H. Krude, H. Biebermann, W. Luck, R. Horn, G. Brabant, A. Grüters, Severe early-onset obesity, adrenal insufficiency and red hair pigmentation caused by POMC mutations in humans. *Nat. Genet.* **19**, 155–157 (1998).
41. H. Krude, A. Grüters, Implications of proopiomelanocortin (POMC) mutations in humans: the POMC deficiency syndrome. *Trends Endocrinol. Metab.* **11**, 15–22 (2000).
42. E. L. Air, S. C. Benoit, K. A. Blake Smith, D. J. Clegg, S. C. Woods, Acute third ventricular administration of insulin decreases food intake in two paradigms. *Pharmacol. Biochem. Behav.* **72**, 423–429 (2002).
43. D. J. Clegg, C. A. Riedy, K. A. B. Smith, S. C. Benoit, S. C. Woods, Differential sensitivity to central leptin and insulin in male and female rats. *Diabetes.* **52**, 682–687 (2003).
44. C. Benedict, W. Kern, B. Schultes, J. Born, M. Hallschmid, Differential sensitivity of men and women to anorexigenic and memory-improving effects of intranasal insulin. *J. Clin. Endocrinol. Metab.* **93**, 1339–1344 (2008).
45. S. Obici, B. B. Zhang, G. Karkanias, L. Rossetti, Hypothalamic insulin signaling is required for inhibition of glucose production. *Nat. Med.* **8**, 1376–1382 (2002).
46. J. C. Erickson, K. E. Clegg, R. D. Palmiter, Sensitivity to leptin and susceptibility to seizures of mice lacking neuropeptide Y. *Nature.* **381**, 415–421 (1996).

47. K. W. Williams, M. M. Scott, J. K. Elmquist, Modulation of the central melanocortin system by leptin, insulin, and serotonin: co-ordinated actions in a dispersed neuronal network. *Eur. J. Pharmacol.* **660**, 2–12 (2011).
48. Y. Xu, J. K. Elmquist, M. Fukuda, Central nervous control of energy and glucose balance: focus on the central melanocortin system. *Ann. N. Y. Acad. Sci.* **1243**, 1–14 (2011).
49. R. D. Cone, Studies on the physiological functions of the melanocortin system. *Endocr. Rev.* **27**, 736–749 (2006).
50. C. Girardet, A. A. Butler, Neural melanocortin receptors in obesity and related metabolic disorders. *Biochim. Biophys. Acta.* **1842**, 482–494 (2014).
51. P. Chen, S. M. Williams, K. L. Grove, M. S. Smith, Melanocortin 4 receptor-mediated hyperphagia and activation of neuropeptide Y expression in the dorsomedial hypothalamus during lactation. *J. Neurosci.* **24**, 5091–5100 (2004).
52. D. A. Morgan, L. N. McDaniel, T. Yin, M. Khan, J. Jiang, M. R. Acevedo, S. A. Walsh, L. L. B. Ponto, A. W. Norris, M. Lutter, K. Rahmouni, H. Cui, Regulation of glucose tolerance and sympathetic activity by MC4R signaling in the lateral hypothalamus. *Diabetes.* **64**, 1976–1987 (2015).
53. Y.-H. Choi, T. Fujikawa, J. Lee, A. Reuter, K. W. Kim, Revisiting the ventral medial nucleus of the hypothalamus: the roles of SF-1 neurons in energy homeostasis. *Front. Neurosci.* **7**, 71 (2013).
54. T. Klöckener, S. Hess, B. F. Belgardt, L. Paeger, L. A. W. Verhagen, A. Husch, J.-W. Sohn, B. Hampel, H. Dhillon, J. M. Zigman, B. B. Lowell, K. W. Williams, J. K. Elmquist, T. L. Horvath, P. Kloppenburg, J. C. Brüning, High-fat feeding promotes obesity via insulin receptor/PI3K-dependent inhibition of SF-1 VMH neurons. *Nat. Neurosci.* **14**, 911–918 (2011).
55. J.-W. Sohn, Y. Oh, K. W. Kim, S. Lee, K. W. Williams, J. K. Elmquist, Leptin and insulin engage specific PI3K subunits in hypothalamic SF1 neurons. *Mol. Metab.* **5**, 669–679 (2016).
56. S. M. Sternson, G. M. G. Shepherd, J. M. Friedman, Topographic mapping of VMH --> arcuate nucleus microcircuits and their reorganization by fasting. *Nat. Neurosci.* **8**, 1356–1363 (2005).
57. K. W. Kim, J. J. Donato, E. D. Berglund, Y.-H. Choi, D. Kohno, C. F. Elias, R. A. Depinho, J. K. Elmquist, FOXO1 in the ventromedial hypothalamus regulates energy balance. *J. Clin. Invest.* **122**, 2578–2589 (2012).
58. Y. J. Kim, S. Bi, Knockdown of neuropeptide Y in the dorsomedial hypothalamus reverses high-fat diet-induced obesity and impaired glucose tolerance in rats. *Am. J. Physiol. J. Physiol.* **310**, R134–42 (2016).
59. F. Zheng, Y. J. Kim, P.-T. Chao, S. Bi, Overexpression of neuropeptide Y in the dorsomedial hypothalamus causes hyperphagia and obesity in rats. *Obes. (Silver Spring).* **21**, 1086–1092 (2013).
60. P.-T. Chao, L. Yang, S. Aja, T. H. Moran, S. Bi, Knockdown of NPY expression in the dorsomedial hypothalamus promotes development of brown adipocytes and prevents diet-induced obesity. *Cell Metab.* **13**, 573–583 (2011).
61. J. A. Brown, H. L. Woodworth, G. M. Leinninger, To ingest or rest? Specialized roles of lateral hypothalamic area neurons in coordinating energy balance. *Front. Syst. Neurosci.* **9**, 9 (2015).
62. D. M. Nance, R. A. Gorski, Similar effects of estrogen and lateral hypothalamic lesions on feeding behavior of female rats. *Brain Res. Bull.* **3**, 549–553 (1978).
63. S. W. Corbett, R. E. Keesey, Energy balance of rats with lateral hypothalamic lesions. *Am. J. Physiol.* **242**, E273–9 (1982).
64. K. M. Milam, J. S. Stern, L. H. Storlien, R. E. Keesey, Effect of lateral hypothalamic lesions on regulation of body weight and adiposity in rats. *Am. J. Physiol.* **239**, R337–43 (1980).
65. J. R. Charles, M. A. Duva, G. J. Ramirez, R. L. Lara, C. R. Yang, B. G. Stanley, Activation of lateral hypothalamic mGlu1 and mGlu5 receptors elicits feeding in rats. *Neuropharmacology.* **79**, 59–65 (2014).
66. K. Thongkhao-on, D. Wirtshafter, S. A. Shippy, Feeding specific glutamate surge in the rat lateral hypothalamus revealed by low-flow push-pull perfusion. *Pharmacol. Biochem. Behav.* **89**, 591–597 (2008).
67. A. C. Hausen, J. Ruud, H. Jiang, S. Hess, H. Varbanov, P. Kloppenburg, J. C. Brüning, Insulin-dependent activation of MCH neurons impairs locomotor activity and insulin sensitivity in obesity. *Cell Rep.* **17**, 2512–2521 (2016).
68. N. Weiss, F. Miller, S. Cazaubon, P.-O. Couraud, The blood-brain barrier in brain homeostasis and neurological diseases. *Biochim. Biophys. Acta.* **1788**, 842–857 (2009).

69. R. C. Janzer, M. C. Raff, Astrocytes induce blood-brain barrier properties in endothelial cells. *Nature.* **325**, 253–257 (1987).
70. A. M. Fernandez, E. Hernandez-Garzón, P. Perez-Domper, A. Perez-Alvarez, S. Mederos, T. Matsui, A. Santi, A. Trueba-Saiz, L. García-Guerra, J. Pose-Utrilla, J. Fielitz, E. N. Olson, R. Fernandez de la Rosa, L. Garcia Garcia, M. A. Pozo, T. Iglesias, A. Araque, H. Soya, G. Perea, E. D. Martin, I. Torres Aleman, Insulin regulates astrocytic glucose handling through cooperation with IGF-I. *Diabetes.* **66**, 64–74 (2017).
71. C. García-Cáceres, C. Quarta, L. Varela, Y. Gao, T. Gruber, B. Legutko, M. Jastroch, P. Johansson, J. Ninkovic, C.-X. Yi, O. Le Thuc, K. Szigeti-Buck, W. Cai, C. W. Meyer, P. T. Pfluger, A. M. Fernandez, S. Luquet, S. C. Woods, I. Torres-Alemán, C. R. Kahn, M. Götz, T. L. Horvath, M. H. Tschöp, Astrocytic insulin signaling couples brain glucose uptake with nutrient availability. *Cell.* **166**, 867–880 (2016).
72. C. J. Garwood, L. E. Ratcliffe, S. V Morgan, J. E. Simpson, H. Owens, I. Vazquez-Villaseñor, P. R. Heath, I. A. Romero, P. G. Ince, S. B. Wharton, Insulin and IGF1 signalling pathways in human astrocytes in vitro and in vivo; characterisation, subcellular localisation and modulation of the receptors. *Mol. Brain.* **8**, 51 (2015).
73. I. González-García, T. Gruber, C. García-Cáceres, Insulin action on astrocytes: from energy homeostasis to behaviour. *J. Neuroendocrinol.* **33**, e12953 (2021).
74. H. Takahashi, F. Takata, J. Matsumoto, T. Machida, A. Yamauchi, S. Dohgu, Y. Kataoka, Brain pericyte-derived soluble factors enhance insulin sensitivity in GT1–7 hypothalamic neurons. *Biochem. Biophys. Res. Commun.* **457**, 532–537 (2015).
75. P. A. Wilden, C. R. Kahn, K. Siddle, M. F. White, Insulin receptor kinase domain autophosphorylation regulates receptor enzymatic function. *J. Biol. Chem.* **267**, 16660–16668 (1992).
76. A. Belfiore, R. Malaguarnera, V. Vella, M. C. Lawrence, L. Sciacca, F. Frasca, A. Morrione, R. Vigneri, Insulin receptor isoforms in physiology and disease: an updated view. *Endocr. Rev.* **38**, 379–431 (2017).
77. K. Siddle, Signalling by insulin and IGF receptors: supporting acts and new players. *J. Mol. Endocrinol.* **47**, R1–10 (2011).
78. J. Havrankova, J. Roth, M. Brownstein, Insulin receptors are widely distributed in the central nervous system of the rat. *Nature.* **272**, 827–829 (1978).
79. M. A. Abbott, D. G. Wells, J. R. Fallon, The insulin receptor tyrosine kinase substrate p58/53 and the insulin receptor are components of CNS synapses. *J. Neurosci.* **19**, 7300–7308 (1999).
80. J. Unger, T. H. McNeill, R. T. 3rd Moxley, M. White, A. Moss, J. N. Livingston, Distribution of insulin receptor-like immunoreactivity in the rat forebrain. *Neuroscience.* **31**, 143–157 (1989).
81. A. Adem, S. S. Jossan, R. d'Argy, P. G. Gillberg, A. Nordberg, B. Winblad, V. Sara, Insulin-like growth factor 1 (IGF-1) receptors in the human brain: quantitative autoradiographic localization. *Brain Res.* **503**, 299–303 (1989).
82. J. De Keyser, N. Wilczak, J. P. De Backer, L. Herroelen, G. Vauquelin, Insulin-like growth factor-I receptors in human brain and pituitary gland: an autoradiographic study. *Synapse.* **17**, 196–202 (1994).
83. E. Blázquez, E. Velázquez, V. Hurtado-Carneiro, J. M. Ruiz-Albusac, Insulin in the brain: Its pathophysiological implications for states related with central insulin resistance, type 2 diabetes and alzheimer's disease. *Front. Endocrinol. (Lausanne).* **5**, 1–21 (2014).
84. J. J. Boucher, A. A. Kleinridders, C. R. Kahn, Insulin Receptor Signaling in Normal and Insulin-Resistant States. *Cold Spring Harb. Perspect. Biol.* 2014. **6**, a009191 (2014).
85. H. A. Metz HE, Insulin Receptor Substrate Regulation of Phosphoinositide 3-Kinase. *Clin. Cancer Res.* **17**, 206–212 (2011).
86. S. Hanke, M. Mann, The phosphotyrosine interactome of the insulin receptor family and its substrates IRS-1 nad IRS-2. *Mol. Cell. Proteomics.* **8**, 519–534 (2009).
87. T. F. Franke, D. R. Kaplan, L. C. Cantley, A. Tokert, Direct regulation of the Akt proto-oncogene product by phosphatidylinositol-3, 4- bisphosphate. *Am. Assoc. Adv. Sci. Stable.* **275**, 665–668 (2016).
88. D. D. Sarbassov, D. A. Guertin, S. M. Ali, D. M. Sabatini, Phosphorylation and regulation of Akt/PKB by the rictor-mTOR complex. *Science (80-.).* **307**, 1098–1101 (2005).
89. B. D. Manning, A. Toker, AKT/PKB signaling : navigating the network. *Cell.* **169**, 381–405 (2017).

90. B. A. Hemmings, D. F. Restuccia, PI3K-PKB/Akt pathway. *Cold Spring Harb. Perspect. Biol.* **4**, a011189 (2012).
91. B. D. Hopkins, M. D. Goncalves, Insulin–PI3K signalling: an evolutionarily insulated metabolic driver of cancer. *Nat. Rev. Endocrinol.* **16**, 276–283 (2020).
92. A. M. Dieterle, P. Böhler, H. Keppeler, S. Alers, N. Berleth, S. Drießen, N. Hieke, S. Pietkiewicz, A. S. Löffler, C. Peter, A. Gray, N. R. Leslie, H. Shinohara, T. Kurosaki, M. Engelke, J. Wienands, M. Bonin, S. Wesselborg, B. Stork, PDK1 controls upstream PI3K expression and PIP 3 generation. *Oncogene.* **33**, 3043–3053 (2014).
93. R. A. Haeusler, T. E. McGraw, D. Accili, Biochemical and cellular properties of insulin receptor signalling. *Nat. Rev. Mol. Cell Biol.* **19**, 31–44 (2018).
94. Y. Jeon, S. Aja, G. V. Ronnett, E. K. Kim, D-chiro-inositol glycan reduces food intake by regulating hypothalamic neuropeptide expression via AKT-FoxO1 pathway. *Biochem. Biophys. Res. Commun.* **470**, 818–823 (2016).
95. M. S. Kim, Y. K. Pak, P. G. Jang, C. Namkoong, Y. S. Choi, J. C. Won, K. S. Kim, S. W. Kim, H. S. Kim, J. Y. Park, Y. B. Kim, K. U. Lee, Role of hypothalamic Foxo1 in the regulation of food intake and energy homeostasis. *Nat. Neurosci.* **9**, 901–906 (2006).
96. T. Kitamura, Y. Feng, Y. I. Kitamura, S. C. Chua, A. W. Xu, G. S. Barsh, L. Rossetti, D. Accili, Forkhead protein FoxO1 mediates Agrp-dependent effects of leptin on food intake. *Nat. Med.* **12**, 534–540 (2006).
97. T. Saneyasu, S. Fujita, A. Kitashiro, S. Fukuzo, K. Honda, H. Kamisoyama, Hypothalamic Akt-mediated signaling regulates food intake in chicks. *Neurosci. Lett.* **670**, 48–52 (2018).
98. K. D. Niswender, C. D. Morrison, D. J. Clegg, R. Olson, D. G. Baskin, M. G. Myers, R. J. Seeley, M. W. Schwartz, Insulin activation of phosphatidylinositol 3-kinase in the hypothalamic arcuate nucleus a key mediator of insulin-induced anorexia. *Diabetes.* **52**, 227–231 (2003).
99. Y. Li, T. J. Soos, X. Li, J. Wu, M. DeGennaro, X. Sun, D. R. Littman, M. J. Birnbaum, R. D. Polakiewicz, Protein kinase C θ inhibits insulin signaling by phosphorylating IRS1 at Ser1101. *J. Biol. Chem.* **279**, 45304–45307 (2004).
100. J. E. Chin, F. Liu, R. A. Roth, Activation of protein kinase Cα inhibits insulin-stimulated tyrosine phosphorylation of insulin receptor substrate-1. *Mol. Endocrinol.* **8**, 51–58 (1994).
101. M. Li, S. G. Vienberg, O. Bezy, B. T. O'Neill, C. R. Kahn, Role of PKCδ in insulin sensitivity and skeletal muscle metabolism. *Diabetes.* **64**, 4023–4032 (2015).
102. K. Shi, K. Egawa, H. Maegawa, T. Nakamura, S. Ugi, Y. Nishio, A. Kashiwagi, Protein-tyrosine phosphatase 1B associates with insulin receptor and negatively regulates insulin signaling without receptor internalization. *J. Biochem.* **136**, 89–96 (2004).
103. S. Galic, C. Hauser, B. B. Kahn, F. G. Haj, B. G. Neel, N. K. Tonks, T. Tiganis, Coordinated regulation of insulin signaling by the protein tyrosine phosphatases PTP1B and TCPTP. *Mol. Cell. Biol.* **25**, 819–829 (2005).
104. L. J. Holt, K. Siddle, Grb10 and Grb14: enigmatic regulators of insulin action - And more? *Biochem. J.* **388**, 393–406 (2005).
105. F. Liu, R. A. Roth, Grb-IR: A SH2-domain-containing protein that binds to the insulin receptor and inhibits its function. *PNAS.* **92**, 10287–10291 (1995).
106. A. Yoshimura, T. Naka, M. Kubo, SOCS proteins, cytokine signalling and immune regulation. *Nat. Rev. Immunol.* **7**, 454–465 (2007).
107. J. Feng, S. Lu, B. Ou, Q. Liu, J. Dai, C. Ji, H. Zhou, H. Huang, Y. Ma, The role of JNK signaling pathway in obesity-driven insulin resistance. *Diabetes, Metab. Syndr. Obes. Targets Ther.* **13**, 1399–1406 (2020).
108. S. Karina, M. Flores-le, E. Avila-muñoz, N. Rodr, C. Arias, K. Sánchez-Alegría, M. Flores-León, E. Avila-muñoz, N. Rodríguez-Corona, C. Arias, PI3K signaling in neurons: a central node for the control of multiple functions. *Int. J. Mol. Sci.* **19** (2018).
109. L. Wang, K. Zhou, Z. Fu, D. Yu, H. Huang, X. Zang, Brain Development and Akt signaling : the crossroads of signaling pathway and neurodevelopmental diseases. *J. Mol. Neurosci.* **61**, 379–384 (2017).
110. L. A. Jansen, G. M. Mirzaa, G. E. Ishak, B. J. O. Roak, J. B. Hiatt, W. H. Roden, S. A. Gunter, S. L. Christian, S. Collins, J. St-onge, J. G. Ojemann, C. Adams, J. Rivie, J. Shendure, R. F. Hevner, W. B. Dobyns, PI3K / AKT pathway mutations cause a spectrum of brain malformations from megalencephaly to focal cortical dysplasia. *CNS Neurosci. Ther.* **138**, 1613–1628 (2015).
111. F. Xu, L. Na, Y. Li, L. Chen, Roles of the PI3K/AKT/mTOR signalling pathways in neurodegenerative diseases and tumours. *Cell Biosci.* **10**, 1–12 (2020).

112. Z. Xiao, J. Peng, L. Yang, H. Kong, F. Yin, Interleukin-1 β plays a role in the pathogenesis of mesial temporal lobe epilepsy through the PI3K/Akt/mTOR signaling pathway in hippocampal neurons. *J. Neuroimmunol.* **282**, 110–117 (2015).
113. Y. Xu, J. W. Hill, M. Fukuda, L. Gautron, J. Sohn, K. Kim, C. E. Lee, M. J. Choi, D. A. Lauzon, H. Dhillon, B. B. Lowell, J. M. Zigman, J. J. Zhao, Short article PI3K signaling in the ventromedial hypothalamic nucleus is required for normal energy homeostasis. *Cell Metab.* **12**, 88–95 (2010).
114. S. Obici, Z. Feng, G. Karkanias, D. G. Baskin, L. Rossetti, Decreasing hypothalamic insulin receptors causes hyperphagia and. *Nat. Neurosci.* **5** (2002), doi:10.1038/nn861.
115. A. I. Choudhury, H. Heffron, M. A. Smith, H. Al-Qassab, A. W. Xu, C. Selman, M. Simmgen, M. Clements, M. Claret, G. MacColl, D. C. Bedford, K. Hisadome, I. Diakonov, V. Moosajee, J. D. Bell, J. R. Speakman, R. L. Batterham, G. S. Barsh, M. L. J. J. Ashford, D. J. Withers, The role of insulin receptor substrate 2 in hypothalamic and beta cell function. *J. Clin. Invest.* **115**, 940–950 (2005).
116. D. J. Withers, J. S. Gutierrez, H. Towery, D. J. Burks, J. M. Ren, S. Previs, Y. Zhang, D. Bernal, S. Pons, G. I. Shulman, S. Bonner-Weir, M. F. White, Disruption of IRS-2 causes type 2 diabetes in mice. *Nature.* **391**, 900–904 (1998).
117. X. Lin, A. Taguchi, S. Park, J. A. Kushner, F. Li, Y. Li, M. F. White, Dysregulation of insulin receptor substrate 2 in beta cells and brain causes obesity and diabetes. *J. Clin. Invest.* **114**, 908–916 (2004).
118. C. E. Battú, D. Rieger, S. Loureiro, G. V Furtado, H. Bock, R. Pessoa-pureur, C. Gonçalves, M. S. Perry, Alterations of PI3K and Akt signaling pathways in the hippocampus and hypothalamus of Wistar rats treated with highly palatable food. *Nutr. Neurosci.* **15**, 10–18 (2012).
119. T. P. Flagg, D. Enkvetchakul, J. C. Koster, C. G. Nichols, Muscle KATP channels: recent insights to energy sensing and myoprotection. *Physiol. Rev.* **90**, 799–829 (2010).
120. J. E. Brayden, Functional roles of KATP channels in vascular smooth muscle. *Clin. Exp. Pharmacol. Physiol.* **29**, 312–316 (2002).
121. K. Bennett, C. James, K. Hussain, Pancreatic β-cell KATP channels: hypoglycaemia and hyperglycaemia. *Rev. Endocr. Metab. Disord.* **11**, 157–163 (2010).
122. K. Ballanyi, Protective role of neuronal KATP channels in brain hypoxia. *J. Exp. Biol.* **207**, 3201–3212 (2004).
123. K. Minami, T. Miki, T. Kadowaki, S. Seino, Roles of ATP-sensitive K+ channels as metabolic sensors: Studies of Kir6.x null mice. *Diabetes.* **53**, 1–5 (2004).
124. M. L. Zhuo, Y. Huang, D. P. Liu, C. C. Liang, KATP channel: relation with cell metabolism and role in the cardiovascular system. *Int. J. Biochem. Cell Biol.* **37**, 751–764 (2005).
125. N. Ibrahim, M. A. Bosch, J. L. Smart, J. Qiu, M. Rubinstein, O. K. Rønnekleiv, M. J. Low, M. J. Kelly, Hypothalamic proopiomelanocortin neurons are glucose responsive and express KATP channels. *Endocrinology.* **144**, 1331–1340 (2003).
126. A. Pocal, T. K. T. Lam, R. Gutierrez-Juarez, S. Obici, G. J. Schwartz, J. Bryan, L. Aguilar-Bryan, L. Rossetti, Hypothalamic KATP channels control hepatic glucose production. *Nature.* **434**, 1026–1031 (2005).
127. M. Acosta-Martínez, J. E. Levine, Regulation of KATP channel subunit gene expression by hyperglycemia in the mediobasal hypothalamus of female rats. *Am. J. Physiol.* **292**, E1801–E1807 (2007).
128. T. Miki, K. Minami, L. Zhang, M. Morita, T. Gonoi, T. Shiuchi, Y. Minokoshi, J. M. Renaud, S. Seino, ATP-sensitive potassium channels participate in glucose uptake in skeletal muscle and adipose tissue. *Am. J. Physiol. - Endocrinol. Metab.* **283**, 1178–1184 (2002).
129. G. G. Macgregor, K. Dong, C. G. Vanoye, L. Tang, G. Giebisch, S. C. Hebert, Nucleotides and phospholipids compete for binding to the C terminus of K ATP channels. *Proc. Natl. Acad. Sci. U. S. A.* (2002).
130. L. Plum, X. Ma, B. Hampel, N. Balthasar, R. Coppari, H. Münzberg, M. Shanabrough, D. Burdakov, E. Rother, R. Janoschek, J. Alber, B. F. Belgardt, L. Koch, J. Seibler, F. Schwenk, C. Fekete, A. Suzuki, T. W. Mak, W. Krone, T. L. Horvath, Enhanced PIP 3 signaling in POMC neurons causes K ATP channel activation and leads to diet-sensitive obesity. *J. Clin. Invest.* **116** (2006).
131. H. Yang, X. Jiang, B. Li, H. J. Yang, M. Miller, A. Yang, A. Dhar, P. Nikola, Mechanisms of mTORC1 activation by RHEB and inhibition by PRAS40. *Nature.* **552**, 368–373 (2017).
132. M. Jhanwar-Uniyal, J. V Wainwright, A. L. Mohan, M. E. Tobias, R. Murali, C. D. Gandhi, M. H. Schmidt, Diverse signaling mechanisms of mTOR complexes: mTORC1 and mTORC2

in forming a formidable relationship. *Adv. Biol. Regul.* **72**, 51–62 (2019).
133. A. Astrinidis, E. P. Henske, Tuberous sclerosis complex : linking growth and energy signaling pathways with human disease. *Oncogene.* **24**, 7475–7481 (2005).
134. R. Zoncu, A. Efeyan, D. M. Sabatini, mTOR: from growth signal integration to cancer, diabetes and ageing. *Nat. Rev. Mol. Cell Biol.* **12**, 21–35 (2011).
135. N. Takei, H. Nawa, mTOR signaling and its roles in normal and abnormal brain development. *Front. Mol. Neurosci.* **7**, 1–12 (2014).
136. F. Hu, Y. Xu, F. Liu, Hypothalamic roles of mTOR complex I : integration of nutrient and hormone signals to regulate energy homeostasis. *Am. J. Physiol.* **310**, 994–1002 (2016).
137. M. Mannaa, S. Krämer, M. Boschmann, M. Gollasch, mTOR and regulation of energy homeostasis in humans. *J. Mol. Med.* **91**, 1167–1175 (2013).
138. H. E. Kocalis, S. L. Hagan, L. George, M. K. Turney, M. A. Siuta, G. N. Laryea, L. C. Morris, L. J. Muglia, R. L. Printz, G. D. Stanwood, Rictor/mTORC2 facilitates central regulation of energy and glucose homeostasis. *Mol. Metab.* **3**, 394–407 (2014).
139. D. W. Lamming, L. Ye, P. Katajisto, M. D. Goncalves, M. Saitoh, D. M. Stevens, J. G. Davis, A. B. Salmon, A. Richardson, R. S. Ahima, D. A. Guertin, D. M. Sabatini, J. A. Baur, Rapamycin-induced insulin resistance is mediated by mTORC2 loss and uncoupled from longevity. *Science.* **335**, 1638–1643 (2012).
140. L. Martins, D. Fernández-Mallo, M. G. Novelle, M. J. Vázquez, M. Tena-Sempere, R. Nogueiras, M. López, C. Diéguez, Hypothalamic mTOR signaling mediates the orexigenic action of ghrelin. *PLoS One.* **7**, e46923 (2012).
141. O. Kwon, K. W. Kim, M. Kim, Leptin signalling pathways in hypothalamic neurons. *Cell. Mol. Life Sci.* **73**, 1457–1477 (2016).
142. K. Muta, D. A. Morgan, K. Rahmouni, The role of hypothalamic mTORC1 signaling in insulin regulation of food intake, body weight, and sympathetic nerve activity in male mice. *Endocrinology.* **156**, 1398–1407 (2015).
143. H. Mori, K. Inoki, H. Mu, D. Opland, M. Faouzi, E. C. Villanueva, T. Ikenoue, D. Kwiatkowski, O. A. Macdougald, M. G. Myers, K. Guan, Critical role for hypothalamic mTOR activity in energy balance. *Cell Metab.* **9**, 362–374 (2009).
144. H. Ono, G. J. Schwartz, L. Rossetti, H. Ono, A. Pocai, Y. Wang, H. Sakoda, T. Asano, J. M. Backer, G. J. Schwartz, L. Rossetti, Activation of hypothalamic S6 kinase mediates diet-induced hepatic insulin resistance in rats. *J. Clin. Invest.* **118**, 2959–2968 (2008).
145. M. M. Mihaylova, R. J. Shaw, The AMPK signalling pathway coordinates cell growth, autophagy and metabolism. *Nat. Publ. Gr.* **13**, 1016–1023 (2011).
146. J. Kim, G. Yang, Y. Kim, J. Kim, J. Ha, AMPK activators : mechanisms of action and physiological activities. *Exp. Mol. Med.* **48**, e224–12 (2016).
147. H. My Khanh Q, A. W. Kinyua, D. J. Yang, K. W. Kim, hypothalamic AMPK as a regulator of energy homeostasis. *Neural. Plast.* (2016).
148. M. Claret, M. A. Smith, R. L. Batterham, C. Selman, A. I. Choudhury, L. G. D. Fryer, M. Clements, H. Al-qassab, H. Heffron, A. W. Xu, J. R. Speakman, G. S. Barsh, B. Viollet, S. Vaulont, M. L. J. Ashford, D. Carling, D. J. Withers, AMPK is essential for energy homeostasis regulation and glucose sensing by POMC and AgRP neurons. *J. Clin. Invest.* **117** (2007).
149. D. Kohno, H. Sone, S. Tanaka, H. Kurita, D. Gantulga, T. Yada, Neuroscience letters AMP-activated protein kinase activates neuropeptide Y neurons in the hypothalamic arcuate nucleus to increase food intake in rats. *Neurosci. Lett.* **499**, 194–198 (2011).
150. T. S. Oh, H. Cho, J. H. Cho, S. Yu, Hypothalamic AMPK-induced autophagy increases food intake by regulating NPY and POMC expression. *Autophagy.* **12**, 2009–2025 (2016).
151. C. S. Yang, C. K. L. Lam, M. Chari, G. W. C. Cheung, A. Kokorovic, S. Gao, I. Leclerc, G. A. Rutter, T. K. T. Lam, Hypothalamic AMP-activated protein kinase regulates glucose production. *Diabetes.* **59**, 2435–2443 (2010).
152. H. Liu, Y. Xu, F. Hu, AMPK in the ventromedial nucleus of the hypothalamus : a key regulator for thermogenesis. *Front. Endocrinol. (Lausanne).* **11**, 1–13 (2020).
153. A. D. Van Dam, S. Kooijman, M. Schilperoort, P. C. N. Rensen, R. Boon, Regulation of brown fat by AMP-activated protein kinase. *Trends Mol. Med.*, 1–9 (2015).
154. J. Mccain, The MAPK (ERK) pathway investigational combinations for the treatment of BRAF-mutated metastatic melanoma. *Pharm. Ther.* **38** (2013).

155. W. Zhang, H. T. Liu, MAPK signal pathways in the regulation of cell proliferation in mammalian cells. *Cell Res.* **12**, 9–18 (2002).
156. P. Jiao, B. Feng, Y. Li, Q. He, H. Xu, Hepatic ERK activity plays a role in energy metabolism. *Mol. Cell. Endocrinol.* **375**, 157–166 (2013).
157. T. Horie, G. Park, Y. Inaba, E. Hashiuchi, T. Iezaki, K. Tokumura, K. Fukasawa, T. Yamada, M. Hiraiwa, Y. Kitaguchi, H. Kamada, K. Kaneda, T. Tanaka, H. Inoue, E. Hinoi, MAPK Erk5 in leptin receptor–expressing neurons controls body weight and systemic energy homeostasis in female mice. *Endocrinology.* **160**, 2837–2848 (2019).
158. F. Liu, X. Yang, M. Geng, M. Huang, Targeting ERK, an Achilles ' Heel of the MAPK pathway, in cancer therapy. *Acta Pharm. Sin. B.* **8**, 552–562 (2018).
159. A. De Luca, M. R. Maiello, A. D'Alessio, M. Pergameno, N. Normanno, The RAS/RAF/MEK/ERK and the PI3K/AKT signalling pathways: role in cancer pathogenesis and implications for therapeutic approaches. *Expert Opin. Ther. Targets.* **16 Suppl 2**, S17–27 (2012).
160. C. M. Mayer, D. D. Belsham, Insulin directly regulates NPY and AgRP gene expression via the MAPK MEK/ERK signal transduction pathway in mHypoE-46 hypothalamic neurons. *Mol. Cell. Endocrinol.* **307**, 99–108 (2009).
161. K. Rahmouni, C. D. Sigmund, W. G. Haynes, A. L. Mark, Hypothalamic ERK Mediates the anorectic and thermogenic sympathetic effects of leptin. *Diabetes.* **58**, 536–542 (2009).
162. K. Rahmouni, D. A. Morgan, G. M. Morgan, X. Liu, C. D. Sigmund, A. L. Mark, W. G. Haynes, Hypothalamic PI3K and MAPK differentially mediate regional sympathetic activation to insulin. *J. Clin. Invest.* **114** (2004), doi:10.1172/JCI200421737.652.
163. A. Kleinridders, H. A. Ferris, W. Cai, C. R. Kahn, Insulin action in brain regulates systemic metabolism and brain function. *Diabetes.* **63**, 2232–2243 (2014).
164. S. Kullmann, A. Kleinridders, D. M. Small, A. Fritsche, H.-U. Häring, H. Preissl, M. Heni, Central nervous pathways of insulin action in the control of metabolism and food intake. *Lancet. Diabetes Endocrinol.* **8**, 524–534 (2020).
165. C. D. Chapman, W. H. 2nd Frey, S. Craft, L. Danielyan, M. Hallschmid, H. B. Schiöth, C. Benedict, Intranasal treatment of central nervous system dysfunction in humans. *Pharm. Res.* **30**, 2475–2484 (2013).
166. M. C. Petersen, G. I. Shulman, Mechanisms of insulin action and insulin resistance. *Physiol. Rev.* **98**, 2133–2223 (2018).
167. C.-X. Yi, S. E. la Fleur, E. Fliers, A. Kalsbeek, The role of the autonomic nervous liver innervation in the control of energy metabolism. *Biochim. Biophys. acta.* **1802**, 416–431 (2010).
168. H. Inoue, W. Ogawa, A. Asakawa, Y. Okamoto, A. Nishizawa, M. Matsumoto, K. Teshigawara, Y. Matsuki, E. Watanabe, R. Hiramatsu, K. Notohara, K. Katayose, H. Okamura, C. R. Kahn, T. Noda, K. Takeda, S. Akira, A. Inui, M. Kasuga, Role of hepatic STAT3 in brain-insulin action on hepatic glucose production. *Cell Metab.* **3**, 267–275 (2006).
169. K. Kimura, M. Tanida, N. Nagata, Y. Inaba, H. Watanabe, M. Nagashimada, T. Ota, S. Asahara, Y. Kido, M. Matsumoto, K. Toshinai, M. Nakazato, T. Shibamoto, S. Kaneko, M. Kasuga, H. Inoue, Central insulin action activates kupffer cells by suppressing hepatic vagal activation via the nicotinic alpha 7 acetylcholine receptor. *Cell Rep.* **14**, 2362–2374 (2016).
170. H. V Lin, L. Plum, H. Ono, R. Gutiérrez-Juárez, M. Shanabrough, E. Borok, T. L. Horvath, L. Rossetti, D. Accili, Divergent regulation of energy expenditure and hepatic glucose production by insulin receptor in agouti-related protein and POMC neurons. *Diabetes.* **59**, 337–346 (2010).
171. M. Heni, R. Wagner, S. Kullmann, R. Veit, H. Mat Husin, K. Linder, C. Benkendorff, A. Peter, N. Stefan, H.-U. Häring, H. Preissl, A. Fritsche, Central insulin administration improves whole-body insulin sensitivity via hypothalamus and parasympathetic outputs in men. *Diabetes.* **63**, 4083–4088 (2014).
172. D. S. Edgerton, M. C. Moore, J. J. Winnick, M. Scott, B. Farmer, H. Naver, C. B. Jeppesen, P. Madsen, T. B. Kjeldsen, E. Nishimura, C. L. Brand, A. D. Cherrington, Changes in glucose and fat metabolism in response to the administration of a hepato-preferential insulin analog. *Diabetes.* **63**, 3946–3954 (2014).
173. T. D. Farmer, E. C. Jenkins, T. P. O'Brien, G. A. McCoy, A. E. Havlik, E. R. Nass, W. E. Nicholson, R. L. Printz, M. Shiota, Comparison of the physiological relevance of systemic vs. portal insulin delivery to evaluate whole body glucose flux during an insulin clamp. *Am. J. Physiol. Endocrinol. Metab.* **308**, E206–22 (2015).

174. D. K. Sindelar, J. H. Balcom, C. A. Chu, D. W. Neal, A. D. Cherrington, A comparison of the effects of selective increases in peripheral or portal insulin on hepatic glucose production in the conscious dog. *Diabetes*. **45**, 1594–1604 (1996).
175. E. Fabbrini, B. S. Mohammed, F. Magkos, K. M. Korenblat, B. W. Patterson, S. Klein, Alterations in adipose tissue and hepatic lipid kinetics in obese men and women with nonalcoholic fatty liver disease. *Gastroenterology*. **134**, 424–431 (2008).
176. T. Scherer, C. Lindtner, J. O'Hare, M. Hackl, E. Zielinski, A. Freudenthaler, S. Baumgartner-Parzer, K. Tödter, J. Heeren, M. Krššák, L. Scheja, C. Fürnsinn, C. Buettner, Insulin regulates hepatic triglyceride secretion and lipid content via signaling in the brain. *Diabetes*. **65**, 1511–1520 (2016).
177. G. I. Smith, M. Shankaran, M. Yoshino, G. G. Schweitzer, M. Chondronikola, J. W. Beals, A. L. Okunade, B. W. Patterson, E. Nyangau, T. Field, C. B. Sirlin, S. Talukdar, M. K. Hellerstein, S. Klein, Insulin resistance drives hepatic de novo lipogenesis in nonalcoholic fatty liver disease. *J. Clin. Invest.* **130**, 1453–1460 (2020).
178. M. Lafontan, D. Langin, Lipolysis and lipid mobilization in human adipose tissue. *Prog. Lipid Res.* **48**, 275–297 (2009).
179. A. C. Shin, N. Filatova, C. Lindtner, T. Chi, S. Degann, D. Oberlin, C. Buettner, Insulin receptor signaling in POMC, but not AgRP, neurons controls adipose tissue insulin action. *Diabetes*. **66**, 1560–1571 (2017).
180. T. Scherer, J. O'Hare, K. Diggs-Andrews, M. Schweiger, B. Cheng, C. Lindtner, E. Zielinski, P. Vempati, K. Su, S. Dighe, T. Milsom, M. Puchowicz, L. Scheja, R. Zechner, S. J. Fisher, S. F. Previs, C. Buettner, Brain insulin controls adipose tissue lipolysis and lipogenesis. *Cell Metab.* **13**, 183–194 (2011).
181. L. Koch, F. T. Wunderlich, J. Seibler, A. C. Könner, B. Hampel, S. Irlenbusch, G. Brabant, C. R. Kahn, F. Schwenk, J. C. Brüning, Central insulin action regulates peripheral glucose and fat metabolism in mice. *J. Clin. Invest.* **118**, 2132–2147 (2008).
182. S. Kullmann, M. Heni, M. Hallschmid, A. Fritsche, H. Preissl, H.-U. Häring, Brain insulin resistance at the crossroads of metabolic and cognitive disorders in humans. *Physiol. Rev.* **96**, 1169–1209 (2016).
183. K. A. Iwen, T. Scherer, M. Heni, F. Sayk, T. Wellnitz, F. Machleidt, H. Preissl, H.-U. Häring, A. Fritsche, H. Lehnert, C. Buettner, M. Hallschmid, Intranasal insulin suppresses systemic but not subcutaneous lipolysis in healthy humans. *J. Clin. Endocrinol. Metab.* **99**, E246–51 (2014).
184. S. Dash, C. Xiao, C. Morgantini, K. Koulajian, G. F. Lewis, Intranasal insulin suppresses endogenous glucose production in humans compared with placebo in the presence of similar venous insulin concentrations. *Diabetes*. **64**, 766–774 (2015).
185. S. Gancheva, C. Koliaki, A. Bierwagen, P. Nowotny, M. Heni, A. Fritsche, H.-U. Häring, J. Szendroedi, M. Roden, Effects of intranasal insulin on hepatic fat accumulation and energy metabolism in humans. *Diabetes*. **64**, 1966–1975 (2015).
186. C. Brøns, C. B. Jensen, H. Storgaard, N. J. Hiscock, A. White, J. S. Appel, S. Jacobsen, E. Nilsson, C. M. Larsen, A. Astrup, B. Quistorff, A. Vaag, Impact of short-term high-fat feeding on glucose and insulin metabolism in young healthy men. *J. Physiol.* **587**, 2387–2397 (2009).
187. M.-A. Cornier, B. C. Bergman, D. H. Bessesen, The effects of short-term overfeeding on insulin action in lean and reduced-obese individuals. *Metabolism*. **55**, 1207–1214 (2006).
188. B. Cannon, J. Nedergaard, Brown adipose tissue: function and physiological significance. *Physiol. Rev.* **84**, 277–359 (2004).
189. A. Vitali, I. Murano, M. C. Zingaretti, A. Frontini, D. Ricquier, S. Cinti, The adipose organ of obesity-prone C57BL/6J mice is composed of mixed white and brown adipocytes. *J. Lipid Res.* **53**, 619–629 (2012).
190. K. Ikeda, P. Maretich, S. Kajimura, The common and distinct features of brown and beige adipocytes. *Trends Endocrinol. Metab.* **29**, 191–200 (2018).
191. C. Müller, M. J. Voirol, N. Stefanoni, J. F. Surmely, E. Jéquier, R. C. Gaillard, L. Tappy, Effect of chronic intracerebroventricular infusion of insulin on brown adipose tissue activity in fed and fasted rats. *Int. J. Obes. Relat. Metab. Disord. J. Int. Assoc. Study Obes.* **21**, 562–566 (1997).
192. G. T. Dodd, S. Decherf, K. Loh, S. E. Simonds, F. Wiede, E. Balland, T. L. Merry, H. Münzberg, Z.-Y. Zhang, B. B. Kahn, B. G. Neel, K. K. Bence, Z. B. Andrews, M. A. Cowley, T. Tiganis, Leptin and insulin act on POMC neurons to promote the browning of white fat. *Cell*. **160**, 88–104 (2015).

193. C. M. Chen W, Balland E, Hypothalamic insulin resistance in obesity: effects on glucose homeostasis. *Neuroendocrinology.* **104**, 364–381 (2014).
194. H. Ono, Molecular mechanisms of hypothalamic insulin resistance. *Int. J. Mol. Sci.* **20** (2019).
195. V. J. Hamburg NM, McMackin CJ, Huang AL, Shenouda SM, Widlansky ME, Schulz E, Gokce N, Ruderman NB, Keaney JF Jr, Physical inactivity rapidly induces insulin resistance and microvascular dysfunction in healthy volunteers. *Arter. Thromb Vasc Biol.* **23**, 1–7 (2008).
196. K. S. Kim, S. J. Kim, S. Kim, D. W. Choi, Y. J. Ju, E. C. Park, Association of self-reported sedentary time with insulin resistance among Korean adults without diabetes mellitus: a cross-sectional study. *BMC Public Health.* **18**, 1–8 (2018).
197. D. P. Belsham DD, Insulin signalling in hypothalamic neurones. *J. Neuroendocrinol.* **33**, E12919 (2020).
198. I. P. J. P. Costa-nunes, V. Kasatkin, T. Strekalova, E. Veniaminova, A. Demchenko, A. Lyundup, K. Lesch, E. D. Ponomarev, Insulin receptor in the brain: mechanisms of activation and the role in the CNS pathology and treatment. *CNS Neurosci. Ther.* **24**, 763–774 (2018).
199. K. Timper, J. C. Brüning, Hypothalamic circuits regulating appetite and energy homeostasis: Pathways to obesity. *DMM Dis. Model. Mech.* **10**, 679–689 (2017).
200. C. T. De Souza, E. P. Araujo, S. Bordin, R. Ashimine, R. L. Zollner, A. C. Boschero, M. J. A. Saad, L. A. Velloso, Consumption of a fat-rich diet activates a proinflammatory response and induces insulin resistance in the hypothalamus. *Endocrinology.* **146**, 4192–4199 (2005).
201. P. S. Dalvi, J. A. Chalmers, V. Luo, D. Y. Han, L. Wellhauser, Y. Liu, D. Q. Tran, J. Castel, S. Luquet, M. B. Wheeler, D. D. Belsham, High fat induces acute and chronic inflammation in the hypothalamus: Effect of high-fat diet, palmitate and TNF-α on appetite-regulating NPY neurons. *Int. J. Obes.* **41**, 149–158 (2017).
202. L. S. Santos, G. S. Cordeiro, R. J. B. Matos, G. S. Perez, R. T. Silva, G. T. Boaventura, J. M. Barreto-Medeiros, High-fat diet promotes hypothalamic inflammation in animal models: a systematic review. *Nutr. Rev.* **00**, 1–8 (2021).
203. S. B. Wheatcroft, I. L. Williams, A. M. Shah, M. T. Kearney, Pathophysiological implications of insulin resistance on vascular endothelial function. *Diabet. Med.* **20**, 255–268 (2003).
204. T. Scherer, K. Sakamoto, C. Buettner, Brain insulin signalling in metabolic homeostasis and disease. *Nat. Rev. Endocrinol.* **17**, 468–483 (2021).
205. J. Wang, S. Obici, K. Morgan, N. Barzilai, Z. Feng, L. Rossetti, Overfeeding rapidly induces leptin and insulin resistance. *Diabetes.* **50**, 2786–2791 (2001).
206. E. W. Kraegen, P. W. Clark, A. B. Jenkins, E. A. Daley, D. J. Chisholm, L. H. Storlien, Development of muscle insulin resistance after liver insulin resistance in high-fat-fed rats. *Diabetes.* **40**, 1397–1403 (1991).
207. S. N. van der Crabben, G. Allick, M. T. Ackermans, E. Endert, J. A. Romijn, H. P. Sauerwein, Prolonged fasting induces peripheral insulin resistance, which is not ameliorated by high-dose salicylate. *J. Clin. Endocrinol. Metab.* **93**, 638–641 (2008).
208. M. A. Torsoni, J. B. Carvalheira, M. Pereira-Da-Silva, M. A. de Carvalho-Filho, M. J. A. Saad, L. A. Velloso, Molecular and functional resistance to insulin in hypothalamus of rats exposed to cold. *Am. J. Physiol. Endocrinol. Metab.* **285**, E216–23 (2003).
209. R. F. Rifkin, M. Potgieter, J.-B. Ramond, D. A. Cowan, Ancient oncogenesis, infection and human evolution. *Evol. Appl.* **10**, 949–964 (2017).

CHAPTER SEVEN

Insulin and Brain Reward Systems

Brian C. Liu, Qingchen Zhang, and Emmanuel N. Pothos

CONTENTS

INTRODUCTION

Insulin is historically known to regulate metabolism in muscle, liver, and adipose tissue. But in recent years, the effect of insulin signaling in the brain has been the subject of a more thorough investigation. We are now aware that central insulin modulates not only several functions, including appetite, reproduction, body temperature, and glucose and lipid metabolism but also higher-order cognitive processing. Of particular clinical interest, insulin resistance in the brain is implicated in mood disorders.

DOI: 10.1201/9781003079927-7

HISTORICAL PERSPECTIVE

In 1921, Frederick Banting and Charles Best, at the University of Toronto, set out to identify a substance produced by the islets of Langerhans that could regulate glucose metabolism. They were able to extract this substance, known as insulin, and demonstrated that it could reduce hyperglycemia in dogs. Next, with the help of biochemist James Collip, they were able to extract a more concentrated form of insulin from the beef pancreas. In 1922, Leonard Thompson became the first person with type I diabetes to be treated with insulin (1).

Insulin is a pancreatic peptide hormone. It is responsible for maintaining healthy blood glucose levels, specifically through regulating protein, lipid and carbohydrate metabolism, promoting glucose uptake, and cell division and growth. Insulin is a dipeptide with A and B chains that are linked by disulfide bridges. It contains 51 amino acids, with the A chain made up of 21 amino acids and the B chain containing 30 amino acids. Insulin has an iso-electric point of pH 5.5 and a molecular weight of 5,802 (2).

The gene for insulin is found on the short arm of chromosome 11 in humans (3). First, pre-proinsulin is synthesized. Then, it is converted into proinsulin when signal peptidases remove the signal peptide, leaving the A and B chains and a connecting (C) peptide. This process occurs in the rough endoplasmic reticulum, and secretory vesicles transport the proinsulin to the Golgi apparatus. Here enzymes outside of the Golgi separate the C peptide from the proinsulin to create insulin (4). When insulin is released in response to glucose, the secretion response can be characterized as biphasic. There is an initial rapid secretion, and then a longer, sustained secretion of insulin at a lower level (5).

PERIPHERAL INSULIN

Traditionally, insulin has been associated with regulating glucose homeostasis (6). In the periphery, insulin is known to regulate body weight and food intake (7). Produced by beta cells in the pancreas, insulin is also responsible for lowering glucose levels in various organs. Classically, insulin targets the liver, skeletal muscle, and adipose tissue to regulate metabolism in the body (8). Insulin resistance occurs when there are deficits in one or more of these sites of action (9).

THE INSULIN RECEPTOR

In 1970, House and Weidemann demonstrated that insulin was binding to cell membranes in the liver of rats (10). The insulin receptor (IR) was shown to be a dimer that had two alpha and two beta subunits that are connected by a disulfide bond (11). In 1982 the IR was identified as a receptor tyrosine kinase. When insulin binds to the receptor, it activates its beta subunit (12, 13). Insulin receptor substrate (IRS)-1 was the first intra-cellular substrate of the IR that was discovered in 1985 (14). Also in 1985, two groups independently determined the cDNA sequence for the human IR. There are two isoforms of the receptor, IR-A and IR-B, that contrast by 12 residues (15, 16). Downstream of the IR is the signaling pathway that controls metabolism in the body, which includes protein kinase B, mitogen-activated protein kinase (MAPK), phosphoinositide 3-kinase, Ras, Phosphatase and tensin homolog, protein tyrosine phosphatases, and many other proteins (17).

When insulin binds to the alpha subunit of the IR extracellularly, this enables ATP to bind to the beta subunit of the IR, a signaling cascade that appears to be relevant to the action of insulin in the central nervous system (CNS) as well. This causes the phosphorylation of the beta subunit, which in turn phosphorylates the downstream IRS (2). IRS is then important for binding to downstream signaling molecules. There are four different IRS proteins, each having distinct functions in the cellular cascade of insulin. IRS 1 mediates glucose detection and insulin secretion from these cells. IRS 2 is primarily responsible for pancreatic beta cell growth. Although IRS 1 and IRS 2 are found in many of the same tissues, IRS 1 is thought to be the main IRS in skeletal muscles, while IRS 2 is the primary IRS of the liver (18). Cells in the liver, adipose tissue, and pancreatic beta cells are the only ones who express IRS 3, while cells in the kidney, thymus, and brain express IRS 4 (19, 20). Downstream, phosphatidylinositol 3-kinase (PI 3-kinase) is one of the most important enzymes that the IRS binds to. PI 3-kinase acts through Akt/protein kinase B, protein kinase C and PI-dependent protein kinases 1 and 2, which are serine or threonine kinases. This action results in most of insulin's metabolic effects such as glucose uptake and lipid, glycogen, and protein synthesis. The cell growth and mitogenic effects of insulin are mediated by a different pathway involving the protein Ras (19).

TARGETED DELETION OF INSULIN RECEPTORS IN THE CNS VERSUS PERIPHERAL SITES

Mice with IRs deleted in the whole brain have shown abnormal behavior including increased food intake and obesity (21). Insulin infusions in healthy men also result in a decreased appetite (22). Those findings demonstrate that the brain is an insulin-sensitive organ (23). When IRs are activated, the phosphorylation of the subunit of the IR modifies the cell function through distinct secondary messengers of different signaling pathways (24). The functional changes include glucose uptake, glucose synthesis, gluconeogenesis, cell growth, differentiation, and protein synthesis (25).

The knockout of IRs using the Cre/loxP recombination system in various tissues in mouse models has been an important tool for understanding the role of insulin. This is done by using a targeting vector with the mouse IR gene with a selection cassette surrounded by loxP sites upstream of exon 4 and another loxP site downstream of exon 4. Embryonic stem cells are then transfected with this targeting vector and then with a plasmid containing Cre cDNA, which results in the removal of the selection cassette. Mouse blastocytes are subsequently injected with these clones, and these mice are bred with C57BL/6J mice. When the offspring of this breeding is injected with the Cre recombinase, exon 4 of the IRlox allele is deleted, resulting in an IR-specific knockout (26).

Mice born with ubiquitous IR knockout do not appear to have any metabolic problems at first (27). However, this quickly changes, as insulin levels increase 1,000 times more than normal after feeding, and glucose levels rise at supraphysiological levels. Within a few days after birth, beta-cell failure occurs when insulin storage granules in the cytoplasms of the beta cells collapse. Eventually, the mice die from diabetic ketoacidosis (28).

IR knockout in different tissues is associated with several other phenotypes. First, the knockout of IRs in skeletal muscle leads to decreased insulin signaling and glucose transport but does not induce insulin resistance (29). Mice with IR knockout in both white and brown adipose tissue exhibit decreased fat pad mass and triglyceride content. These mice also do not develop glucose intolerance when overfed and do not gain weight during aging or when hyperphagia is induced (30). Mice with liver-specific IR knockout demonstrate glucose intolerance, insulin resistance and hyperinsulinemia (31).

INSULIN IN THE CENTRAL NERVOUS SYSTEM

The central nervous system (CNS) used to be thought of as an insulin-independent system. This changed in the late 1970s when Havrankova et al. became the first to detect insulin in the rat brain using radioimmunoassays (32). Furthermore, they discovered that the levels of insulin in the CNS were independent of the levels of peripheral insulin (33).

There were two main hypotheses for the origin of insulin in the brain. First, Margolis and Altzuler postulated that insulin could cross the blood-brain barrier. They were able to demonstrate that cerebrospinal fluid insulin levels increased after peripheral injections of insulin (34). Similar results were also seen in a study with dogs (35). The transport of insulin from the periphery to the brain has also been shown to be regulated by multiple factors. This includes obesity, in which dogs fed with a high-fat diet exhibited reduced insulin transport into the brain (36). Insulin transport into the CNS was also inhibited by dexamethasone, suggesting that glucocorticoids may also regulate this process (37).

The second hypothesis suggests that insulin is produced locally in the CNS, as opposed to being transported across the blood-brain barrier. High concentrations of insulin and C-peptide, which are released by the formation of insulin, were identified in the brains of human cadavers so insulin could, in part, be synthesized in the brain (38). Second, insulin mRNA was found in the rat paraventricular nucleus of the hypothalamus through in situ hybridization (39) and in the olfactory bulbs, dentate gyrus, and hippocampus of neonatal rabbits (40). Rat and mouse CNS neuronal cultures feature insulin-like immunoreactivity (41, 42). Further immunohistochemistry studies demonstrated that fetal rabbit brain cell cultures secrete insulin-like mRNA and an insulin-like substance that could not be distinguished from real insulin (42).

BRAIN INSULIN RECEPTORS AND INSULIN SIGNALING

The discovery of the IRs in the brain further demonstrated that the CNS is indeed an insulin-sensitive organ. IRs are found throughout the brain, but in the cerebellum, hypothalamus and cortex, expression of IRs is higher (43). In addition, neurons and neuroglial cells both express IRs. Neurons express the IR-A isoform of the alpha subunit, whereas glial cells express IR-B. The IR-B isoform includes exon 11, while the IR-A isoform does not (44, 45).

Insulin signaling in the brain is largely similar to insulin signaling in the periphery. The main difference seems to involve the downstream glucose transporters. In the periphery, the insulin signaling pathway results in the translocation of glucose transporter 4 (GLUT4) to the membrane in order to allow for glucose uptake from the bloodstream (46). In the brain, GLUT4 is not widely expressed and is primarily found in regions such as the hypothalamus and hippocampus (47, 48). Glucose uptake in the brain occurs through GLUT1, which is insulin-insensitive, in astrocytes and endothelial cells in the blood-brain barrier (49, 50).

Neuronal-specific IR knockouts have also helped us develop a better understanding of the role of insulin in the brain. In a seminal paper, Bruning et al. demonstrated that mice with IR knockout in neurons had increased food intake, developed obesity, increased body fat, increased leptin levels, and insulin resistance. These mice, however, did not have any impairment in neuronal development and survival (51). Furthermore, these mice had decreased adrenaline and norepinephrine responses to hypoglycemia, suggesting that brain insulin is important for responding to hypoglycemia (21).

LEPTIN

Leptin is a hormone that is important for regulating metabolism and inhibiting food intake while it significantly impacts central pathways coding for food and drug reward. Mutations in leptin and in the leptin receptor cause hyperphagia and obesity (52, 53). When leptin receptors expressed on POMC neurons were knocked out in mice, the mice were obese and had an abnormal expression of neuropeptides within the hypothalamus (54). In rodents with mutated leptin receptors, reactivation of these leptin receptors in the arcuate resulted in smaller improvements in body weight, but significant improvements in the level of blood glucose and hyperinsulinemia (55–57). At the same time, leptin receptors are expressed on VTA dopamine cell bodies and leptin-deficient mice exhibit not only obesity but also significant deficits in central dopamine exocytosis (58). All of this evidence suggests that hypothalamic arcuate POMC neurons are essential in the pathway for regulating glucose homeostasis via leptin, while the role of leptin and insulin in modulating reward pathways appears to be synergistic at the very least.

GHRELIN

Ghrelin is often known as the "hunger hormone" and is secreted from the stomach (59). The growth hormone secretagogue receptor (GHSR), the receptor for ghrelin, is found in the hypothalamus as well (60, 61). Mice with a GHSR knockout fed with a high-fat diet store fewer calories and have lower body weight and adipose tissue compared to control mice. These mice also demonstrate improved glucose homeostasis. The site of action by which ghrelin increases food intake and adipose tissue appears to be the hypothalamus (62). Indeed, it has been hypothesized that the effects of ghrelin may oppose and regulate the effects of leptin and insulin in the hypothalamus (61). There is additional evidence that ghrelin, like leptin, not only impacts on homeostatic processing of energy balance but also on hedonic properties of food intake and underlying neurotransmission (62–65) so much so that it plays a significant role in the neurochemical 'signatures" of obesity and eating disorders. Although the synergistic or additive role of central insulin to ghrelin effects on food reward has not been adequately investigated, one can only assume that such interaction is at play and likely constitutes a significant part of the neurochemical phenotype of disorders that impact both metabolism and hedonic/compulsive food intake.

INSULIN, DOPAMINE, FEEDING BEHAVIOR AND THE HYPOTHALAMUS

Levels of insulin have been shown to rise in anticipation of feeding, and this effect may be due to the animals learning which environmental cues predict feeding (66, 67). Furthermore, injections of insulin into the cerebral ventricles of baboons decrease food intake and body weight (68). Insulin applied intra-nasally in humans has also been shown to decrease food intake (22). Central insulin decreased endogenous glucose production in mice and humans, as well (69, 70). PI 3-kinase is activated downstream of IRs in the periphery. When PI 3-kinase inhibitors are injected into the ventricles, then insulin no longer decreases food intake or glucose production (71, 72), demonstrating that PI 3-kinase is indeed involved in the central insulin signaling pathway as well.

Beta-cells in the pancreas produce insulin in order to send information regarding peripheral fat to the hypothalamus (73–75), and the hypothalamus is known to be an important structure for controlling energy homeostasis (76). Brain response to intranasal insulin was decreased in the hypothalamus in patients with high visceral adipose tissue compared to those with lower visceral adipose tissue, as shown in one magnetic resonance imaging study (77). Furthermore, IRs are indeed expressed in the hypothalamus (78). Decreasing the number of IRs in the hypothalamus caused insulin resistance and hyperphagia (an unusually strong sense of hunger) in rats (79). Within the hypothalamus, insulin and leptin act on the arcuate nucleus (ARC) to modulate feeding behavior (60). The effects of insulin on feeding behavior in the hypothalamus involve the PI 3-kinase signaling pathway, specifically in arcuate POMC neurons (6, 80). Using Phosphatase and tensin homolog (PTEN), a negative regulator of PI 3-kinase, it has been shown that phosphatidylinositol (3,4,5) triphosphate (PIP3), which is produced by PI 3-kinase, in the hypothalamus of rats is also important for modulating glucose production and food intake (79). Specifically, insulin is involved in reducing food intake through the regulation of POMC and agouti-related peptide (AgRP) neurons in the hypothalamus (81). Interestingly, the knockout of IRs in POMC neurons does not affect glucose homeostasis. Instead, the loss of IRs in POMC neurons causes insulin resistance and diabetes when paired with the deletion of leptin receptors as well (82). Injections of dopamine bilaterally into the hypothalamus of obese Zucker rats were also able to reduce food intake, therefore indicating that insulin and dopamine may interact in the hypothalamus to regulate feeding behavior (83).

CENTRAL INSULIN AND MOOD DISORDERS

Diabetes is a metabolic disease that is often characterized by insulin resistance, in which the body does not produce or does not respond to insulin. Epidemiological studies have demonstrated a prevalence rate of depression that is three times higher for patients with type 1 diabetes and nearly two times higher for patients with type 2 diabetes compared to those without diabetes (84). Even in previously diabetic and prediabetic patients, there are elevated rates of depression (85). One meta-analysis revealed that those with depression have a 60% increased risk of getting type 2 diabetes (86). There appears to be an association between diabetes and anxiety as well, as 14% of patients with diabetes report having generalized anxiety disorder (GAD) and 40% report having elevated anxiety symptoms (87). With regards to the mechanisms linking diabetes and mood disorders, insulin resistance in the brain is a common irregularity between diabetes and depression (88) and has emerged as a potential target for treatments for anhedonic symptoms in depression (89).

The relationship between diabetes and mood behaviors can arise as early as childhood. Children with the greatest insulin resistance demonstrated increased depressive symptoms, as well as reduced volumes in the anterior cingulate and hippocampus (90). One longitudinal study also found that those with depressive symptoms at baseline had greater insulin resistance when measured 6 years later (91).

Inflammation may be one of the mediators between depression and diabetes. Firstly, inflammation has been linked to insulin resistance (92). Likewise, depression has also been associated with inflammation in both the peripheral and central nervous systems (93). Therefore, a potential shared mechanism between the two diseases is the inflammatory response (86, 94). Another possible mechanism mediating the relationship between depression and diabetes is stress. Psychological stress is associated with increased depressive-like symptoms (95) as well as the development of insulin resistance (96). One common contributor to both inflammation and the insulin signaling pathway is macrophages. Specifically, macrophages secrete tumor necrosis factor (TNF) alpha, which is involved in insulin signaling (97). Similarly, adipose tissue also secretes TNF alpha in obesity, which can contribute to insulin resistance and depression (98). Mice treated with infliximab, a TNF alpha antibody, were protected against the development of depressive-like behaviors (99). One study using bupropion, an antidepressant medication, showed that when given to rats, bupropion reduced TNF alpha and insulin resistance (100). Pro-inflammatory cytokines, which are also activated by macrophages, are indeed also involved in the development of depression (101).

Research with diabetic or obese mouse models also suggests a linkage between depression and diabetes. Mice fed with a high-fat diet and *db/db* mice (a mouse model of type 2 diabetes lacking leptin receptors) demonstrated increased depressive- and anxiety-like behaviors through several behavioral paradigms, including the forced swimming test, elevated plus maze test, and open field test (102–105). Some of the most used behavioral paradigms to assess depressive- and anxiety-like behaviors, as well as reward-associated behaviors, are depicted in Table 7.1. High-fat diet-induced obesity also resulted in an increased stress response, with higher corticosterone levels following acute restraint stress. Furthermore, mice fed with a high-fat diet demonstrated changes in the neural reward circuitry that is associated with depressive-like behavior (103). Treatments that have traditionally been used for diabetes, pioglitazone and rosiglitazone, have been shown to reverse the effects of diabetes or high-fat diet-induced depression (106, 107).

Though the direction of the relationship between depression and diabetes is not clear, one potential mechanism that could explain this association is the reward system. One popular theory on the etiology of depression is the monoamine deficiency hypothesis, which suggests that decreased monoamine neurotransmitters (serotonin, dopamine, and norepinephrine) can contribute to depressive disorders (108). The literature on the relationship between dopamine and depression is especially strong, with several studies demonstrating that dopamine dysregulation can contribute to depressive disorders and behaviors (109), specifically dopamine neurons in the ventral tegmental area (VTA) (110). The habenula (Hb) has also emerged as an important regulator of the dopaminergic neurons in the VTA (111, 112). In addition, the Hb has a high density of IRs (113); and studies in humans demonstrate Hb impairments in people with depression (114).

Knockouts of IRs in various areas of the brain also result in increased depressive- and anxiety-like behavior through dopamine signaling. Neuron-specific IR knockout induced depressive- and anxiety-like behaviors in mice. Furthermore, these mice showed decreased dopamine signaling in the dorsal striatum and NAcc compared to controls. IR knockout was associated with increased levels of Monoamine Oxidase (MAO) A and B, which are important for the degradation of monoamines (78). Astrocyte-specific IR knockout also resulted in increased depressive- and anxiety-like behavior in mice. This effect was due to decreased ATP exocytosis, which in turn decreased purinergic signaling on dopaminergic neurons in the nucleus accumbens, medial prefrontal cortex, and dorsal caudate putamen (115). Similarly, the lack of IRs and the related insulin-like growth factor-1 receptor (IGF1R) in the hippocampus and amygdala resulted in increased anxiety-like behavior (116).

CENTRAL INSULIN AS A DRIVER OF REWARD-ASSOCIATED BEHAVIOR

The Reward System

Reward-associated behavior was first studied by Olds J. and Milner P. in the early 1950s (117). Through electrical stimulation, rodents tend to seek strong and repeated activation of specific brain sites including the fibers of passage in the medial forebrain bundle. The intensity and repeating pattern of stimulation is coding for "pleasure". Further research since then has shown that reward-associated behavior is relevant not only to one type of emotion but also a series of reward-related cognitive processes, including associative learning, incentive salience, and positively valenced emotions (118, 119). Thus, reward-associated behavior can be defined as the enhancement of behavior by positive stimulations that are related to the difference between gain and expectation (120).

The neurobiology of reward behavior is complicated since different brain regions are involved. The basal ganglia, in particular, seem to play an important role in the reward circuitry (120).

The basal ganglia can be separated into the caudate nucleus, the putamen, and the globus pallidus. The substantia nigra and ventral tegmental area are in the core of the midbrain and along with the subthalamic nucleus in the ventral thalamus have very close functional relationships with the basal ganglia (121). The functions of the basal ganglia can be summarized into two major parts: (i) regulation of movement by modulating motor systems and (ii) formation of emotions.

The functional units in the basal ganglia can be separated into 'loops' (Figure 7.1). The primary loops in the basal ganglia include the body movement loop, oculomotor loop, prefrontal loop, and limbic loop. The body movement loop and oculomotor loop are crucial to controlling the

TABLE 7.1
Rodent Behavioral Tests to Assess Mood- and Stress Disorder-Like Behaviors, as well as Reward-Seeking Behaviors, and Distinguish Them from Malaise

Test	Function	Description
Forced swimming	Evaluate anxiety- and depression-like behaviors	Rodent is placed in a water tank and forced to swim. Time spent moving and attempting to escape onto a rescue platform and time spent immobile are measured.
Elevated plus maze	Anxiety-related behaviors	Rodents are placed in the center of a plus sign-shaped maze. Two of the arms of the maze are enclosed, and two are open. The amount of time the rodent spends in the open versus enclosed areas of the maze is measured.
Open field	Anxiety-related behaviors	Rodents are placed in a large box (open field) and allowed to move freely around the box. The amount of time the rodent spends in the middle area of the open field versus the sides of the box is measured.
Tail suspension	Despair, depressive-like behaviors	Test measures the length of time it takes for the animal to stop resisting suspension.
Overcrowding	Anxiety-related behaviors	Overcrowding the home cage induces a form of chronic stress in rodents.
No bedding, wet bedding, cage tilting	Anxiety-related behaviors	Lack of bedding or damp bedding induces a form of chronic stress in rodents.
Light cycle shifts	Anxiety-related behaviors	Abrupt and unpredictable changes in light/dark cycle induce chronic stress in rodents.
Variable stressors (thermal, noise, light etc.)	Anxiety-related behaviors	Random exposure to different stressors induces stress in rodents.
Restraint, immobilization	Anxiety-related behaviors	Use of restraint devices for rodents is usually employed for induction of acute stress.
Social defeat or predator stress	Anxiety- and depression-like behaviors, despair	Controlled but repeated exposure without physical contact or harm to an aggressor animal or its scent is inducing chronic stress in rodents and eventually leads to lack of resisting behaviors.
Learned helplessness	Anxiety- and depression-like behaviors, despair	Mild, inescapable aversive stimuli delivered repeatedly to a rodent will induce this form of depression-like behavior.
Grooming	Flat mood, apathy, malaise	Time spent grooming is measured after animal is sprayed with water mist. Grooming is also used to evaluate compulsive behavior in rodents.
Fur coat state	Flat or negative mood, apathy, malaise	Overall condition of fur is evaluated under baseline conditions as indicative of normal grooming.
Nest building	Flat or negative mood, apathy, malaise	Nest-building activity is standard measure of emotional and physical health in rodents.
Feeding changes	Anxiety- and depression-like behaviors, malaise	Feeding suppression is indicative of mood disorders and/or malaise.
Sucrose preference test	Anhedonia, reward-related deficits	Healthy rodents consistently and strongly prefer sweet solutions over plain water.
LiCl taste test	Malaise	The lithium chloride taste aversion test is among the most reliable behavioral tests for demonstrating a state of malaise in rodents through failure to exhibit aversion.

(Continued)

TABLE 7.1 (*CONTINUED*)

Rodent Behavioral Tests to Assess Mood- and Stress Disorder-Like Behaviors, as well as Reward-Seeking Behaviors, and Distinguish Them from Malaise

Test	Function	Description
Electrical self-stimulation	Reward-associated behavior	Stimulating voluntarily through bar pressing the medial forebrain bundle fibers of passage in the CNS with an implanted electrode induces reward. To be distinguished from deep brain stimulation, which is used therapeutically for PD.
Conditioned place preference	Reward-associated behavior	The rodent learns to increase time spent in a compartment where they have previously been trained to receive a rewarding stimulus versus a neutral compartment.
Self-administration of drugs of misuse (typically intravenous)	Reward-associated behavior	Through instrumental learning and different schedules of reinforcement, this test precisely evaluates the capacity of the animal to acquire, maintain, extinguish and relapse to self-administration of a drug of misuse, typically through bar pressing in a Skinner box or equivalent.
Vertical, horizontal locomotion	Reward-associated behavior	Locomotor stereotypy characterizes some drugs of misuse (e.g. d-amphetamine) and is used as evidence of pharmacologically relevant dose regimen.

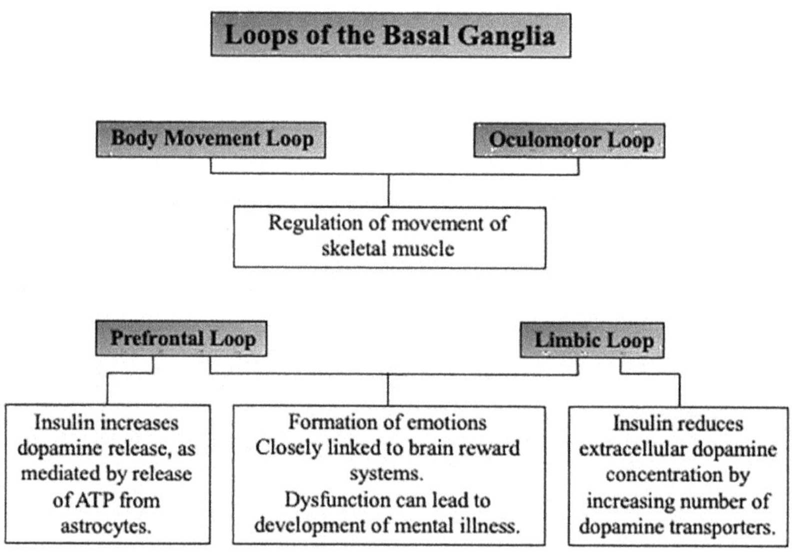

Figure 7.1 Functional units of the basal ganglia loops linked to reward and most impactful mechanisms of dopamine modulation by central insulin.

movement of the skeletal muscle, while the prefrontal loop and limbic loop contribute to the formation of memory and emotion (122). In general, each loop in the basal ganglia can be divided into 4 parts: the cortex input, the striatum, the pallidum, and the thalamus or the cortex output (123). The limbic loop is crucial to forming emotions, and the dysfunction of the limbic loop is linked to many mental disorders, while prefrontal loops may relate to cognitive processes. Both loops are closely associated with the reward system.

The cortical input of the limbic loop is from the amygdala, hippocampus, orbitofrontal cortex, anterior cingulate cortex, temporal cortex, projecting to the ventral striatum, then ventral pallidum and the mediodorsal nucleus of the thalamus. The cortical input of the prefrontal loop consists of the pathway from the dorsolateral prefrontal cortex to the striatum; anterior caudate to pallidum; globus pallidus to thalamus; mediodorsal and ventral anterior nuclei back to the prefrontal cortex. Dysfunction of the prefrontal loops is associated with mental disorders. The indirect pathway in basal ganglia is considered to be an important part of the functional loops. It involves regulation sites projecting to the striatum with an inhibitory signal, then the striatum projecting to the external segment of pallidum with the inhibitory signal, then the to the subthalamic nucleus. The subthalamic nucleus projects to the internal segment of the pallidum (121). Interestingly, the inhibitory signal is processed in the striatum via dopaminergic neurons, but the relevant dopamine receptor for inhibitory signals at the postsynaptic neurons is the D2 receptor rather than the D1 (123). Thus, the increase in dopamine release in the ventral and dorsal striatum may have a different effect on the downstream signaling pathway. At the same time, IRs are distributed in different regions of these functional loops likely affecting signaling pathways in a specific manner depending on which neurotransmitter and neurotransmitter receptors are involved (23).

The function of the distinct loops in basal ganglia focuses on the modulation of hyperkinetic cortical activation. The associated cortex is stimulated to form a wanting and learning process, while the pallidum sends an inhibitory signal to the cortex to inhibit over-activation. The striatum projects to the pallidum and sends an inhibitory signal to cancel the inhibition of the pallidum to the cortex. In other words, the function of the striatum is a 'disinhibition' of the forming of 'wanting', 'learning'. The striatum is positively regulated by cortical input from the forebrain and other structures in the midbrain (123). The ventral tegmental area (VTA) regulates the limbic loop by projecting to the ventral striatum and medial prefrontal cortex (mesoaccumbens and mesocortical pathways), and the substantial nigra pars compacta (SNpc) partly regulates all basal ganglia loops by projecting to the dorsal striatum (nigrostriatal pathway) (124).

Overall, at least three neurotransmitters have been identified to be important in the brain reward system: dopamine, glutamate and GABA.

Dopamine

A key neurotransmitter of the basal ganglia loops is dopamine. The midbrain contains approximately 80% of all dopamine neurons in the CNS in a very compact area (Figure 7.2) projecting forward in distinct pathways. This is unlike other neurotransmitters, such as norepinephrine and serotonin, that diffusely project from the hindbrain throughout the cerebrum and cerebellum. Neurons from SNpc and VTA (Figure 7.2) to the ventral and dorsal striatum are dopaminergic and the dopamine receptor

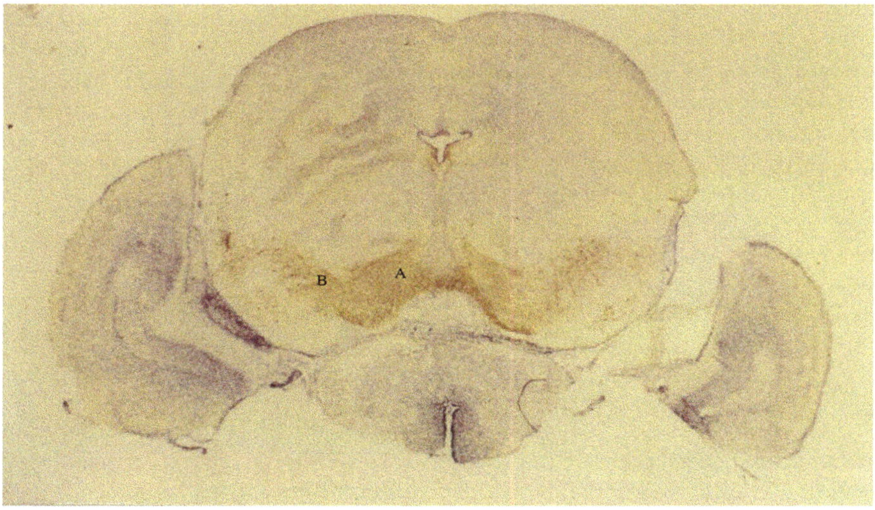

Figure 7.2 Tyrosine hydroxylase immunostaining of dopamine cell bodies in the rodent midbrain (coronal section). (A) Ventral Tegmental Area, origin of the mesoaccumbens and mesocortical projections. (B) Substantia Nigra Pars Compacta, origin of the nigrostriatal projection. The anatomy of dopaminergic systems is substantially preserved in the CNS of higher mammals like primates and humans.

at the postsynaptic neurons is primarily the D1-type receptor. Dopaminergic terminals are also present in the prefrontal cortex, and dysregulation of the dopaminergic neurons was established in patients with depression and anxiety (125, 126).

In the reward circuitry, the mesoaccumbens projection from the VTA to the nucleus accumbens (NAcc), the mesocortical projection from the VTA to the medial prefrontal cortex (and, in part, the nigrostriatal projection from SNpc to dorsal striatum) have positive regulation functions. Therefore, loss of dopamine neurons in these pathways will not necessarily cause loss of the normal desiring function (127). The dopamine function in this process is 'liking' rather than 'wanting'. A sample equation can clearly explain the function of dopamine: Dopamine response = Reward occurred − Reward predicted.

It is worth mentioning that the dysfunction of the reward system mediated by dopamine is closely associated with substance use disorders and obesity (128, 129), while central dopaminergic neurotransmission in the NAcc, prefrontal cortex, and striatum is sensitive to metabolic and energy balance stages (130–133), as well as changes in insulin signaling (134). Continuous firing of the dopaminergic neuron causes the downregulation of the expression of the D1 dopamine receptor in the postsynaptic neuron, and the dopamine release in baseline conditions will also be downregulated to preserve homeostasis (120). Further, insulin binds directly to IRs expressed on dopamine neurons (135, 136) and regulates dopamine signaling, reward processing, and reward behavior by increasing dopamine transporter (DAT) density and function in the striatum (78, 137, 138). Insulin also interacts with D2-like dopamine receptors to affect dopamine-linked reward behaviors in humans and animals (139, 140), significantly modifies brain reward thresholds (141), and attenuates preference for food intake, and hedonic food intake under satiety (137, 142).

Glutamate

The glutamate system primarily projects from neuronal cell bodies in the prefrontal cortex to the hypothalamus, VTA, amygdala, and NAcc following activation of their dopamine receptors by the mesocortical pathway. The function of the glutamate neuron has been associated with the formation of memories through long-term potentiation and synaptic facilitation at the N-methyl-D-aspartate (NMDA) receptor. The interaction between the NMDA receptor and dopamine D1 receptor may influence receptor internalization and relate to the neurochemistry of substance use disorders (143).

GABA

Gabaergic neurons in the reward system exist mainly in the form of medium spiny (inter)neurons in the nucleus accumbens – ventral pallidum – VTA - NAcc –hippocampus – VTA pathways. Gabaergic neurons serve as inhibitory of the target area, and they are very important in forming the disinhibition loops (144).

THE MESOLIMBIC DOPAMINE SYSTEM AND CENTRAL INSULIN

Dopamine neurons in the VTA and SNpc express IRs (135), as well as the insulin receptor substrate-2 (IRS-2) and phosphatidylinositol (3,4,5)-trisphosphate (PIP3) (145). In addition, insulin can induce long-term depression (LTD) on excitatory synapses on dopamine neurons, even in very low concentrations, and this effect is mediated by endocannabinoids (146). Insulin action in the VTA has been shown to have important effects on dopamine transporters as well as dopamine release. Insulin injected intraventricularly increased mRNA of the dopamine transporter (DAT) in the VTA and SNpc (147). Insulin applied to the VTA reduced dopamine concentrations, and this effect was reversed in the presence of a DAT inhibitor. Therefore, insulin may have reduced the dopamine extracellular concentration by increasing the number of DATs (137).

NUCLEUS ACCUMBENS

IRs are expressed in the NAcc on neurons (148) and on glial cells (115). Knockout of IR on astrocytes in the NAcc decreased dopamine release (115). Insulin applied to striatal slices increased dopamine release in the NAcc (149). The effect of insulin on dopamine release was apparently mediated by exocytosis of ATP from astrocytes (115). Other studies have reported cholinergic interneurons as the mediator (149).

MECHANISMS OF INSULIN REGULATION OF DOPAMINE RELEASE

As a peptide, it is difficult for insulin to cross the blood-brain barrier. The illustration of the whole process has not been clearly established. One potential mechanism is that circulating insulin interacts with the IRs on the luminal membrane of the microvascular brain endothelial cells following an endocytotic process. It is released there and can permeate the Virchow-Robin space, which at the capillary level is filled with the basement membrane material. Once across the endothelium it may interact with pericytes, astrocytes, or enter the blood-cerebrospinal fluid (BCSF) barrier and act directly on neurons (150).

Three mechanisms appear to play an important role in insulin-regulating dopamine release.

First, insulin modulates dopamine release through the vesicular nucleotide transporter in astrocytes. Astrocytes potentially regulate dopamine release by modulating ATP release. When active, the ATP receptor in the presynaptic neuron will open Ca^{2+} channels and increase the release of dopamine. In astrocytes, ATP is released by the formation of N -ethylmaleimide-sensitive factor attachment protein receptors (SNAREs) (151). Munc-18 is an inhibitory protein that binds to the SNAREs to prevent their formation. The tyrosine phosphorylation of Munc-18 will increases the formation of SNAREs (115, 152). Insulin signaling has been confirmed to catalyze the tyrosine phosphorylation of Munc-18. Loss of insulin will cause a decrease in SNAREs and a decrease in ATP release, followed by a decrease in dopamine release (115).

Second, insulin modulates dopamine release through MAO A and MAO B in the neurons and glial cells. The downregulation of MAO-A and MAO-B was confirmed to be directly induced by insulin stimulation (78). Furthermore, the dopamine reuptake transporter is also attenuated by insulin.

Third, studies have also shown that the morphology of the mitochondria is impacted by insulin signaling. The change can induce oxidative stress in neurons. Dopamine reuptake in neurons can therefore be potentially inhibited as a consequence of the metabolic changes in neurons (78).

The possible role of central insulin and IRs in the release of other neurotransmitters and endogenous signals like endogenous opioids and cannabinoids still needs to be investigated in-depth to better understand the impact of brain insulin signaling on reward systems.

INSULIN AND SUBSTANCE USE DISORDERS

Given what is known about the link between insulin and dopamine release, it is not surprising that insulin deficiencies are often linked to addictive disorders. Substance use disorders involve the compulsive use of a misused substance, such as alcohol, despite significant adverse consequences, physical, psychological, family, professional, and otherwise, for the patient. Those who misuse substances have an increased risk of developing diabetes (153). The prevalence of substance use disorders in adults with diabetes is 17%, compared to just 8% in adults without diabetes (154). In diabetics, the use of illicit substances can increase disease complications and exacerbate symptoms. One study found that patients with diabetes who were also addicted to opioids experienced increased serum glucose levels and decreased high-density lipoprotein cholesterol, which further contributed to the metabolic syndrome (155). The development of the metabolic syndrome in patients who misuse substances may be due to increased damage of cells caused by nutritional deficiency (156). Recently, intranasal insulin has emerged as a potential treatment for substance use disorders (157).

EATING DISORDERS

Several investigators and clinical scientists have argued that eating disorders, such as anorexia nervosa, bulimia nervosa, or binge eating disorder, are part of the addictive disorder spectrum (158–161). Studies have also demonstrated increased insulin sensitivity in individuals with anorexia nervosa (162). On the other hand, bulimia nervosa and binge eating disorder were associated with decreased insulin sensitivity (163). In patients with eating disorders, the insulin response to food was diminished as well (162). Anorexia nervosa and bulimia nervosa are also common complications seen in patients with Type I diabetes (164). Furthermore, women with type I diabetes may have up to 2.4 times an increased risk of developing an eating disorder compared to women of the same age without diabetes (165). These results point to insulin as an important modulator in the development of eating disorders and may be an important clinical target for the treatment of eating disorders. Nonetheless, further research on the mechanisms in which insulin is involved in eating disorders is still necessary.

INSULIN AND PARKINSON'S DISEASE

It is often overlooked that Parkinson's disease (PD) is a neurodegenerative disease, in which patients have not only severe motor defects, such as tremors and rigidity of movement but also significant anhedonia. The loss of dopamine neurons in the midbrain, specifically the SNpc, is one of the main neurological markers of the disease (166). Recently, there has been a growing amount of evidence linking PD and diabetes. Studies have suggested that patients with diabetes have an increased risk of developing PD (167–170). Even in PD patients who have not been diagnosed with diabetes, there is a high percentage (58.4%) of patients who have insulin resistance (171). PD patients with comorbid diabetes also appear to have more severe PD symptoms, including postural instability and gait rigidity (172). Furthermore, in one study, insulin resistance was present in 62% of patients with PD with dementia, suggesting an association between insulin resistance and dementia (173). Impaired glucose metabolism has also been seen in the early stages of sporadic PD (174).

One of the biggest risk factors for PD is age, with about 1% of the population over the age of 60 having PD (175). mRNA levels of the IR in the brain also decrease with age (176). Specifically, in the substantia nigra, patients with PD display significantly lower levels of IR mRNA compared to normal controls (177). In one immunohistochemistry study, anti-IR antibody did not stain the neurons of the SNpc in the brains of patients who had PD, indicating that the loss of IRs is implicated in the development of PD (178). IGF-1 has also been implicated in PD. The level of IGF-1 in the serum of patients with PD was higher compared to healthy controls, and this increased level of IGF-1 was correlated with impaired motor function in the patients with PD (179). Levels of IGF-1 were also higher in the cerebrospinal fluid (CSF) of patients with PD compared to normal controls (180). Such an array of findings in PD patients invites the question of whether PD-related anhedonia is causatively linked to insulin and IR deficiencies and points to the need of investigating this issue further in animal models of PD and in human patients, likely with imaging technology when patients engage in reward-oriented tasks.

CONCLUDING REMARKS

It is perhaps too early to claim a sufficient understanding of the several roles of central insulin in physiology and disease. Nevertheless, in this chapter, we have attempted to focus on the central interactions of insulin and its receptors with dopaminergic neurotransmission and mesolimbic pathways, a system that is apparently coding for the reward value and reward prediction of stimuli like food and drugs of misuse. And the possible impact of such interactions on the phenotype of disorders with signature deficits of dopaminergic neurotransmission like addictive disorders, mood disorders, and Parkinson's disease. The direct impact of insulin on major regulators of dopamine exocytosis, like the dopamine transporter, as well as the modulation of dopaminergic signaling by cellular and molecular mechanisms that reside with IRs in both neurons and astrocytes provide a good first insight into how central insulin functions and on possible new therapeutic approaches for these disorders that target central IRs.

Interestingly, direct evidence for a possible therapeutic role of insulin in substance use disorders is still scarce but some initial work points to possibly significant benefits of intranasal insulin treatment as stated above while the majority of studies in this area so far focus more on the perspective of food addiction and eating disorders. The recent increase in federal funding through the HEAL initiative in the United States with emphasis, among other things, on possible alternative (non-opioid) treatments for opioid use disorder may further help illustrate the role of signals like insulin in the treatment of addiction. Furthermore, there is increasing evidence that appetite-regulating peptides (e.g. GLP-1, ghrelin) play a role in addictive disorders and insulin is likely to confirm such possibilities rather than constitute an exception. The lack of appropriate pharmacological interventions for most substance use disorders makes it imperative to actively investigate more in-depth the impact of insulin on this field.

ACKNOWLEDGEMENTS

Part of the original research discussed in this chapter was funded through RO1 DK065872 (Pothos), a Pilot and Feasibility Award through Joslin Diabetes Center and Boston University (Pothos) and the Graduate Programs in Pharmacology and Experimental Therapeutics and Pharmacology and Drug Development, Graduate School of Biomedical Sciences, Tufts University, Boston, MA, United States.

REFERENCES

1. L. Rosenfeld, Insulin: discovery and controversy. *Clin Chem* **48**, 2270–2288 (2002).
2. G. Wilcox, Insulin and insulin resistance. *Clin Biochem Rev* **26**, 19–39 (2005).
3. D. Schroder, H. Zuhlke, Gene technology, characterization of insulin gene and the relationship to diabetes research. *Endokrinologie* **79**, 197–209 (1982).
4. G. Dodson, D. Steiner, The role of assembly in insulin's biosynthesis. *Curr Opin Struct Biol* **8**, 189–194 (1998).
5. T. K. Bratanova-Tochkova et al., Triggering and augmentation mechanisms, granule pools, and biphasic insulin secretion. *Diabetes* **51 Suppl 1**, S83–90 (2002).
6. K. W. Williams, M. M. Scott, J. K. Elmquist, Modulation of the central melanocortin system by leptin, insulin, and serotonin: co-ordinated actions in a dispersed neuronal network. *Eur J Pharmacol* **660**, 2–12 (2011).
7. J. Rodin, J. Wack, E. Ferrannini, R. A. DeFronzo, Effect of insulin and glucose on feeding behavior. *Metabolism* **34**, 826–831 (1985).
8. V. Baron, E. Van Obberghen, Mechanism of insulin action. *C R Seances Soc Biol Fil* **189**, 25–41 (1995).
9. T. M. Batista, N. Haider, C. R. Kahn, Defining the underlying defect in insulin action in type 2 diabetes. *Diabetologia* **64**, 994–1006 (2021).
10. P. D. House, M. J. Weidemann, Characterization of an [125 I]-insulin binding plasma membrane fraction from rat liver. *Biochem Biophys Res Commun* **41**, 541–548 (1970).
11. J. Massague, P. F. Pilch, M. P. Czech, Electrophoretic resolution of three major insulin receptor structures with unique subunit stoichiometries. *Proc Natl Acad Sci U S A* **77**, 7137–7141 (1980).
12. M. Kasuga, F. A. Karlsson, C. R. Kahn, Insulin stimulates the phosphorylation of the 95,000-dalton subunit of its own receptor. *Science* **215**, 185–187 (1982).
13. M. Kasuga et al., Insulin stimulation of phosphorylation of the beta subunit of the insulin receptor. Formation of both phosphoserine and phosphotyrosine. *J Biol Chem* **257**, 9891–9894 (1982).
14. M. F. White, R. Maron, C. R. Kahn, Insulin rapidly stimulates tyrosine phosphorylation of a Mr-185,000 protein in intact cells. *Nature* **318**, 183–186 (1985).
15. Y. Ebina et al., The human insulin receptor cDNA: the structural basis for hormone-activated transmembrane signalling. *Cell* **40**, 747–758 (1985).
16. A. Ullrich et al., Human insulin receptor and its relationship to the tyrosine kinase family of oncogenes. *Nature* **313**, 756–761 (1985).
17. K. Siddle, Signalling by insulin and IGF receptors: supporting acts and new players. *J Mol Endocrinol* **47**, R1–10 (2011).
18. Y. Kido, J. Nakae, D. Accili, Clinical review 125: the insulin receptor and its cellular targets. *J Clin Endocrinol Metab* **86**, 972–979 (2001).
19. D. J. Burks, M. F. White, IRS proteins and beta-cell function. *Diabetes* **50 Suppl 1**, S140–145 (2001).
20. D. J. Withers, M. White, Perspective: the insulin signaling system--a common link in the pathogenesis of type 2 diabetes. *Endocrinology* **141**, 1917–1921 (2000).
21. S. J. Fisher, J. C. Bruning, S. Lannon, C. R. Kahn, Insulin signaling in the central nervous system is critical for the normal sympathoadrenal response to hypoglycemia. *Diabetes* **54**, 1447–1451 (2005).
22. C. Benedict, W. Kern, B. Schultes, J. Born, M. Hallschmid, Differential sensitivity of men and women to anorexigenic and memory-improving effects of intranasal insulin. *J Clin Endocrinol Metab* **93**, 1339–1344 (2008).
23. A. Kleinridders, H. A. Ferris, W. Cai, C. R. Kahn, Insulin action in brain regulates systemic metabolism and brain function. *Diabetes* **63**, 2232–2243 (2014).
24. P. De Meyts, The Insulin Receptor and Its Signal Transduction Network in *Endotext*, K. R. Feingold et al., eds. South Dartmouth, MA, MDText.com, Inc., 2000.
25. C. M. Taniguchi, B. Emanuelli, C. R. Kahn, Critical nodes in signalling pathways: insights into insulin action. *Nat Rev Mol Cell Biol* **7**, 85–96 (2006).
26. R. N. Kulkarni et al., Tissue-specific knockout of the insulin receptor in pancreatic beta cells

creates an insulin secretory defect similar to that in type 2 diabetes. *Cell* **96**, 329–339 (1999).
27. A. Louvi, D. Accili, A. Efstratiadis, Growth-promoting interaction of IGF-II with the insulin receptor during mouse embryonic development. *Dev Biol* **189**, 33–48 (1997).
28. J. Nakae, Y. Kido, D. Accili, Distinct and overlapping functions of insulin and IGF-I receptors. *Endocr Rev* **22**, 818–835 (2001).
29. J. C. Bruning et al., A muscle-specific insulin receptor knockout exhibits features of the metabolic syndrome of NIDDM without altering glucose tolerance. *Mol Cell* **2**, 559–569 (1998).
30. M. Bluher et al., Adipose tissue selective insulin receptor knockout protects against obesity and obesity-related glucose intolerance. *Dev Cell* **3**, 25–38 (2002).
31. M. D. Michael et al., Loss of insulin signaling in hepatocytes leads to severe insulin resistance and progressive hepatic dysfunction. *Mol Cell* **6**, 87–97 (2000).
32. J. Havrankova, D. Schmechel, J. Roth, M. Brownstein, Identification of insulin in rat brain. *Proc Natl Acad Sci U S A* **75**, 5737–5741 (1978).
33. J. Havrankova, J. Roth, M. J. Brownstein, Concentrations of insulin and insulin receptors in the brain are independent of peripheral insulin levels. Studies of obese and streptozotocin-treated rodents. *J Clin Invest* **64**, 636–642 (1979).
34. R. U. Margolis, N. Altszuler, Insulin in the cerebrospinal fluid. *Nature* **215**, 1375–1376 (1967).
35. S. C. Woods, D. Porte, Jr., Relationship between plasma and cerebrospinal fluid insulin levels of dogs. *Am J Physiol* **233**, E331–334 (1977).
36. K. J. Kaiyala, R. L. Prigeon, S. E. Kahn, S. C. Woods, M. W. Schwartz, Obesity induced by a high-fat diet is associated with reduced brain insulin transport in dogs. *Diabetes* **49**, 1525–1533 (2000).
37. G. D. Baura et al., Insulin transport from plasma into the central nervous system is inhibited by dexamethasone in dogs. *Diabetes* **45**, 86–90 (1996).
38. A. Dorn, A. Rinne, H. G. Bernstein, H. J. Hahn, M. Ziegler, Insulin and C-peptide in human brain neurons (insulin/C-peptide/brain peptides/immunohistochemistry/radioimmunoassay). *J Hirnforsch* **24**, 495–499 (1983).
39. W. S. Young, 3rd, Periventricular hypothalamic cells in the rat brain contain insulin mRNA. *Neuropeptides* **8**, 93–97 (1986).
40. S. U. Devaskar et al., Insulin gene expression and insulin synthesis in mammalian neuronal cells. *J Biol Chem* **269**, 8445–8454 (1994).
41. M. K. Raizada, Localization of insulin-like immunoreactivity in the neurons from primary cultures of rat brain. *Exp Cell Res* **143**, 351–357 (1983).
42. R. Schechter et al., Immunohistochemical and in situ hybridization study of an insulin-like substance in fetal neuron cell cultures. *Brain Res* **636**, 9–27 (1994).
43. D. F. Hopkins, G. Williams, Insulin receptors are widely distributed in human brain and bind human and porcine insulin with equal affinity. *Diabet Med* **14**, 1044–1050 (1997).
44. M. Adamo, M. K. Raizada, D. LeRoith, Insulin and insulin-like growth factor receptors in the nervous system. *Mol Neurobiol* **3**, 71–100 (1989).
45. A. Kleinridders, Deciphering brain insulin receptor and insulin-like growth factor 1 receptor signalling. *J Neuroendocrinol* **28**, (2016).
46. D. Leto, A. R. Saltiel, Regulation of glucose transport by insulin: traffic control of GLUT4. *Nat Rev Mol Cell Biol* **13**, 383–396 (2012).
47. C. Leloup et al., Discrete brain areas express the insulin-responsive glucose transporter GLUT4. *Brain Res Mol Brain Res* **38**, 45–53 (1996).
48. C. M. Reno et al., Brain GLUT4 knockout mice have impaired glucose tolerance, decreased insulin sensitivity, and impaired hypoglycemic counterregulation. *Diabetes* **66**, 587–597 (2017).
49. B. V. Zlokovic, The blood-brain barrier in health and chronic neurodegenerative disorders. *Neuron* **57**, 178–201 (2008).
50. I. A. Simpson, A. Carruthers, S. J. Vannucci, Supply and demand in cerebral energy metabolism: the role of nutrient transporters. *J Cereb Blood Flow Metab* **27**, 1766–1791 (2007).
51. J. C. Bruning et al., Role of brain insulin receptor in control of body weight and reproduction. *Science* **289**, 2122–2125 (2000).
52. H. Chen et al., Evidence that the diabetes gene encodes the leptin receptor: identification of a mutation in the leptin receptor gene in db/db mice. *Cell* **84**, 491–495 (1996).
53. D. L. Coleman, Effects of parabiosis of obese with diabetes and normal mice. *Diabetologia* **9**, 294–298 (1973).
54. N. Balthasar et al., Leptin receptor signaling in POMC neurons is required for normal body weight homeostasis. *Neuron* **42**, 983–991 (2004).

55. R. Coppari et al., The hypothalamic arcuate nucleus: a key site for mediating leptin's effects on glucose homeostasis and locomotor activity. *Cell Metab* **1**, 63–72 (2005).
56. G. J. Morton et al., Leptin regulates insulin sensitivity via phosphatidylinositol-3-OH kinase signaling in mediobasal hypothalamic neurons. *Cell Metab* **2**, 411–420 (2005).
57. G. J. Morton et al., Arcuate nucleus-specific leptin receptor gene therapy attenuates the obesity phenotype of Koletsky (fa(k)/fa(k)) rats. *Endocrinology* **144**, 2016–2024 (2003).
58. S. Fulton et al., Leptin regulation of the mesoaccumbens dopamine pathway. *Neuron* **51**, 811–822 (2006).
59. E. Arvat et al., Preliminary evidence that Ghrelin, the natural GH secretagogue (GHS)-receptor ligand, strongly stimulates GH secretion in humans. *J Endocrinol Invest* **23**, 493–495 (2000).
60. A. Abizaid, T. L. Horvath, Brain circuits regulating energy homeostasis. *Regul Pept* **149**, 3–10 (2008).
61. T. L. Horvath, S. Diano, P. Sotonyi, M. Heiman, M. Tschop, Minireview: ghrelin and the regulation of energy balance--a hypothalamic perspective. *Endocrinology* **142**, 4163–4169 (2001).
62. J. M. Zigman et al., Mice lacking ghrelin receptors resist the development of diet-induced obesity. *J Clin Invest* **115**, 3564–3572 (2005).
63. O. Al Massadi, R. Nogueiras, C. Dieguez, J. A. Girault, Ghrelin and food reward. *Neuropharmacology* **148**, 131–138 (2019).
64. V. B. Bogdanov et al., Reward-related brain activity and behavior are associated with peripheral ghrelin levels in obesity. *Psychoneuroendocrinology* **112**, 104520 (2020).
65. E. Micioni Di Bonaventura et al., Assessing the role of ghrelin and the enzyme ghrelin O-acyltransferase (GOAT) system in food reward, food motivation, and binge eating behavior. *Pharmacol Res* **172**, 105847 (2021).
66. D. L. Drazen, T. P. Vahl, D. A. D'Alessio, R. J. Seeley, S. C. Woods, Effects of a fixed meal pattern on ghrelin secretion: evidence for a learned response independent of nutrient status. *Endocrinology* **147**, 23–30 (2006).
67. S. C. Woods et al., Conditioned insulin secretion and meal feeding in rats. *J Comp Physiol Psychol* **91**, 128–133 (1977).
68. S. C. Woods, E. C. Lotter, L. D. McKay, D. Porte, Jr., Chronic intracerebroventricular infusion of insulin reduces food intake and body weight of baboons. *Nature* **282**, 503–505 (1979).
69. M. Heni et al., Hypothalamic and striatal insulin action suppresses endogenous glucose production and may stimulate glucose uptake during hyperinsulinemia in lean but not in overweight men. *Diabetes* **66**, 1797–1806 (2017).
70. K. Kimura et al., Central insulin action activates Kupffer cells by suppressing hepatic vagal activation via the nicotinic alpha 7 acetylcholine receptor. *Cell Rep* **14**, 2362–2374 (2016).
71. R. W. Gelling et al., Insulin action in the brain contributes to glucose lowering during insulin treatment of diabetes. *Cell Metab* **3**, 67–73 (2006).
72. K. D. Niswender et al., Insulin activation of phosphatidylinositol 3-kinase in the hypothalamic arcuate nucleus: a key mediator of insulin-induced anorexia. *Diabetes* **52**, 227–231 (2003).
73. R. C. Frederich et al., Leptin levels reflect body lipid content in mice: evidence for diet-induced resistance to leptin action. *Nat Med* **1**, 1311–1314 (1995).
74. J. Panksepp, D. M. Nance, Insulin, glucose and hypothalamic regulation of feeding. *Physiol Behav* **9**, 447–451 (1972).
75. S. C. Woods, D. Porte, Jr., Neural control of the endocrine pancreas. *Physiol Rev* **54**, 596–619 (1974).
76. G. Williams et al., The hypothalamus and the control of energy homeostasis: different circuits, different purposes. *Physiol Behav* **74**, 683–701 (2001).
77. S. Kullmann et al., Selective insulin resistance in homeostatic and cognitive control brain areas in overweight and obese adults. *Diabetes Care* **38**, 1044–1050 (2015).
78. A. Kleinridders et al., Insulin resistance in brain alters dopamine turnover and causes behavioral disorders. *Proc Natl Acad Sci U S A* **112**, 3463–3468 (2015).
79. S. Obici, Z. Feng, G. Karkanias, D. G. Baskin, L. Rossetti, Decreasing hypothalamic insulin receptors causes hyperphagia and insulin resistance in rats. *Nat Neurosci* **5**, 566–572 (2002).
80. K. W. Williams et al., Segregation of acute leptin and insulin effects in distinct populations of arcuate proopiomelanocortin neurons. *J Neurosci* **30**, 2472–2479 (2010).

81. B. F. Belgardt, J. C. Bruning, CNS leptin and insulin action in the control of energy homeostasis. *Ann N Y Acad Sci* **1212**, 97–113 (2010).

82. J. W. Hill et al., Direct insulin and leptin action on pro-opiomelanocortin neurons is required for normal glucose homeostasis and fertility. *Cell Metab* **11**, 286–297 (2010).

83. Z. J. Yang, M. M. Meguid, J. K. Chai, C. Chen, A. Oler, Bilateral hypothalamic dopamine infusion in male Zucker rat suppresses feeding due to reduced meal size. *Pharmacol Biochem Behav* **58**, 631–635 (1997).

84. T. Roy, C. E. Lloyd, Epidemiology of depression and diabetes: a systematic review. *J Affect Disord* **142 Suppl**, S8–21 (2012).

85. S. Chen et al., Association of depression with pre-diabetes, undiagnosed diabetes, and previously diagnosed diabetes: a meta-analysis. *Endocrine* **53**, 35–46 (2016).

86. B. Mezuk, W. W. Eaton, S. Albrecht, S. H. Golden, Depression and type 2 diabetes over the lifespan: a meta-analysis. *Diabetes Care* **31**, 2383–2390 (2008).

87. A. B. Grigsby, R. J. Anderson, K. E. Freedland, R. E. Clouse, P. J. Lustman, Prevalence of anxiety in adults with diabetes: a systematic review. *J Psychosom Res* **53**, 1053–1060 (2002).

88. E. S. N. M. Lyra et al., Insulin resistance as a shared pathogenic mechanism between depression and type 2 diabetes. *Front Psychiatry* **10**, 57 (2019).

89. J. A. Hamer et al., Brain insulin resistance: a treatment target for cognitive impairment and anhedonia in depression. *Exp Neurol* **315**, 1–8 (2019).

90. M. K. Singh et al., Brain and behavioral correlates of insulin resistance in youth with depression and obesity. *Horm Behav* **108**, 73–83 (2019).

91. L. B. Shomaker et al., Longitudinal study of depressive symptoms and progression of insulin resistance in youth at risk for adult obesity. *Diabetes Care* **34**, 2458–2463 (2011).

92. P. Dandona, A. Aljada, A. Bandyopadhyay, Inflammation: the link between insulin resistance, obesity and diabetes. *Trends Immunol* **25**, 4–7 (2004).

93. C. H. Lee, F. Giuliani, The role of inflammation in depression and fatigue. *Front Immunol* **10**, 1696 (2019).

94. M. J. Stuart, B. T. Baune, Depression and type 2 diabetes: inflammatory mechanisms of a psychoneuroendocrine co-morbidity. *Neurosci Biobehav Rev* **36**, 658–676 (2012).

95. L. Yang et al., The effects of psychological stress on depression. *Curr Neuropharmacol* **13**, 494–504 (2015).

96. L. Li, X. Li, W. Zhou, J. L. Messina, Acute psychological stress results in the rapid development of insulin resistance. *J Endocrinol* **217**, 175–184 (2013).

97. G. S. Hotamisligil, Mechanisms of TNF-alpha-induced insulin resistance. *Exp Clin Endocrinol Diabetes* **107**, 119–125 (1999).

98. G. S. Hotamisligil, N. S. Shargill, B. M. Spiegelman, Adipose expression of tumor necrosis factor-alpha: direct role in obesity-linked insulin resistance. *Science* **259**, 87–91 (1993).

99. Y. N. Liu et al., TNFalpha mediates stress-induced depression by upregulating indoleamine 2,3-dioxygenase in a mouse model of unpredictable chronic mild stress. *Eur Cytokine Netw* **26**, 15–25 (2015).

100. M. Ahmed, W. M. El-Bakly, A. M. Zaki, L. F. abd Alzez, O. El serafi, Bupropion effects on high-fat diet-induced steatohepatitis and endothelial dysfunction in rats: role of tumour necrosis factor-alpha. *J Phar Pharmacol* **66**, 793–801 (2014).

101. J. C. Felger, F. E. Lotrich, Inflammatory cytokines in depression: neurobiological mechanisms and therapeutic implications. *Neuroscience* **246**, 199–229 (2013).

102. A. N. Sharma, K. M. Elased, T. L. Garrett, J. B. Lucot, Neurobehavioral deficits in db/db diabetic mice. *Physiol Behav* **101**, 381–388 (2010).

103. S. Sharma, S. Fulton, Diet-induced obesity promotes depressive-like behaviour that is associated with neural adaptations in brain reward circuitry. *Int J Obesity (2005)* **37**, 382–389 (2013).

104. J. Deslauriers, M. Toth, A. Der-Avakian, V. B. Risbrough, Current status of animal models of posttraumatic stress disorder: behavioral and biological phenotypes, and future challenges in improving translation. *Biol Psychiatry* **83**, 895–907 (2018).

105. B. Planchez, A. Surget, C. Belzung, Animal models of major depression: drawbacks and challenges. *J Neural Transm (Vienna)* **126**, 1383–1408 (2019).

106. Y. Kurhe, R. Mahesh, Pioglitazone, a PPARγ agonist rescues depression associated with obesity using chronic unpredictable mild stress model

in experimental mice. *Neurobiol Stress* **3**, 114–121 (2016).
107. A. N. Sharma, K. M. Elased, J. B. Lucot, Rosiglitazone treatment reversed depression- but not psychosis-like behavior of db/db diabetic mice. *J Psychopharmacol (Oxford, England)* **26**, 724–732 (2012).
108. P. L. Delgado, Depression: the case for a monoamine deficiency. *J Clin Psychiatry* **61 Suppl 6**, 7–11 (2000).
109. P. Belujon, A. A. Grace, Dopamine system dysregulation in major depressive disorders. *Int J Neuropsychopharmacol* **20**, 1036–1046 (2017).
110. D. Chaudhury et al., Rapid regulation of depression-related behaviours by control of midbrain dopamine neurons. *Nature* **493**, 532–536 (2013).
111. G. R. Christoph, R. J. Leonzio, K. S. Wilcox, Stimulation of the lateral habenula inhibits dopamine-containing neurons in the substantia nigra and ventral tegmental area of the rat. *J Neurosci* **6**, 613–619 (1986).
112. J. Shumake, A. Ilango, H. Scheich, W. Wetzel, F. W. Ohl, Differential neuromodulation of acquisition and retrieval of avoidance learning by the lateral habenula and ventral tegmental area. *J Neurosci* **30**, 5876–5883 (2010).
113. J. Unger et al., Distribution of insulin receptor-like immunoreactivity in the rat forebrain. *Neuroscience* **31**, 143–157 (1989).
114. L. J. Boulos, E. Darcq, B. L. Kieffer, Translating the habenula-from rodents to humans. *Biol Psychiatry* **81**, 296–305 (2017).
115. W. Cai et al., Insulin regulates astrocyte gliotransmission and modulates behavior. *J Clin Invest* **128**, 2914–2926 (2018).
116. M. Soto, W. Cai, M. Konishi, C. R. Kahn, Insulin signaling in the hippocampus and amygdala regulates metabolism and neurobehavior. *Proc Natl Acad Sci U S A* **116**, 6379–6384 (2019).
117. J. Olds, P. Milner, Positive reinforcement produced by electrical stimulation of septal area and other regions of rat brain. *J Comp Physiol Psychol* **47**, 419–427 (1954).
118. K. C. Berridge, M. L. Kringelbach, Pleasure systems in the brain. *Neuron* **86**, 646–664 (2015).
119. W. Schultz, Neuronal reward and decision signals: from theories to data. *Physiol Rev* **95**, 853–951 (2015).
120. H. Hu, Reward and aversion. *Annu Rev Neurosci* **39**, 297–324 (2016).
121. D. Purves, *Neuroscience*. (Oxford University Press, New York, Sixth edition., 2018), pp. 1 volume (various pagings).
122. P. J. Morgane, J. R. Galler, D. J. Mokler, A review of systems and networks of the limbic forebrain/limbic midbrain. *Prog Neurobiol* **75**, 143–160 (2005).
123. G. E. Alexander, M. R. DeLong, P. L. Strick, Parallel organization of functionally segregated circuits linking basal ganglia and cortex. *Annu Rev Neurosci* **9**, 357–381 (1986).
124. A. Kleinridders, E. N. Pothos, Impact of brain insulin signaling on dopamine function, food intake, reward, and emotional behavior. *Curr Nutr Rep* **8**, 83–91 (2019).
125. S. Ikemoto, Dopamine reward circuitry: two projection systems from the ventral midbrain to the nucleus accumbens-olfactory tubercle complex. *Brain Res Rev* **56**, 27–78 (2007).
126. M. Watabe-Uchida, L. Zhu, S. K. Ogawa, A. Vamanrao, N. Uchida, Whole-brain mapping of direct inputs to midbrain dopamine neurons. *Neuron* **74**, 858–873 (2012).
127. K. C. Berridge, T. E. Robinson, What is the role of dopamine in reward: hedonic impact, reward learning, or incentive salience? *Brain Res Brain Res Rev* **28**, 309–369 (1998).
128. P. J. Allen et al., Rationale and consequences of reclassifying obesity as an addictive disorder: neurobiology, food environment and social policy perspectives. *Physiol Behav* **107**, 126–137 (2012).
129. E. L. Gardner, Addiction and brain reward and antireward pathways. *Adv Psychosom Med* **30**, 22–60 (2011).
130. E. N. Pothos, L. Hernandez, B. G. Hoebel, Chronic food deprivation decreases extracellular dopamine in the nucleus accumbens: implications for a possible neurochemical link between weight loss and drug abuse. *Obes Res* **3 Suppl 4**, 525S–529S (1995).
131. E. N. Pothos, I. Creese, B. G. Hoebel, Restricted eating with weight loss selectively decreases extracellular dopamine in the nucleus accumbens and alters dopamine response to amphetamine, morphine, and food intake. *J Neurosci* **15**, 6640–6650 (1995).
132. B. M. Geiger et al., Deficits of mesolimbic dopamine neurotransmission in rat dietary obesity. *Neuroscience* **159**, 1193–1199 (2009).

133. B. M. Geiger et al., Evidence for defective mesolimbic dopamine exocytosis in obesity-prone rats. *FASEB J* **22**, 2740–2746 (2008).
134. E. A. Schur, J. Tong, Insulin action to inhibit food intake: is it all in your head? *J Clin Endocrinol Metab*, (2021).
135. D. P. Figlewicz, S. B. Evans, J. Murphy, M. Hoen, D. G. Baskin, Expression of receptors for insulin and leptin in the ventral tegmental area/substantia nigra (VTA/SN) of the rat. *Brain Res* **964**, 107–115 (2003).
136. D. P. Figlewicz, Insulin, food intake, and reward. *Semin Clin Neuropsychiatry* **8**, 82–93 (2003).
137. D. M. Mebel, J. C. Wong, Y. J. Dong, S. L. Borgland, Insulin in the ventral tegmental area reduces hedonic feeding and suppresses dopamine concentration via increased reuptake. *Eur J Neurosci* **36**, 2336–2346 (2012).
138. S. J. Russo et al., IRS2-Akt pathway in midbrain dopamine neurons regulates behavioral and cellular responses to opiates. *Nat Neurosci* **10**, 93–99 (2007).
139. S. A. Eisenstein et al., Insulin, central dopamine D2 receptors, and monetary reward discounting in obesity. *PLoS One* **10**, e0133621 (2015).
140. A. J. Sipols, G. D. Stuber, S. N. Klein, M. S. Higgins, D. P. Figlewicz, Insulin and raclopride combine to decrease short-term intake of sucrose solutions. *Peptides* **21**, 1361–1367 (2000).
141. K. D. Carr, G. Kim, S. Cabeza de Vaca, Hypoinsulinemia may mediate the lowering of self-stimulation thresholds by food restriction and streptozotocin-induced diabetes. *Brain Res* **863**, 160–168 (2000).
142. E. L. Air, S. C. Benoit, K. A. Blake Smith, D. J. Clegg, S. C. Woods, Acute third ventricular administration of insulin decreases food intake in two paradigms. *Pharmacol Biochem Behav* **72**, 423–429 (2002).
143. A. E. Kelley, K. C. Berridge, The neuroscience of natural rewards: relevance to addictive drugs. *J Neurosci* **22**, 3306–3311 (2002).
144. T. R. Stratford, A. E. Kelley, Evidence of a functional relationship between the nucleus accumbens shell and lateral hypothalamus subserving the control of feeding behavior. *J Neurosci* **19**, 11040–11048 (1999).
145. A. W. Pardini et al., Distribution of insulin receptor substrate-2 in brain areas involved in energy homeostasis. *Brain Res* **1112**, 169–178 (2006).
146. G. Labouèbe et al., Insulin induces long-term depression of ventral tegmental area dopamine neurons via endocannabinoids. *Nat Neurosci* **16**, 300–308 (2013).
147. D. P. Figlewicz, P. Szot, M. Chavez, S. C. Woods, R. C. Veith, Intraventricular insulin increases dopamine transporter mRNA in rat VTA/substantia nigra. *Brain Res* **644**, 331–334 (1994).
148. S. Kar, J. G. Chabot, R. Quirion, Quantitative autoradiographic localization of [125I]insulin-like growth factor I, [125I]insulin-like growth factor II, and [125I]insulin receptor binding sites in developing and adult rat brain. *J Comp Neurol* **333**, 375–397 (1993).
149. M. A. Stouffer et al., Insulin enhances striatal dopamine release by activating cholinergic interneurons and thereby signals reward. *Nat Commun* **6**, 8543 (2015).
150. S. M. Gray, E. J. Barrett, Insulin transport into the brain. *Am J Physiol Cell Physiol* **315**, C125–C136 (2018).
151. X. Cao et al., Astrocyte-derived ATP modulates depressive-like behaviors. *Nat Med* **19**, 773–777 (2013).
152. J. Jiao et al., Munc18-1 catalyzes neuronal SNARE assembly by templating SNARE association. *Elife* **7**, (2018).
153. P. Sharma, Y. Balhara, Opioid use and diabetes: an overview. *J Soc Health Diabetes* **4**, 6 (2016).
154. L.-T. Wu et al., Substance use and mental diagnoses among adults with and without type 2 diabetes: Results from electronic health records data. *Drug Alcohol Dependence* **156**, 162–169 (2015).
155. G. A. Karam et al., Effects of opium addiction on some serum factors in addicts with non-insulin-dependent diabetes mellitus. *Add Biol* **9**, 53–58 (2004).
156. A. Virmani, Z. K. Binienda, S. F. Ali, F. Gaetani, Metabolic syndrome in drug abuse. *Ann N Y Acad Sci* **1122**, 50–68 (2007).
157. B. Kashyap, L. R. Hanson, W. H. Frey Ii, Intranasal insulin: a treatment strategy for addiction. *Neurotherapeutics* **17**, 105–115 (2020).
158. N. C. Barbarich-Marsteller, R. W. Foltin, B. T. Walsh, Does anorexia nervosa resemble an addiction? *Curr Drug Abuse Rev* **4**, 197–200 (2011).
159. S. E. Cassin, K. M. von Ranson, Is binge eating experienced as an addiction? *Appetite* **49**, 687–690 (2007).
160. A. N. Gearhardt, M. A. White, M. N. Potenza, Binge eating disorder and food addiction. *Curr Drug Abuse Rev* **4**, 201–207 (2011).

161. J. H. Lacey, Self-damaging and addictive behaviour in bulimia nervosa. A catchment area study. *Br J Psychiatry* **163**, 190–194 (1993).
162. A. C. Prince, S. J. Brooks, D. Stahl, J. Treasure, Systematic review and meta-analysis of the baseline concentrations and physiologic responses of gut hormones to food in eating disorders. *Am J Clin Nutr* **89**, 755–765 (2009).
163. A. Ilyas et al., The metabolic underpinning of eating disorders: a systematic review and meta-analysis of insulin sensitivity. *Mol Cell Endocrinol* **497**, 110307 (2019).
164. S. J. Crow, P. K. Keel, D. Kendall, Eating disorders and insulin-dependent diabetes mellitus. *Psychosomatics* **39**, 233–243 (1998).
165. J. M. Jones, M. L. Lawson, D. Daneman, M. P. Olmsted, G. Rodin, Eating disorders in adolescent females with and without type 1 diabetes: cross sectional study. *BMJ (Clinical Research ed.)* **320**, 1563–1566 (2000).
166. J. Lotharius, P. Brundin, Pathogenesis of Parkinson's disease: dopamine, vesicles and alpha-synuclein. *Nat Rev Neurosci* **3**, 932–942 (2002).
167. J. A. Driver et al., Prospective cohort study of type 2 diabetes and the risk of Parkinson's disease. *Diabetes Care* **31**, 2003–2005 (2008).
168. G. Hu, P. Jousilahti, S. Bidel, R. Antikainen, J. Tuomilehto, Type 2 diabetes and the risk of Parkinson's disease. *Diabetes Care* **30**, 842–847 (2007).
169. Q. Xu et al., Diabetes and risk of Parkinson's disease. *Diabetes Care* **34**, 910–915 (2011).
170. G. Pagano et al., Diabetes mellitus and Parkinson disease. *Neurology* **90**, e1654–e1662 (2018).
171. E. Hogg et al., High prevalence of undiagnosed insulin resistance in non-diabetic subjects with Parkinson's disease. *J Parkinsons Dis* **8**, 259–265 (2018).
172. V. Kotagal et al., Diabetes is associated with postural instability and gait difficulty in Parkinson disease. *Parkinsonism Relat Disord* **19**, 522–526 (2013).
173. D. Bosco et al., Dementia is associated with insulin resistance in patients with Parkinson's disease. *J Neurol Sci* **315**, 39–43 (2012).
174. L. Dunn et al., Dysregulation of glucose metabolism is an early event in sporadic Parkinson's disease. *Neurobiol Aging* **35**, 1111–1115 (2014).
175. A. Reeve, E. Simcox, D. Turnbull, Ageing and Parkinson's disease: why is advancing age the biggest risk factor? *Ageing Res Rev* **14**, 19–30 (2014).
176. J. A. Kushner, The role of aging upon beta cell turnover. *J Clin Invest* **123**, 990–995 (2013).
177. M. Takahashi et al., Insulin receptor mRNA in the substantia nigra in Parkinson's disease. *Neurosci Lett* **204**, 201–204 (1996).
178. I. Moroo et al., Loss of insulin receptor immunoreactivity from the substantia nigra pars compacta neurons in Parkinson's disease. *Acta Neuropathol* **87**, 343–348 (1994).
179. J. Godau et al., Serum insulinlike growth factor 1 as possible marker for risk and early diagnosis of Parkinson disease. *Arch Neurol* **68**, 925–931 (2011).
180. F. Mashayekhi, E. Mirzajani, M. Naji, M. Azari, Expression of insulin-like growth factor-1 and insulin-like growth factor binding proteins in the serum and cerebrospinal fluid of patients with Parkinson's disease. *J Clin Neurosci* **17**, 623–627 (2010).

CHAPTER EIGHT

The Impact of Insulin on Brain Serotonergic System: Consequences on Diabetes-Associated Mood Disorders[1]

Hugo Martin
Sebastian Bullich and Bruno P. Guiard
Xavier Fioramonti

CONTENTS

INTRODUCTION: THE COMORBIDITY DIABETES-DEPRESSION

The focus of the present chapter is to describe the effects of insulin on the brain's serotonergic system and the implications for the regulation of mood. The literature showing that insulin is able to modulate the activity of serotonin (5-HT) neurons in the dorsal and/or median raphe nuclei (DRN/MRN) is limited, as described below. The main evidence that insulin may modulate this neuronal population comes from extensive literature reporting that the risk of developing mood disorders is significantly increased in patients with type 1 or type 2 diabetes (T1D or T2D).

Over the past two decades, epidemiological studies have suggested a strong link between diabetes and depression. Indeed, it appears that the prevalence of depression is twice as high, if not higher, in patients with T1D or T2D (1–4). Based on Anderson's meta-analysis (1), the prevalence of depression appears to be the same for both diabetes subtypes. Nevertheless, studies emphasize that the risk of developing mood disorders is higher in patients with T2D than in T1D (5). These differences might be related to different inflammatory states between T1D and T2D patients, as there is evidence that inflammatory biomarkers correlate with depressive symptoms only in T2D patients (6). Alternatively, because one of the most obvious

[1] All the authors have contributed equally.

DOI: 10.1201/9781003079927-8

differences between T1D and T2D is insulin resistance, we hypothesize in this review that impaired insulin signaling at the peripheral and/or central level contributes, at least in part, the development of mood disorders. Although beyond the scope of this review, it is noteworthy that a reciprocal link also exists, albeit less documented. Accordingly, several studies have suggested that the prevalence of diabetes may also be twice as high in depressed patients (7, 8). Overall, these data highlight the existence of a bidirectional relationship between metabolic and psychiatric disorders.

Several interconnected biological mechanisms have been suggested to explain the comorbidity diabetes/psychiatric disorders including oxidative stress, inflammation, and insulin resistance. For this chapter, we only address the potential involvement of brain insulin resistance in the development of diabetes-associated mood disorders and the eventuality that the serotonin system is impacted by such pathological conditions. Supporting this hypothesis, a clinical study recently reported that depressive symptoms correlate with insulin resistance (8). Preclinical studies also consolidate this concept, as high-fat diet (HFD)-induced insulin resistance has been repeatedly shown to cause anxiety and depressive-like phenotypes in mice or rats (9–12). For instance, depressive-like symptoms are observed in genetic models of T2D such as in leptin-receptor deficient mice (db/db) or Zucker fa/fa rats (13, 14). These behavioral abnormalities are not limited to T2D, as streptozotocin (STZ) injection-induced T1D in mice also promotes anxiety and depressive-like symptoms (15–17). Conversely, animal studies showed that chronic mild stress, social defeat, or dexamethasone-induced depressive-like symptoms are associated with peripheral insulin resistance (17–19). Nevertheless, in these models, it has never been shown whether metabolic impairments are a cause or a consequence of mood disorders. Of particular note, several preclinical studies have examined the potential therapeutic effects of insulin or antidiabetic drugs in these models of T1D or T2D. They concur to demonstrate their beneficial effects on emotionality in association with enhancement of peripheral insulin signaling and/or anti-inflammatory properties. For example, in a rat model of T2D, chronic intraperitoneal administration of insulin elicits a comparable antidepressant-like response than the selective serotonin reuptake inhibitor (SSRI) sertraline (20). An important question that remains unresolved is whether the impairment of peripheral and/or central insulin signaling contributes to modifying brain circuits involved in depression. The observation that subchronic intraperitoneal insulin treatment reverses freezing behavior without attenuating hyperglycemia in an STZ-T1D mouse model (15), strongly suggests that the stimulation of insulin signaling may promote a therapeutic independently from an improvement of peripheral metabolism. This review summarizes the arguments in favor of the hypothesis that insulin resistance in the brain is key to these deleterious effects of diabetes on behavior.

OVERVIEW OF BRAIN INSULIN SIGNALING

Insulin signaling in the brain has not been fully characterized but has been an important area of research for decades. Because of the molecular weight of insulin (5,800 g/mol) and the failure of previous studies to detect this hormone in cerebrospinal fluid (CSF) (21, 22), it has long been assumed that insulin cannot cross the blood-brain barrier (BBB). In the 1970s, Margolis and Altszuler were the first to detect insulin in CSF, and this discovery has been further confirmed with different approaches (23–25). Today, it is well accepted that peripheral insulin readily crosses the BBB through a saturable receptor-mediated process at the level of the choroid plexus (26–28). This transcytosis may occur through many different proteins that are capable of binding and internalizing insulin. It could therefore be mediated by a direct pathway that recruits the insulin receptor (IR) itself. Accordingly, a recent study reports that the IR antagonist S-961 blunts insulin uptake in brain endothelial cells in culture (29), while another study shows that the same antagonist slightly reduces insulin transport in the mouse brain, suggesting that insulin transport may occur independently from its receptor (30). Interestingly, recent evidence shows that insulin can be produced directly by choroid plexus (31) and may be delivered to brain structures closed to the ventricles. Finally, it is important to take into consideration that insulin growth factor-1 (IGF1) is synthesized in the brain and displays a potent effect on brain functions commonly associated with insulin in the brain such as its ability to stimulate plasticity, learning, and memory (32).

In the brain, in situ hybridization and immunohistochemistry analyzes identified the IR in several discrete areas with different expression levels: the olfactory bulbs, the cortex, the hippocampus, and the hypothalamus are examples of structures endowed with a high density of IR (33–35). Alternative splicing of IR mRNA results in two different forms of IR namely IR-A and IR-B. Interestingly, the IR-A isoform has a stronger affinity for IGF-1, suggesting inevitable crosstalk (36). Similarly, insulin can interact with IGF1-receptor (IGF1-R) at high doses, although its distribution appears to be different from that of IR (34), suggesting well-dissociated signaling and action between IGF-1 and insulin. Figure 8.1 summarizes the main signaling pathways activated downstream the IR (Figure 8.1). All in all, these studies strongly suggest that peripheral insulin reaches the brain where it may play a critical role in higher brain functions including mood regulation.

EFFECTS OF BRAIN INSULIN ON EMOTIONALITY

Both human and animal studies have used intranasal delivery of insulin (IDI) to study the effect of insulin in the brain without affecting peripheral metabolism (i.e., with its ability to induce hypoglycemia). In clinical studies, a single IDI treatment has been shown to attenuate the hypothalamic-pituitary-adrenal (HPA) axis hyperactivity in response to psychosocial stress (36). In addition, chronic IDI treatment has been reported to improve self-confidence and well-being, reduce anger, and reduce anxiety in healthy individuals (37, 38). Considering that IDI also shows pro-cognitive

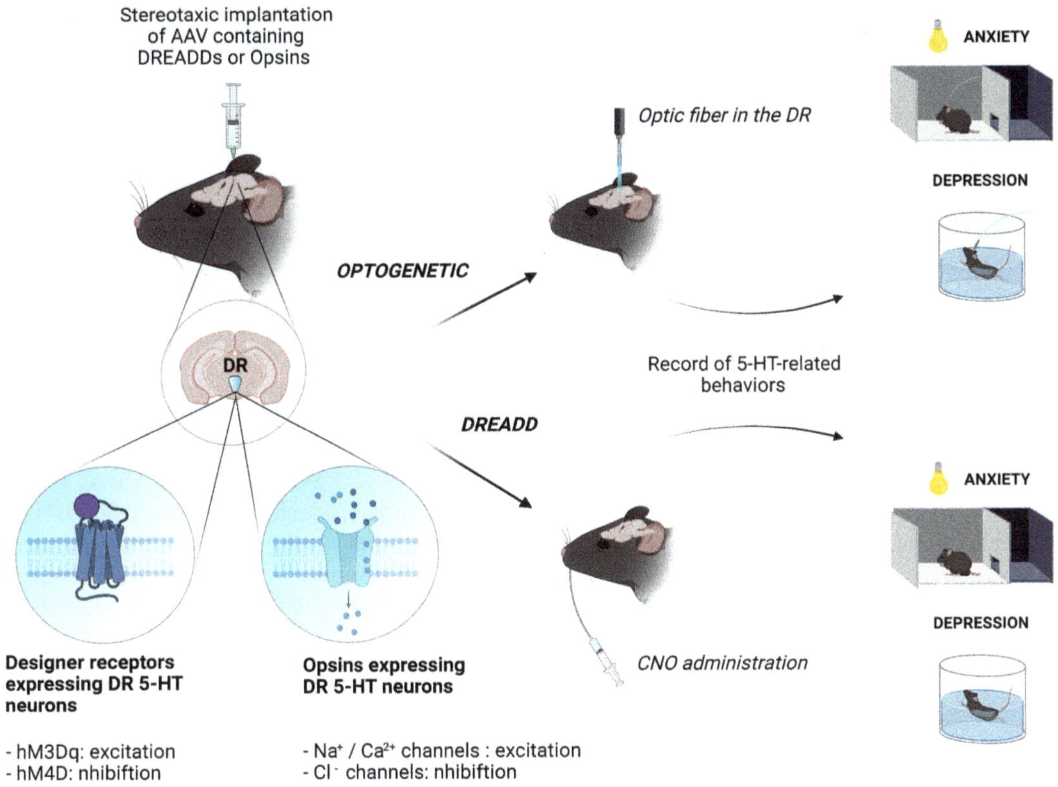

Figure 8.1 Schematic representation for the use of optogenetics and pharmacogenetics in the study of 5-HT system-dependent emotional behavior. Animals are injected with a viral vector expressing either an opsin or a DREADD which will promote the activation or inhibition of selective neuronal networks (DR 5-HT neurons in this diagram). Modulation of these neurons will be performed in response to photoactivation/inhibition by an optic fiber placed beforehand by stereotaxy for the optogenetics approach. For the pharmacogenetics approach, modulation of neurons will be performed in response to CNO injected intraperitoneally or given in the drinking water. During the photo-modulation for the optogenetics approach or after the CNO administration, emotional behaviors will be evaluated through appropriate behavioral tests monitoring either anxiety or resignation.

effects and that cognitive impairment is recognized as a core symptom of depression, this pharmacological intervention may represent a new opportunity for an "add-on" strategy with currently available antidepressants (ADs). For example, in light of the memory impairments related to the antimuscarinic effect of the tricyclics (TCAs), we assume that IDI could attenuate the cognitive side-effects of this class of antidepressant. Beyond these considerations, it has been recently demonstrated that the insulin-sensitizing agent metformin is a promising compound to be used as an adjunct in major depression (39). Studies conducted in animal models also support a role for insulin in the brain in regulating emotional behaviors. Subchronic IDI treatments resulted in an apparent decrease in anxiety along with an improvement in memory in rodents (40, 41). Robust antidepressant-like activity has also been described in response to a single intracerebroventricular (ICV) injection of insulin in naive rats.

Although the association between insulin resistance and depressive symptoms has already been described (42), the possibility that such a process occurs in the brain to influence neurobehavior has been raised for a decade. Kleinridders and coworkers were the first to investigate whether insulin resistance in the brain, selectively modeled in neurons by genetic deletion of IR, causes behavioral changes in mice (43). Thus, NIRKO mice (for Neuronal Insulin Receptor Knock-Out) spontaneously exhibit pro-depressive- and anxiety-like phenotypes across the lifespan. These changes parallel impairments in brain dopaminergic activity but this study did not investigate the putative involvement of other monoaminergic neuronal populations. More recently, studies have shown that selective knockdown of IR in the hippocampus or hypothalamus using lentiviral approaches also leads to the same conclusion that the genetic depletion of IR indiscrete brain areas favors emotional and cognitive impairments (44, 45). Going one step further, knock-down of both IR and IGF1-R in the hippocampus or amygdala has been shown to exaggerate anxiety. More studies are needed to shed light on the nature of the specific neuronal circuits underlying the effects of insulin in the brain. It is important to note that the expression of IR is not restricted to neurons and has also been demonstrated in astrocytes (46). Constitutive or inducible knock-out of IR in astrocytes also induces a depression-like phenotype in GIRKO mice (Glial Insulin Receptor Knock-Out mice) (47). Therefore, a deficit in the action of insulin selectively in the brain is critical for the regulation of emotional states. Based on these observations, we can assume that impaired insulin signaling in the brain, resulting from insulin resistance (T2D) or impaired insulin secretion (T1D), may be the putative cornerstone for the comorbidity between diabetes and depression. An important challenge in the future will be to identify the neural circuits impacted by the disruption of insulin signaling that is relevant to the physiopathology of depressive disorders.

EVIDENCE OF THE IMPACT OF INSULIN ONTO SEROTONERGIC NEUROTRANSMISSION

Overview of the Role of DRN 5-HT Neurons in Mood

Serotonin neurons originating from the DRN and MRN innervate the entire forebrain and midbrain and as such are considered important in modulating various biological functions, particularly emotional states (48). Although pharmacological and neuroanatomical/functional studies have been widely used to gain insight into the role of DRN and MRN and their respective neuronal projections in mood disorders (49), the poor selectivity of these approaches has not allowed for an accurate and comprehensive insight into the role of the various serotonergic circuits. The recent development of opto- and chemo-genetic approaches has helped to delineate these 5-HT pathways more precisely. In particular, direct optogenetic activation of DRN 5-HT neurons has been shown to elicit antidepressant-like effects, whereas activation of MRN 5-HT neurons promotes anxiety associated with an increase in 5-HT levels in the hippocampus (50, 51). Although optogenetic approaches are useful for acute control of specific 5-HT neurons, they are not suitable for mimicking long-term modulation of 5-HT neuronal activity. Therefore, chemogenetic approaches have been introduced to overcome this problem, and hM3Dq DREADD (Designer Receptor Exclusively Activated Designer Drug) coupled to the Gαq pathway has been successfully used to activate 5-HT neuron subpopulations. Under these conditions, short- and long-term activation of DRN 5-HT neurons by clozapine N-oxides (CNO) induces behavioral responses reminiscent of the effects of antidepressants (52, 53), thus establishing a causal link between serotonergic DRN activity and emotion. In marked contrast, activation of 5-HT neurons in the MRN mediates anxiety-like and depression-like behaviors (54). Overall, these findings likely reinforce the role of 5-HT neurons in the control of emotional behavior by DRN and MRN in opposite directions.

At the molecular level, it is well known that DRN and MRN 5-HT neuronal activity is regulated by auto- and hetero-receptors. Regarding the auto-regulation of this neuronal population, a very important element controlling 5-HT release throughout the brain is the 5-HT1A autoreceptor present on the soma and dendrites of 5-HT neurons. Pharmacological and genetic studies demonstrated that 5-HT1A autoreceptors exert a negative feedback influence on 5-HT firing activity (55–57). Similarly, 5-HT1B autoreceptors are located at the nerve terminal and also play a critical role by attenuating 5-HT output in projections areas such as the frontal cortex and the hippocampus (58–60). Finally, recent evidence suggests the existence of a positive 5-HT autoreceptor, namely the 5-HT2B subtype whose activation has been shown to enhance serotonergic tone (61). All these receptors represent putative targets to allow insulin-regulating 5-HT neurotransmission.

Technical approaches: Optogenetics and pharmacogenetics - *Principles and uses*

Optogenetics and pharmacogenetics approaches have been designed to remotely and selectively control spatiotemporally the activity of neurons to regulate the behavior and physiological functions (62). Optogenetics relies on the genetic modification of known microbial opsin genes encoding light-sensitive channels, which, once activated by light, either activate or inhibit targeted neurons. The pharmacogenetics approach, also called Designer Receptors Exclusively Activated by Designer Drugs (DREADD), consists of the use of receptor proteins that have been engineered to exclusively respond to inert agonist to activate or inhibit key downstream G protein-dependent signaling pathways (Gs-, Gq- or Gi-signaling pathways). The first DREADD has been designed from a modified human muscarinic acetylcholine receptor. Specific neurons can be targeted by injecting a viral vector (adeno-associated viral (AAV) most of the time) expressing the opsin or the DREADD in a cre-dependent manner in a transgenic mouse expressing the cre-recombinase in the selective neuronal population (TpH2-cre or Pet1-cre in the case of brain 5-HT neurons). The optogenetics approach requires that an optic fiber "lighting-up" the neurons addressing the opsin to be inserted by stereotaxy in the brain, in the area of interest. In opposition, the pharmacogenetics approach is less invasive as it only requires to administer the inert DREADD agonist intraperitoneally or in the drinking water for long term neuronal modulation (in some cases, injecting the inert agonist locally in the brain by stereotaxy could be necessary). Clozapine-N-oxide (CNO) is the first DREADD agonist developed and, so far, the most widely used. CNO selectively activates DREADDs but is not an active ligand for the endogenous muscarinic acetylcholine receptors. The cutting-edge optogenetics and pharmacogenetics are powerful tools in neuroscience that allow selective and bidirectional modulation of the activity of defined populations of neurons with unprecedented specificity. The applications of optogenetics and pharmacogenetics are limitless as these approaches allow the correlation between the activity of a specific neuronal network and a physiological function in a free moving animal. Almost all functions controlled by the brain can be studied inkling feeding behaviors, hormonal regulations, memory or mood. Figure 8.1 summarizes the application of optogenetics and pharmacogenetics in the study of 5-HT system-dependent emotional behaviors (Figure 8.1).

Evidence of Insulin Action on the 5-HT Neurons

Transcriptomic data analysis revealed that 5-HT neurons express IR mRNA (63). However, there is no evidence that 5-HT neurons preferentially express the IR-A or IR-B isoform. However, given that the IR-A is substantially expressed in neurons (64), we might assume that the IR-A isoform is specifically expressed in serotonergic neurons to control their functional activity in response to insulin. Moreover, it is not worth mentioning that IR-A or IR-B expression is not restricted to 5-HT neurons within the DRN, as these isoforms also appear to be expressed in non-5-HT neurons (65). Extracellular single-unit electrophysiological recordings showed that peripheral insulin injection decreased the firing rate of 5- HT neurons recorded in the Raphe pallidus and Raphe obscurus (66). A limitation of using peripheral insulin injections is the induction of iatrogenic hypoglycemia. Therefore, it is not possible to determine whether the effects of insulin on 5-HT neurotransmission are due to its action on the brain or whether they result from reduced plasma glucose levels. To circumvent this limitation, ICV injection of insulin has been used to study the effects of insulin on the central serotonergic system without inducing hypoglycemia. For example, using intracerebral microdialysis *in vivo*, ICV insulin was reported to increase levels of 5-HT and 5-HIAA in Raphe Magnus, suggesting that insulin increases serotonergic neuronal activity (67). These studies suggest that insulin acts on 5-HT neurons, but one might hypothesize that insulin does not act directly on 5-HT neurons but indirectly through the modulation of specific DRN/MRN afferents. The IR is expressed throughout the brain and 5-HT neurons receive projections from many brain areas (68). Ex vivo studies such as patch-clamp recordings on brain slices are needed to address the question of the direct/indirect effects of insulin on 5-HT neurons. The observation that peripheral injection of insulin increased both plasma and brain tryptophan levels (63, 69) suggests that peripheral insulin promotes the uptake of tryptophan into the brain, which in turn may favor 5-HT synthesis. *In vitro* studies also suggested that insulin-mediated increased serotonin levels may be explained by a different mechanism resulting from the inhibition of serotonin-degrading enzymes [i.e., monoamine oxidases A and B (MAO-A/B)] (43). However, conflicting results were obtained, as other data from the literature showed that peripheral administration of insulin increased 5-HT turnover. Thus, peripheral injection of insulin is thought to rise CSF and brain levels of 5-hydroxyindoleacetic acid (5-HIAA), suggesting that this hormone decreases 5-HT levels (70–73).

Indirect evidence also suggests that insulin modulates the serotonergic system in the brain. This arises from studies showing that this specific circuitry is altered in animal models of T1D or T2D. For example, in the STZ-induced T1D rodent model, impaired insulin production has been associated with significant attenuation of brain 5-HT levels throughout the brain (incl. hypothalamus, hippocampus, cortex, and brainstem) (74–77). More recently, it has been proposed that this decrease in brain 5-HT levels in STZ-injected animals may be due to increased MAO-A/MAO-B activities (74, 78). Consistent with these tissue analyses, *in vivo* intracerebral microdialysis studies reported lower hippocampal and hypothalamic extracellular 5-HT levels in STZ-treated rodents (79–81). A higher capacity of 5- HT reuptake by the 5-HT transporter SERT has also been suggested to explain this attenuated serotonergic activity (82). The role of plasma levels of branched-chain amino acids (BCAAs) has also been proposed as one of the mechanisms involved. Indeed, BCAAs are significantly elevated in STZ-injected rodents but also in HFD-fed mice (83). BCAAs negatively affect the serotonergic system, as these amino acids are known to compete with tryptophan to cross the blood-brain barrier through the transporter L-type amino acid transporter 1 (LAT1) (84).

We recently performed in vivo intracerebral microdialysis to demonstrate that HFD-induced insulin resistance decreased extracellular 5 HT levels in the hippocampus (12). As above-mentioned, the activity of 5- HT neurons of the DRN is finely regulated by inhibitory 5-HT1A autoreceptors (55). Thus, our recent results showed that HFD-fed mice displayed a more robust hypothermic response to the 5-HT1A receptor agonist 8-OHDPAT than control animals, providing clear evidence that insulin resistance is associated with hypersensitization of this inhibitory autoreceptor (12). Using ex vivo patch-clamp experiments, we also showed that DRN-5-HT neurons from T2D-HFD-fed mice exhibit decreased firing frequency and altered intrinsic electrophysiological properties (83), confirming the idea that the main brain serotonergic circuit originating from the DRN is blunted in this animal model. In the latter study, we also showed that improving insulin sensitivity with the antidiabetic drug metformin restored DRN 5- HT neuron activity, hippocampal 5-HT levels, and HFD-induced mood disorders (83), strengthening the hypothesis that alteration in the brain 5-HT network may be a critical factor in the development of T2D-induced mood disorders.

Effects of Insulin on 5-HT Sensitive Cells – Crosstalk between Insulin and Serotonin Signaling Pathways

IR expression is enriched in the cortex and the hippocampus. These postsynaptic serotonergic regions have obvious potential for morphological changes. This has prompted researchers to investigate how insulin affects different forms of plasticity and related functions, such as learning, memory, and mood. In the hippocampus, neuronal stem cells (NSCs) proliferate and differentiate, resulting in mature neurons that integrate into the pre-existing neuronal network. Neural stem cells may also produce astrocytes and oligodendrocytes, which are important for regulating emotional states (85–87). A variety of evidence suggests that insulin can regulate all these processes. In particular, insulin has been shown to promote adult hippocampal neurogenesis by stimulating the survival, proliferation, and differentiation of NSC (87–89). Furthermore, decreased expression or function of insulin receptors detected in genetic (db/db) or environmental (HFD) models of metabolic diseases are associated with impaired adult neurogenesis (90–92). Although the molecular determinants of these defects have not been fully elucidated, these findings strongly suggest that insulin sensitivity in the brain plays a crucial role.

As shown in Figure 8.1, insulin was reported to increase the phosphorylation of Akt and GSK3β, both of which are involved in adult hippocampal neurogenesis (93), and 16 weeks of HFD prevented these effects (94). It is worth noting that HFD-induced brain insulin resistance is also related to the weakened activation of the PI3-kinase/Akt/GSK3β pathway in response to 5-HT in this region (94). In addition to this pathway, it is fully demonstrated that mTOR complex 1 (mTORC1) can be activated in response to insulin and serotonin (95). Insulin and serotonin can stimulate postsynaptic PSD-95 through the PI3K/mTOR signaling pathway (96, 97) (Figure 8.2). These findings paved the way for the interaction between insulin and 5-HT, which may underlie the link between metabolic diseases and depressive diseases. It is worth noting that adult hippocampal neurogenesis also involves fine-tuned interactions between NSC and astrocytes (98, 99). These glial cells simultaneously express insulin receptor and serotonin receptor (47, 99). Collectively, these results indicate that insulin and serotonin converge to, among other functions, modulate brain plasticity.

CONCLUDING REMARKS

In this chapter, we tackle the question whether insulin modulates emotional behaviors. The literature suggests that the effect of this hormone

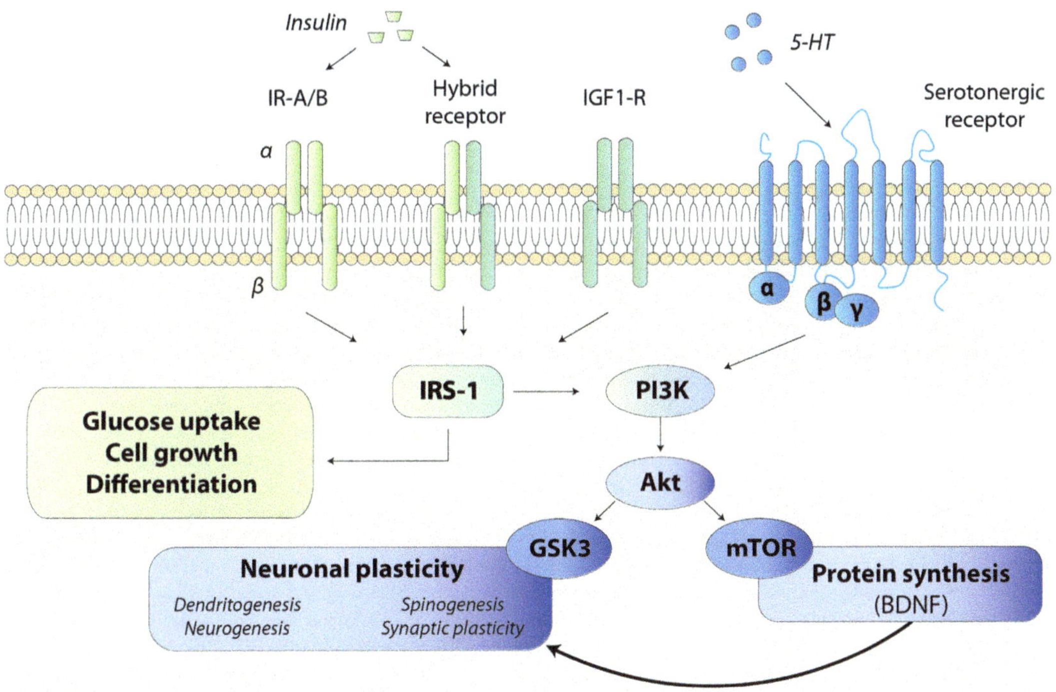

Figure 8.2 Putative crosstalk between insulin and serotonin pathways in relation with neuronal plasticity. Insulin and serotonergic receptors converge towards PI3K which is a node underpinning neuronal plasticity through mTOR and GSK3 signaling. IR-A/B: insulin receptor A/B, IGF1R: insulin-like growth factor 1 receptor, IRS-1: insulin receptor substrate 1, PI3K: phosphoinositide 3-kinase, AKT: protein kinase B, mTOR: mechanistic target of rapamycin, GSK3: glycogen synthase kinase-3, 5-HT: 5-hydroxytryptamine.

on the brain leads to neurobehavioral changes but several points have yet to be solved, notably whether or not 5-HT neurons are involved in the brain action of insulin? Here, we provided evidence that insulin may act at the soma of serotonergic neurons and/or terminals, as summarized in Figure 8.3.

The idea that enhancing or stimulating insulin signaling in the brain, and in particular insulin signaling in 5-HT neurons, may be an interesting

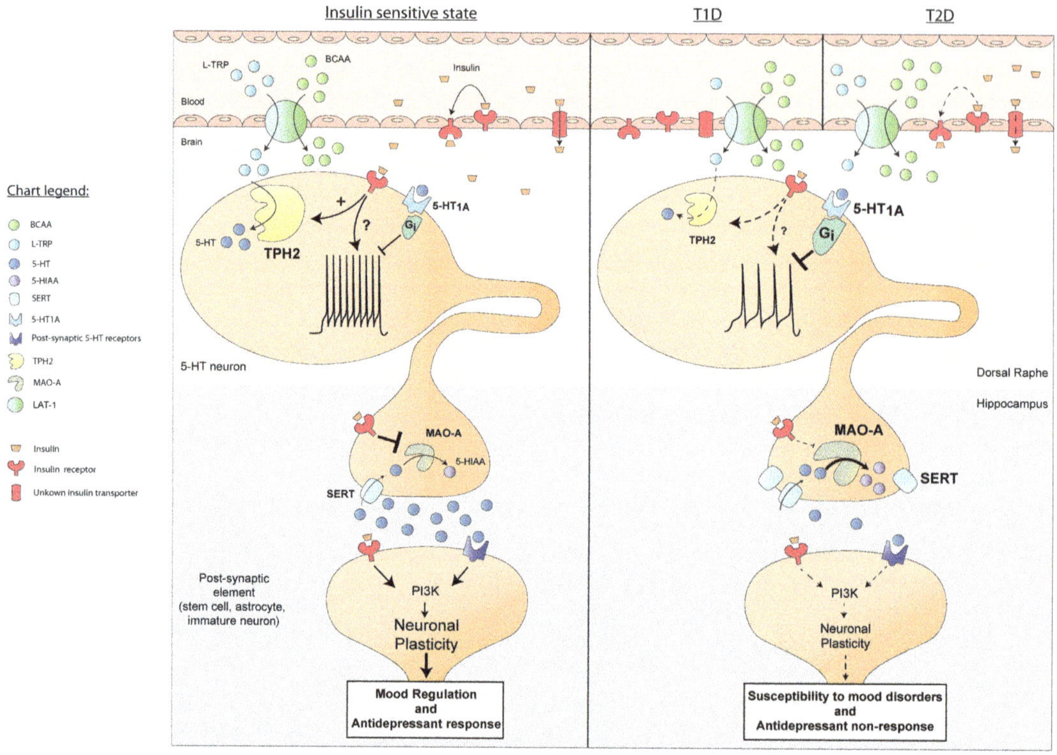

Figure 8.3 Alteration of insulin signaling in the serotonergic system during T1D or T2D. Insulin sensitive state (left panel): L-tryptophan (L-TRP) competes with Branched Chain Amino Acids (BCAAs) to enter the brain through the Large Amino Acid Transporter 1 (LAT-1), although BCAA has greater affinity for LAT-1 than L-TRP. Insulin can cross the blood-brain barrier (BBB) through transcytosis of its own receptor or by other unknown mechanisms. L-TRP is converted to serotonin (5-HT) by the enzyme tryptophan hydroxylase-2 (TPH2) which is positively modulated by insulin, improving serotonin production. At the synapse, after the reuptake of 5-HT by the serotonin transporter (SERT), monoamine oxidase-A (MAO-A) catabolizes serotonin to 5-Hydroxyindoleacetic acid (5-HIAA) at the presynaptic level. MAO-A expression is negatively regulated by insulin receptor signaling. On the post-synaptic element, activation of both serotonergic receptors and insulin receptor induces neuronal plasticity through activation of PI3K which is the common factor of these pathways. Such mechanisms contribute to mood regulation and effective antidepressant response. Pathological conditions (right panels): In T1D, beta-cell loss leads to a lack of insulin secretion with a known deleterious effect on glucose homeostasis. L-TRP levels are significantly decreased in plasma, resulting in diminished tryptophan transport and availability in the brain. Regarding T2D in which blood insulin level varies along the evolution of the pathology, insulin receptor signaling is altered leading to insulin resistance. Moreover, BCAAs levels are increased and thus favor BCAAs transit through the BBB at the expense of L-TRP, resulting, such as in T1D, in a decreased availability of L-TRP in the brain. In both conditions, such lower levels of precursor result in lower levels of brain 5-HT. On top of that, impaired insulin receptor signaling lacks to positively modulate TPH2 activity which also contributes to decrease the level of brain serotonin. Firing rate activity of DRN 5-HT neurons is decreased presumably due to increased sensitivity of inhibitory 5-HT_{1A} autoreceptors that have been reported in such conditions. Decreased serotonin release is a possible consequence of the decreased firing rate of DRN 5-HT neurons but also by increased activity of MAO-A breakdown activity due to blunted insulin action onto 5-HT neurons. At the post-synaptic level, impaired insulin receptor signaling and decreased serotonin levels weakly activate the PI3K pathway and thus fail to induce neuronal plasticity. Such pathological mechanisms contribute to the susceptibility of mood disorders occurrence and antidepressant treatment non-response. Reproduced with permission from (101).

approach to improve depressive symptoms or treatment response, as it has been shown that patients with diabetes are less sensitive to antidepressants (100). Therefore, further animal and clinical studies need to be performed to support this hypothesis and to determine the relevance of insulin sensitization strategies either alone or as an add-on strategy for patients with depressive symptoms and metabolic disorders.

ACKNOWLEDGMENTS

HM is supported by a fellowship INRAE-Région Nouvelle Aquitaine. This work was supported in part by research awards from the Société Française du Diabète (SFD) (XF, GBP), Fondation Bordeaux Université - FMC (XF).

REFERENCES

1. R. J. Anderson, K. E. Freedland, R. E. Clouse, P. J. Lustman, The prevalence of comorbid depression in adults with diabetes: a meta-analysis. *Diabetes Care.* **24**, 1069–1078 (2001).
2. T. Roy, C. E. Lloyd, Epidemiology of depression and diabetes: a systematic review. *J Affect Disord.* **142 Suppl**, S8–21 (2012).
3. S. Ali, M. A. Stone, J. L. Peters, M. J. Davies, K. Khunti, The prevalence of co-morbid depression in adults with Type 2 diabetes: a systematic review and meta-analysis. *Diabet Med* **23**, 1165–1173 (2006).
4. B. Buchberger, H. Huppertz, L. Krabbe, B. Lux, J. T. Mattivi, A. Siafarikas, Symptoms of depression and anxiety in youth with type 1 diabetes: a systematic review and meta-analysis. *Psychoneuroendocrinology.* **70**, 70–84 (2016).
5. J. Silverstein, P. Cheng, K. J. Ruedy, C. Kollman, R. W. Beck, G. J. Klingensmith, J. R. Wood, S. Willi, F. Bacha, J. Lee, E. Cengiz, M. J. Redondo, W. V. Tamborlane, Pediatric diabetes consortium, depressive symptoms in youth with type 1 or type 2 diabetes: results of the pediatric diabetes consortium screening assessment of depression in diabetes study. *Diabetes Care.* **38**, 2341–2343 (2015).
6. C. Herder, A. Schmitt, F. Budden, A. Reimer, B. Kulzer, M. Roden, T. Haak, N. Hermanns, Association between pro- and anti-inflammatory cytokines and depressive symptoms in patients with diabetes-potential differences by diabetes type and depression scores. *Transl Psychiatry.* **7**, 1 (2018).
7. I.-C. Chien, E.-L. Wu, C.-H. Lin, Y.-J. Chou, P. Chou, Prevalence of diabetes in patients with major depressive disorder: a population-based study. *Compr Psychiatry.* **53**, 569–575 (2012).
8. C. M. Phillips, I. J. Perry, Depressive symptoms, anxiety and well-being among metabolic health obese subtypes. *Psychoneuroendocrinology.* **62**, 47–53 (2015).
9. S. Dutheil, K. T. Ota, E. S. Wohleb, K. Rasmussen, R. S. Duman, High-fat diet induced anxiety and anhedonia: impact on brain homeostasis and inflammation. *Neuropsychopharmacology.* **41**, 1874–1887 (2016).
10. A. M. Hassan, G. Mancano, K. Kashofer, E. E. Fröhlich, A. Matak, R. Mayerhofer, F. Reichmann, M. Olivares, A. M. Neyrinck, N. M. Delzenne, S. P. Claus, P. Holzer, High-fat diet induces depression-like behaviour in mice associated with changes in microbiome, neuropeptide Y, and brain metabolome. *Nutr Neurosci.* **22**, 877–893 (2019).
11. J. L. Yang, D. X. Liu, H. Jiang, F. Pan, C. S. Ho, R. C. Ho, The effects of high-fat-diet combined with chronic unpredictable mild stress on depression-like behavior and leptin/LepRb in male rats. *Sci Rep.* **6** (2016), doi:10.1038/srep35239.
12. J. Zemdegs, G. Quesseveur, D. Jarriault, L. Pénicaud, X. Fioramonti, B. P. Guiard, High-fat diet-induced metabolic disorders impairs 5-HT function and anxiety-like behavior in mice. *Br. J. Pharmacol.* **173**, 2095–2110 (2016).
13. A.-L. Dinel, C. André, A. Aubert, G. Ferreira, S. Layé, N. Castanon, Cognitive and emotional alterations are related to hippocampal inflammation in a mouse model of metabolic syndrome. *PLoS ONE.* **6**, e24325 (2011).
14. S. Li, X. Zhai, P. Rong, M. F. McCabe, X. Wang, J. Zhao, H. Ben, S. Wang, Therapeutic effect of Vagus nerve stimulation on depressive-like behavior, hyperglycemia and insulin receptor expression in Zucker fatty rats. *PLoS ONE.* **9**, e112066 (2014).
15. S. Miyata, N. Yamada, S. Hirano, S. Tanaka, J. Kamei, Diabetes attenuates psychological stress-elicited 5-HT secretion in the prefrontal cortex but not in the amygdala of mice. *Brain Res.* **1147**, 233–239 (2007).
16. C.-W. Huang, T.-W. Hong, Y.-J. Wang, K.-C. Chen, J.-C. Pei, T.-Y. Chuang, W.-S. Lai, S.-H. Tsai, R. Chu, W.-C. Chen, L.-Y. Sheen, S. Takahashi, S.-T. Ding, T.-L. Shen, Ophiocordyceps formosana improves hyperglycemia and depression-like behavior in an STZ-induced diabetic mouse model. *BMC Complement Altern Med.* **16**, 310 (2016).

17. Y. Pan, Y. Hong, Q.-Y. Zhang, L.-D. Kong, Impaired hypothalamic insulin signaling in CUMS rats: Restored by icariin and fluoxetine through inhibiting CRF system. *Psychoneuroendocrinology*. **38**, 122–134 (2013).
18. E. Delanogare, R. M. de Souza, G. K. Rosa, F. G. Guanabara, A. Rafacho, E. L. G. Moreira, Enriched environment ameliorates dexamethasone effects on emotional reactivity and metabolic parameters in mice. *Stress*. **23**, 466–473 (2020).
19. H. Tsuneki, E. Tokai, C. Sugawara, T. Wada, T. Sakurai, T. Sasaoka, Hypothalamic orexin prevents hepatic insulin resistance induced by social defeat stress in mice. *Neuropeptides*. **47**, 213–219 (2013).
20. C. C. Sestile, J. C. Maraschin, M. P. Rangel, R. K. N. Cuman, E. A. Audi, Antidepressant-like effect of insulin in streptozotocin-induced type 2 diabetes mellitus rats. *Basic Clin Pharmacol Toxicol*. **119**, 243–248 (2016).
21. N. J. Elgee, R. H. Williams, N. D. Lee, Distribution and degradation studies with insulin I131. *J Clin Invest*. **33**, 1252–1260 (1954).
22. N. Haugaard, M. Vaughan, E. S. Haugaard, W. C. Stadie, Studies of radioactive injected labeled insulin. *J Biol Chem*. **208**, 549–563 (1954).
23. R. U. Margolis, N. Altszuler, Insulin in the cerebrospinal fluid. *Nature*. **215**, 1375–1376 (1967).
24. L. J. Stein, D. M. Dorsa, D. G. Baskin, D. P. Figlewicz, H. Ikeda, S. P. Frankmann, M. R. Greenwood, D. Porte, S. C. Woods, Immunoreactive insulin levels are elevated in the cerebrospinal fluid of genetically obese Zucker rats. *Endocrinology*. **113**, 2299–2301 (1983).
25. S. C. Woods, D. Porte, Relationship between plasma and cerebrospinal fluid insulin levels of dogs. *Am J Physiol*. **233**, E331–334 (1977).
26. W. A. Banks, A. J. Kastin, Differential permeability of the blood-brain barrier to two pancreatic peptides: insulin and amylin. *Peptides*. **19**, 883–889 (1998).
27. G. D. Baura, D. M. Foster, D. Porte, S. E. Kahn, R. N. Bergman, C. Cobelli, M. W. Schwartz, Saturable transport of insulin from plasma into the central nervous system of dogs in vivo. A mechanism for regulated insulin delivery to the brain. *J Clin Invest*. **92**, 1824–1830 (1993).
28. K. R. Duffy, W. M. Pardridge, Blood-brain barrier transcytosis of insulin in developing rabbits. *Brain Res*. **420**, 32–38 (1987).
29. S. M. Gray, K. W. Aylor, E. J. Barrett, Unravelling the regulation of insulin transport across the brain endothelial cell. *Diabetologia*. **60**, 1512–1521 (2017).
30. E. M. Rhea, C. Rask-Madsen, W. A. Banks, Insulin transport across the blood-brain barrier can occur independently of the insulin receptor: insulin transport across the BBB. *J Physiol*. **596**, 4753–4765 (2018).
31. C. H. Mazucanti, Q.-R. Liu, D. Lang, N. Huang, J. F. O'Connell, S. Camandola, J. M. Egan, Release of insulin produced by the choroid plexis is regulated by serotonergic signaling. *JCI Insight*. **4** (2019), doi:10.1172/jci.insight.131682.
32. I. Pomytkin, J. P. Costa-Nunes, V. Kasatkin, E. Veniaminova, A. Demchenko, A. Lyundup, K.-P. Lesch, E. D. Ponomarev, T. Strekalova, Insulin receptor in the brain: mechanisms of activation and the role in the CNS pathology and treatment. *CNS Neurosci Ther*. **24**, 763–774 (2018).
33. J.-T. Dou, M. Chen, F. Dufour, D. L. Alkon, W.-Q. Zhao, Insulin receptor signaling in long-term memory consolidation following spatial learning. *Learn Mem*. **12**, 646–655 (2005).
34. A. M. Fernandez, I. Torres-Alemán, The many faces of insulin-like peptide signalling in the brain. *Nat Rev Neurosci*. **13**, 225–239 (2012).
35. W.-Q. Zhao, H. Chen, M. J. Quon, D. L. Alkon, Insulin and the insulin receptor in experimental models of learning and memory. *Eur J Pharmacol*. **490**, 71–81 (2004).
36. Y. Yamaguchi, J. S. Flier, H. Benecke, B. J. Ransil, D. E. Moller, Ligand-binding properties of the two isoforms of the human insulin receptor. *Endocrinology*. **132**, 1132–1138 (1993).
37. C. Benedict, Intranasal insulin improves memory in humans. *Psychoneuroendocrinology*. **29**, 1326–1334 (2004).
38. D. S. Ferreira de Sá, S. Römer, A. H. Brückner, T. Issler, A. Hauck, T. Michael, Effects of intranasal insulin as an enhancer of fear extinction: a randomized, double-blind, placebo-controlled experimental study. *Neuropsychopharmacology*. **45**, 753–760 (2020).
39. M. S. Abdallah, E. M. Mosalam, A.-A. A. Zidan, K. S. Elattar, S. A. Zaki, A. N. Ramadan, A. M. Ebeid, The antidiabetic metformin as an adjunct to antidepressants in patients with major depressive disorder: a proof-of-concept, randomized, double-blind, placebo-controlled trial. *Neurotherapeutics*. **17**, 1897–1906 (2020).
40. E. Beirami, S. Oryan, S. M. Seyedhosseini Tamijani, A. Ahmadiani, L. Dargahi, Intranasal insulin treatment alleviates methamphetamine

induced anxiety-like behavior and neuroinflammation. *Neurosci Lett*. **660**, 122–129 (2017).
41. D. R. Marks, K. Tucker, M. A. Cavallin, T. G. Mast, D. A. Fadool, Awake intranasal insulin delivery modifies protein complexes and alters memory, anxiety, and olfactory behaviors. *J Neurosci*. **29**, 6734–6751 (2009).
42. N. L. Rasgon, H. A. Kenna, Insulin resistance in depressive disorders and Alzheimer's disease: revisiting the missing link hypothesis. *Neurobiol Aging*. **26 Suppl 1**, 103–107 (2005).
43. A. Kleinridders, W. Cai, L. Cappellucci, A. Ghazarian, W. R. Collins, S. G. Vienberg, E. N. Pothos, C. R. Kahn, Insulin resistance in brain alters dopamine turnover and causes behavioral disorders. *Proc Natl Acad Sci*. **112**, 3463–3468 (2015).
44. C. A. Grillo, G. G. Piroli, K. F. Kaigler, S. P. Wilson, M. A. Wilson, L. P. Reagan, Downregulation of hypothalamic insulin receptor expression elicits depressive-like behaviors in rats. *Behav Brain Res*. **222**, 230–235 (2011).
45. C. A. Grillo, G. G. Piroli, R. C. Lawrence, S. A. Wrighten, A. J. Green, S. P. Wilson, R. R. Sakai, S. J. Kelly, M. A. Wilson, D. D. Mott, L. P. Reagan, Hippocampal insulin resistance impairs spatial learning and synaptic plasticity. *Diabetes*. **64**, 3927–3936 (2015).
46. M. Heni, A. M. Hennige, A. Peter, D. Siegel-Axel, A.-M. Ordelheide, N. Krebs, F. Machicao, A. Fritsche, H.-U. Häring, H. Staiger, Insulin promotes glycogen storage and cell proliferation in primary human astrocytes. *PLoS ONE*. **6**, e21594 (2011).
47. W. Cai, C. Xue, M. Sakaguchi, M. Konishi, A. Shirazian, H. A. Ferris, M. E. Li, R. Yu, A. Kleinridders, E. N. Pothos, C. R. Kahn, Insulin regulates astrocyte gliotransmission and modulates behavior. *J Clin Invest*. **128**, 2914–2926 (2018).
48. A. Muzerelle, S. Scotto-Lomassese, J. F. Bernard, M. Soiza-Reilly, P. Gaspar, Conditional anterograde tracing reveals distinct targeting of individual serotonin cell groups (B5-B9) to the forebrain and brainstem. *Brain Struct Funct*. **221**, 535–561 (2016).
49. J. J. Mann, Role of the serotonergic system in the pathogenesis of major depression and suicidal behavior. *Neuropsychopharmacology*. **21**, 99S–105S (1999).
50. Y. Ohmura, K. F. Tanaka, T. Tsunematsu, A. Yamanaka, M. Yoshioka, Optogenetic activation of serotonergic neurons enhances anxiety-like behaviour in mice. *Int J Neuropsychopharmacol*. **17**, 1777–1783 (2014).
51. A. R. Abela, C. J. Browne, D. Sargin, T. D. Prevot, X. D. Ji, Z. Li, E. K. Lambe, P. J. Fletcher, Median raphe serotonin neurons promote anxiety-like behavior via inputs to the dorsal hippocampus. *Neuropharmacology*. **168**, 107985 (2020).
52. A. Teissier, A. Chemiakine, B. Inbar, S. Bagchi, R. S. Ray, R. D. Palmiter, S. M. Dymecki, H. Moore, M. S. Ansorge, Activity of Raphé serotonergic neurons controls emotional behaviors. *Cell Rep*. **13**, 1965–1976 (2015).
53. D. J. Urban, H. Zhu, C. A. Marcinkiewcz, M. Michaelides, H. Oshibuchi, D. Rhea, D. K. Aryal, M. S. Farrell, E. Lowery-Gionta, R. H. J. Olsen, W. C. Wetsel, T. L. Kash, Y. L. Hurd, L. H. Tecott, B. L. Roth, Elucidation of the behavioral program and neuronal network encoded by dorsal raphe serotonergic neurons. *Neuropsychopharmacology*. **41**, 1404–1415 (2016).
54. S. Li, W.-Q. Yao, Y.-Z. Tao, L. Ma, X. Liu, [Serotonergic neurons in the median raphe nucleus mediate anxiety- and depression-like behavior]. *Sheng Li Xue Bao*. **70**, 228–236 (2018).
55. P. Blier, G. Piñeyro, M. el Mansari, R. Bergeron, C. de Montigny, Role of somatodendritic 5-HT autoreceptors in modulating 5-HT neurotransmission. *Ann N Y Acad Sci*. **861**, 204–216 (1998).
56. A. Ferrés-Coy, N. Santana, A. Castañé, R. Cortés, M. C. Carmona, M. Toth, A. Montefeltro, F. Artigas, A. Bortolozzi, Acute 5-HT1A autoreceptor knockdown increases antidepressant responses and serotonin release in stressful conditions. *Psychopharmacology*. **225**, 61–74 (2013).
57. J. W. Richardson-Jones, C. P. Craige, B. P. Guiard, A. Stephen, K. L. Metzger, H. F. Kung, A. M. Gardier, A. Dranovsky, D. J. David, S. G. Beck, R. Hen, E. D. Leonardo, 5-HT1A autoreceptor levels determine vulnerability to stress and response to antidepressants. *Neuron*. **65**, 40–52 (2010).
58. A. M. Gardier, D. J. David, G. Jego, C. Przybylski, C. Jacquot, S. Durier, B. Gruwez, E. Douvier, P. Beauverie, N. Poisson, R. Hen, M. Bourin, Effects of chronic paroxetine treatment on dialysate serotonin in 5-HT1B receptor knockout mice. *J Neurochem*. **86**, 13–24 (2003).
59. I. Malagié, D. J. David, P. Jolliet, R. Hen, M. Bourin, A. M. Gardier, Improved efficacy of fluoxetine in increasing hippocampal 5-hydroxytryptamine outflow in 5-HT1B receptor knock-out mice. *Eur J Pharmacol*. **443**, 99–104 (2002).

60. E. Dremencov, E. Gur, B. Lerer, M. E. Newman, Subchronic fluoxetine administration to rats: effects on 5-HT autoreceptor activity as measured by in vivo microdialysis. *Eur Neuropsychopharmacol.* **10**, 229–236 (2000).
61. A. Belmer, E. Quentin, S. L. Diaz, B. P. Guiard, S. P. Fernandez, S. Doly, S. M. Banas, P. M. Pitychoutis, I. Moutkine, A. Muzerelle, A. Tchenio, A. Roumier, M. Mameli, L. Maroteaux, Positive regulation of raphe serotonin neurons by serotonin 2B receptors. *Neuropsychopharmacology.* **43**, 1623–1632 (2018).
62. J. Jiang, H. Cui, K. Rahmouni, Optogenetics and pharmacogenetics: principles and applications. *Am J Physiol Regul, Integr Comp Physiol.* **313**, R633–R645 (2017).
63. B. W. Okaty, M. E. Freret, B. D. Rood, R. D. Brust, M. L. Hennessy, D. deBairos, J. C. Kim, M. N. Cook, S. M. Dymecki, Multi-scale molecular deconstruction of the serotonin neuron system. *Neuron.* **88**, 774–791 (2015).
64. C. J. Garwood, L. E. Ratcliffe, S. V. Morgan, J. E. Simpson, H. Owens, I. Vazquez-Villaseñor, P. R. Heath, I. A. Romero, P. G. Ince, S. B. Wharton, Insulin and IGF1 signalling pathways in human astrocytes in vitro and in vivo; characterisation, subcellular localisation and modulation of the receptors. *Mol Brain.* **8**, 51 (2015).
65. J. Ren, A. Isakova, D. Friedmann, J. Zeng, S. M. Grutzner, A. Pun, G. Q. Zhao, S. S. Kolluru, R. Wang, R. Lin, P. Li, A. Li, J. L. Raymond, Q. Luo, M. Luo, S. R. Quake, L. Luo, Single-cell transcriptomes and whole-brain projections of serotonin neurons in the mouse dorsal and median raphe nuclei. *Elife.* **8** (2019), doi:10.7554/eLife.49424.
66. G. A. Smythe, J. E. Bradshaw, M. V. Nicholson, H. S. Grunstein, L. H. Storlien, Rapid bidirectional effects of insulin on hypothalamic noradrenergic and serotoninergic neuronal activity in the rat: role in glucose homeostasis. *Endocrinology.* **117**, 1590–1597 (1985).
67. F. J. Martín-Cora, C. A. Fornal, C. W. Metzler, B. L. Jacobs, Insulin-induced hypoglycemia decreases single-unit activity of serotonergic medullary raphe neurons in freely moving cats: relationship to sympathetic and motor output: medullary 5-HT neuronal responses to glucoregulatory challenges. *Eur J Neurosci.* **16**, 722–734 (2002).
68. S. Balali Dehkordi, J. Sajedianfard, A. A. Owji, The effect of intra-cerebroventricular injection of insulin on the levels of monoamines on the raphe magnus nucleus of non-diabetic and short-term diabetic rats in the formalin test. *Iran J Basic Med Sci.* **22**, 915–921 (2019).
69. J. D. Fernstrom, R. J. Wurtman, Elevation of plasma tryptophan by insulin in rat. *Metabolism.* **21**, 337–342 (1972).
70. R. G. MacKenzie, M. E. Trulson, Effects of insulin and streptozotocin-induced diabetes on brain tryptophan and serotonin metabolism in rats. *J Neurochem.* **30**, 205–211 (1978).
71. J. Danguir, J. L. Elghozi, D. Laude, Increased dopamine and serotonin metabolites in CSF during severe insulin-induced hypoglycemia in freely moving rats. *Neurochem Int.* **6**, 71–75 (1984).
72. M. Juszkiewicz, The effect of insulin on the central serotonergic system of the rat. *Pol J Pharmacol Pharm.* **37**, 591–600 (1985).
73. M. Orosco, S. Nicolaidis, Insulin and glucose-induced changes in feeding and medial hypothalamic monoamines revealed by microdialysis in rats. *Brain Res Bull.* **33**, 289–297 (1994).
74. D. Gupta, Y. Kurhe, M. Radhakrishnan, Antidepressant effects of insulin in streptozotocin induced diabetic mice: modulation of brain serotonin system. *Physiol Behav.* **129**, 73–78 (2014).
75. P. M. Abraham, J. Paul, C. S. Paulose, Down regulation of cerebellar serotonergic receptors in streptozotocin induced diabetic rats: effect of pyridoxine and Aegle marmelose. *Brain Res Bull.* **82**, 87–94 (2010).
76. M. Barber, B. S. Kasturi, M. E. Austin, K. P. Patel, S. M. J. MohanKumar, P. S. MohanKumar, Diabetes-induced neuroendocrine changes in rats: role of brain monoamines, insulin and leptin. *Brain Res.* **964**, 128–135 (2003).
77. M. Sandrini, G. Vitale, A. V. Vergoni, A. Ottani, A. Bertolini, Streptozotocin-induced diabetes provokes changes in serotonin concentration and on 5-HT1A and 5-HT2 receptors in the rat brain. *Life Sci.* **60**, 1393–1397 (1997).
78. C. S. Mayanil, S. M. Kazmi, N. Z. Baquer, Changes in monoamine oxidase activity in rat brain during alloxan diabetes. *J Neurochem.* **38**, 179–183 (1982).
79. M. Kino, T. Yamato, M. Aomine, Simultaneous measurement of nitric oxide, blood glucose, and monoamines in the hippocampus of diabetic rat: an in vivo microdialysis study. *Neurochem Int.* **44**, 65–73 (2004).
80. T. Yamato, Y. Misumi, S. Yamasaki, M. Kino, M. Aomine, Diabetes mellitus decreases

hippocampal release of neurotransmitters: an in vivo microdialysis study of awake, freely moving rats. *Diabetes Nutr Metab.* **17**, 128–136 (2004).

81. N. Ohtani, M. Ohta, T. Sugano, Microdialysis study of modification of hypothalamic neurotransmitters in streptozotocin-diabetic rats. *J Neurochem.* **69**, 1622–1628 (1997).
82. M. S. Petrisić, S. J. Augood, R. J. Bicknell, Monoamine transporter gene expression in the central nervous system in diabetes mellitus. *J Neurochem.* **68**, 2435–2441 (1997).
83. J. Zemdegs, H. Martin, H. Pintana, S. Bullich, S. Manta, M. A. Marqués, C. Moro, S. Layé, F. Ducrocq, N. Chattipakorn, Metformin promotes anxiolytic and antidepressant-like responses in insulin-resistant mice by decreasing circulating branched-chain amino acids. *J Neurosci.* **39**, 5935–5948 (2019).
84. J. D. Fernstrom, Large neutral amino acids: dietary effects on brain neurochemistry and function. *Amino Acids.* **45**, 419–430 (2013).
85. E. Boda, Myelin and oligodendrocyte lineage cell dysfunctions: new players in the etiology and treatment of depression and stress-related disorders. *Eur J Neurosci.* (2019), doi:10.1111/ejn.14621.
86. G. Quesseveur, A. M. Gardier, B. P. Guiard, The monoaminergic tripartite synapse: a putative target for currently available antidepressant drugs. *Curr Drug Targets.* **14**, 1277–1294 (2013).
87. M. A. I. Aberg, N. D. Aberg, T. D. Palmer, A.-M. Alborn, C. Carlsson-Skwirut, P. Bang, L. E. Rosengren, T. Olsson, F. H. Gage, P. S. Eriksson, IGF-I has a direct proliferative effect in adult hippocampal progenitor cells. *Mol Cell Neurosci.* **24**, 23–40 (2003).
88. A. Lindqvist, P. Mohapel, B. Bouter, H. Frielingsdorf, D. Pizzo, P. Brundin, C. Erlanson-Albertsson, High-fat diet impairs hippocampal neurogenesis in male rats. *Eur J Neurol.* **13**, 1385–1388 (2006).
89. H. Lv, L. Tang, C. Guo, Y. Jiang, C. Gao, Y. Wang, C. Jian, Intranasal insulin administration may be highly effective in improving cognitive function in mice with cognitive dysfunction by reversing brain insulin resistance. *Cogn Neurodyn.* **14**, 323–338 (2020).
90. Y. Murata, Y. Narisawa, R. Shimono, H. Ohmori, M. Mori, K. Ohe, K. Mine, M. Enjoji, A high fat diet-induced decrease in hippocampal newly-born neurons of male mice is exacerbated by mild psychological stress using a communication box. *J Affect Disord.* **209**, 209–216 (2017).
91. S. M. Nam, J. W. Kim, H. J. Kwon, D. Y. Yoo, H. Y. Jung, D. W. Kim, I. K. Hwang, J. K. Seong, Y. S. Yoon, Differential effects of low- and high-dose zinc supplementation on synaptic plasticity and neurogenesis in the hippocampus of control and high-fat diet-fed mice. *Neurochem Res.* **42**, 3149–3159 (2017).
92. S. Okuyama, W. Shinoka, K. Nakamura, M. Kotani, A. Sawamoto, K. Sugawara, M. Sudo, M. Nakajima, Y. Furukawa, Suppressive effects of the peel of Citrus kawachiensis (Kawachi Bankan) on astroglial activation, tau phosphorylation, and inhibition of neurogenesis in the hippocampus of type 2 diabetic db/db mice. *Biosci Biotechnol Biochem.* **82**, 1384–1395 (2018).
93. T.-Y. Eom, R. S. Jope, Blocked inhibitory serine-phosphorylation of glycogen synthase kinase-3alpha/beta impairs in vivo neural precursor cell proliferation. *Biol Psychiatry.* **66**, 494–502 (2009).
94. I. K. Papazoglou, A. Jean, A. Gertler, M. Taouis, C.-M. Vacher, Hippocampal GSK3β as a molecular link between obesity and depression. *Mol Neurobiol.* **52**, 363–374 (2015).
95. K.-P. Lesch, N. Araragi, J. Waider, D. van den Hove, L. Gutknecht, Targeting brain serotonin synthesis: insights into neurodevelopmental disorders with long-term outcomes related to negative emotionality, aggression and antisocial behaviour. *Philos Trans R Soc Lond B Biol Sci.* **367**, 2426–2443 (2012).
96. K. A. Jones, D. P. Srivastava, J. A. Allen, R. T. Strachan, B. L. Roth, P. Penzes, Rapid modulation of spine morphology by the 5-HT2A serotonin receptor through kalirin-7 signaling. *Proc Natl Acad Sci USA.* **106**, 19575–19580 (2009).
97. S.-H. Lee, J. M. Zabolotny, H. Huang, H. Lee, Y.-B. Kim, Insulin in the nervous system and the mind: Functions in metabolism, memory, and mood. *Mol Metab.* **5**, 589–601 (2016).
98. F. Cassé, K. Richetin, N. Toni, Astrocytes' contribution to adult neurogenesis in physiology and Alzheimer's disease. *Front Cell Neurosci.* **12**, 432 (2018).
99. G. Quesseveur, D. J. David, M. C. Gaillard, P. Pla, M. V. Wu, H. T. Nguyen, V. Nicolas, G. Auregan, I. David, A. Dranovsky, P. Hantraye, R. Hen, A. M. Gardier, N. Déglon, B. P. Guiard, BDNF overexpression in mouse hippocampal astrocytes promotes local neurogenesis and elicits anxiolytic-like activities. *Transl Psychiatry.* **3**, e253 (2013).
100. Y. S. Woo, R. S. McIntyre, J.-B. Kim, M.-S. Lee, J.-M. Kim, H. W. Yim, T.-Y. Jun, Association

of treatment response with obesity and other metabolic risk factors in adults with depressive disorders: results from a National Depression Cohort study in Korea (the CRESCEND study). *J Affect Disord*. **203**, 190–198 (2016).

101. H. Martin, S. Bullich, B. P. Guiard, X. Fioramonti, The impact of insulin on the serotonergic system and consequences on diabetes-associated mood disorders. *J Neuroendocrinol*. **33**, e12928 (2021).

CHAPTER NINE

Organ Cross-Talk Regulates (Brain) Insulin Action

Thomas Laeger

CONTENTS

INTRODUCTION

A dysregulated communication between the liver, adipose tissue, muscles, and the intestine contributes to the onset and development of metabolic diseases, including obesity and diabetes. Moreover, the communication between the organs relies on insulin action in the brain to maintain systemic homeostasis. In this chapter, the role of endocrines in improving glucose homeostasis, which is partly driven by the central action of insulin and subsequent feedback signals to the periphery, will be discussed.

ADIPOSE TISSUE

Adipose tissue is now recognized as an endocrine organ and not just as a spot for energy storage. As a major site of energy reservoirs and release, the adipose tissue plays an active role in energy balance via signaling the nutritional state of an organism. During fasting, the fatty acids released by the adipose tissue become the dominant oxidative fuel for all major tissues (not the brain) and induce an increase in hepatic glucose production. In addition, the adipose tissue is capable of releasing numerous peptides and hormones that impact the metabolism of the entire body. Some of the adipose-secreted factors signal the CNS to communicate the energy state, while others target different organs, e.g. liver and muscle. On the one hand, insulin acts in the brain to regulate the activity of brown adipose tissue (1), among other things, whereas adipokines influence the action of insulin in the brain. Therefore, disrupted adipose signaling leads to detrimental outcomes in distant tissues.

DOI: 10.1201/9781003079927-9

LEPTIN

One year before the discovery of adiponectin, in 1994, another adipocyte-derived hormone (16 kDa) was identified and named leptin (2). In the early 1950s, leptin's presence had already been predicted with the demonstration of parabiosis studies, which comprised the surgical joining of two mice by skin-to-skin anastomosis (3). Eventually, the circulation of a leptin-deficient *ob/ob* mouse and a leptin receptor-deficient *db/db* mouse was joined. Given the observation that obesity in the *ob/ob* mouse could be cured when connected to the circulation of the *db/db* mouse, it was concluded that the *ob/ob* mouse was missing a circulating factor (4).

Leptin increases energy expenditure and suppresses food intake (2). Interestingly, leptin and insulin share both intracellular and neuronal signaling pathways, the hypothalamic melanocortin system, and interact sub-additively to suppress food intake (5). Leptin signals the information about energy reserves of the body to the brain (6). Therefore, circulating leptin levels are positively correlated with fat mass and reflect the body lipid content (7). Under the conditions of reduced body fat mass, which eventually decreases leptin levels, a mechanism will be triggered leading to adaptive responses to replenish the energy stores in the organism. As a response to a state of starvation, reduced leptin levels decrease energy expenditure which includes the reduction of reproduction and immune responses (8). On the other hand, animal models that are deficient in leptin (*ob/ob* mice) or cannot sense the leptin due to leptin receptor deficiency (*db/db* mice and *fa/fa* rats) become morbidly obese and insulin-resistant (9). Humans with undetectable serum leptin concentrations due to a monogenetic defect display extreme cases of obesity that can be treated with leptin, which causes weight loss due to lowered appetite and increased energy expenditure (10). Interestingly, this leptin sensitivity does not exist in patients with obesity and increased leptin levels (11), pointing to central leptin resistance.

Leptin does not passively cross the BBB because it is too large. Instead, its transport is regulated by a saturable transport system in proportion to circulating levels, and, furthermore, leptin can access the brain by different routes (12). Six leptin receptor isoforms are known (LepRa-f), which differ in the intracellular domains but consist of the same leptin-binding domain. Whereas LepRa-d and f are transmembrane receptors, LepRe does not contain a transmembrane domain but binds circulating leptin and inhibits central leptin transport (12). The most abundant leptin receptor (LepRb) contains an extended intracellular signaling domain and is expressed in the ARC, ventromedial and dorsomedial hypothalamus, and lateral hypothalamic area where it mediates the main effects of leptin in respect to energy homeostasis (12–14).

Besides the limitation of carbohydrate intake due to leptin's anorexic effect, leptin has been shown to act in the brain to lower blood glucose levels via both insulin-dependent and insulin-independent pathways. On the one hand, growing evidence suggests that hypothalamic leptin signaling is beneficial for peripheral insulin sensitivity by affecting primarily the liver's capacity for glucose production via the hepatic branch of the vagus nerve, without affecting food intake (15, 16). On the other hand, leptin also lowers circulating glucose levels independently of insulin. In rodents that are insulin-deficient due to β-cell ablation, it has been demonstrated that an adenovirally induced hyperleptinemia not only normalized the severe hyperglycemia, but leptin also restored terminally ill rodents to full health, despite the absence of insulin (17). In accordance, without increasing circulating leptin, centrally administered leptin lowers blood glucose levels independently of food intake or peripheral insulin elevation by lowering hepatic glucose production, increasing peripheral glucose uptake, and energy expenditure (18, 19). Particularly worthy of mention is that leptin receptor-deficient *db/db* mice not only develop systemic insulin resistance but are brain insulin resistant (20, 21). The fact that leptin can activate insulin signaling via PI3K activation underlines the importance of leptin in the cross-talk between fat tissue and the insulin action in the brain (22).

The observation that high levels of circulating leptin due to obesity lead in most cases to a central and peripheral leptin resistance makes leptin a difficult tool to apply in the clinical treatment of obesity. Unfortunately, the cause of the leptin resistance is far from known, and clinical trials seem to reject the hypothesis that leptin can successfully improve glucose homeostasis in obese humans with type 2 diabetes (23). Interestingly, the detrimental effects of hyperleptinemia can be restored by specifically blocking the excessive leptin with a long-acting leptin antagonist, which improved glucose tolerance during diet-induced obesity in mice (24).

ADIPONECTIN

One important adipokine which attracted much attention because of its antidiabetic effects is adiponectin. It is the most abundant peptide secreted by the adipose tissue and was initially named Acrp30 (adipocyte complement-related protein of 30 kDa) when it was discovered in 1995 (25). Adiponectin was recently shown to be released in humans by the skeletal muscle in obesity and during exercise (26). In 2001, the insulin-sensitizing effect was independently identified by three different groups (27–29). Acting on peripheral tissues, adiponectin is increasing insulin sensitivity, lowering blood glucose through elevated cellular glucose uptake, and reducing hepatic gluconeogenesis; it is involved in the breakdown of fatty acid in the liver and skeletal muscle, as well as in the inhibition of inflammation and oxidative stress (30).

Full-length adiponectin circulates in the blood in a combination of three different major complexes; a low-molecular-weight (LMW) trimer, a middle-molecular-weight (MMW) hexamer, and a high-molecular-weight (HMW) 12- to 18-mer (31). Once synthesized and released, these complexes have distinct biochemical characteristics and do not interconvert between forms in the circulation (32). Interestingly, the three isoforms activate distinct signaling cascades in a tissue-specific manner. Furthermore, adiponectin is circulating as globular adiponectin, a proteolytically cleaved form (29). A sexual dimorphism is expressed in higher levels of total adiponectin and HMW adiponectin in female humans and rodents compared to males (33, 34), indicating that sexual hormones, such as estrogen and testosterone, are involved in the regulation of plasma adiponectin levels. This might contribute to the observation that females are more insulin sensitive than males. In addition, dietary factors, e.g. fish oils, linoleic acid, soy protein, and garlic extract are known to increase plasma adiponectin levels (35–38). In contrast, a carbohydrate-rich diet with a high glycemic load is associated with lower adiponectin concentrations in humans (39).

Adiponectin regulates glucose and lipid homeostasis, targeting the skeletal muscle, liver, and adipose tissue by acting on adiponectin receptors (AdipoRs). Both AdipoR1 and AdipoR2 are ubiquitously expressed and serve as a receptor for globular and full-length adiponectin existing as at least three species of multimers (31). Unlike the classical G-protein-coupled receptors (GPCRs), the AdipoRs have an inverted membrane topology with an internal (cytoplasmatic) NH_2-terminus and an external (extracellular) COOH-terminus (40). In addition, another receptor named T-cadherin is also known to sense hexameric and HMW adiponectin complexes, but not trimeric or globular adiponectin (41). Whereas AdipoR1 has a high affinity for globular adiponectin and is most abundantly expressed in the skeletal muscle, where it activates AMPK and promotes fatty acid oxidation as well as glucose uptake; AdipoR2 is predominantly expressed in the liver and primarily senses full-length adiponectin, leading to AMPK-dependent and -independent suppression of glucose production through a decrease in the expression of gluconeogenic enzymes (e.g. phosphoenolpyruvate carboxylase and glucose-6-phosphatase) (40, 42). In monocyte-derived macrophages from human adipose tissue, adiponectin increases the secretion of the anti-inflammatory cytokine interleukin-10 (43). Consistent with this finding, an adipose tissue-specific overexpression of adiponectin in leptin-deficient (*ob/ob*) mice reduced systemic inflammation and greatly increased subcutaneous fat mass containing smaller adipocytes, leading to increased insulin sensitivity and improved pancreatic β-cell survival (44). It is therefore hypothesized that the ectopic fat accumulation in the liver and the associated insulin resistance is caused by an inability to expand subcutaneous adipose tissue appropriately, with adiponectin as the underlying signal to promote the storage of triglycerides preferentially in the adipose tissue (44).

Besides targeting peripheral tissues, the trimer and hexamer form of adiponectin that have been found in the cerebrospinal fluid (CSF) can act on the brain as well and regulate food intake and energy expenditure (45). Both receptors, AdipoR1 and AdipoR2, are expressed in the arcuate and lateral hypothalamic nuclei, and in the paraventricular hypothalamus (45, 46). However, there is some controversy on whether adiponectin stimulates or suppresses food intake and increases or decreases the energy expenditure in rodents. It has been shown that adiponectin enters the CSF from the circulation, stimulates food intake, and decreases energy expenditure, whereas adiponectin-deficient mice exhibit a decreased food intake and an increased energy expenditure (45). In contrast, an adiponectin treatment in the lateral ventricle of rats elicited an anorexigenic response (46), a slightly reduced food intake without altering the energy expenditure (47), but had no effect on food intake and increased the energy expenditure after an intracerebroventricular administration in mice (48). The exact reason for this discrepancy is not known but leaves room for speculation that besides the animal strain, the animal condition (fasted, fed), a different expression pattern of AdipoRs in the brain, or the different used recombinant adiponectin agents may contribute to the different outcomes.

Both AdipoR1 and AdipoR2 are downregulated in obesity-linked insulin resistance and diabetes (49). Like the receptors, the serum level of adiponectin is also decreased in patients with obesity (50) and type 2 diabetes (34). In addition, lower levels of adiponectin are related to a higher incidence of diabetes, shown in longitudinal and prospective human studies (31). The HMW oligomer of adiponectin plays a major role in improving insulin sensitivity and protecting against type 2 diabetes due to its more active form of the protein compared to the other complexes (42). Impaired adiponectin multimerization and mutations are associated with an increased risk of obesity, insulin resistance, and type 2 diabetes (42, 51). Thus, the HMW adiponectin level might be a more useful marker to evaluate insulin resistance than the total adiponectin levels (52). In consistency, increasing specifically the HMW form of adiponectin in mice by overexpression protected the mice from insulin resistance on a high-fat diet (53). The research, therefore, focuses on understanding the prevention of HMW oligomer adiponectin formation. In addition, peroxisome proliferator-activated receptor gamma (PPARγ) agonists (thiazolidinediones) have insulin-sensitizing effect in patients with type 2 diabetes. These oral antihyperglycemic drugs (e.g. rosiglitazone, pioglitazone, troglitazone) increase the circulating levels of adiponectin in both humans and animals (27, 54, 55). However, troglitazone had to be removed from the market due to serious liver injury including acute liver failure. Interestingly, pioglitazone-induced improvement of insulin sensitivity may occur adiponectin-dependently in the liver and adiponectin-independently in skeletal muscle as demonstrated in adiponectin-deficient mice (56). In the human brain (postmortem frontal cortex), it was demonstrated that insulin and adiponectin signaling interacts (57). Here, adiponectin enhances insulin receptor activation and signaling while insulin inhibits adiponectin-induced signaling.

Furthermore, adiponectin mediates the beneficial glucose homeostasis response of fibroblast growth factor 21 (FGF21) following either acute or chronic FGF21 administration (58, 59). Increased levels of adiponectin in response to administration of FGF21 have been reported in rodents and humans (58–60). Whereas in rodents, the beneficial glucose homeostasis response of acute or chronic FGF21 administration seems to be mediated by adiponectin in a few studies (58, 59), in another study with adiponectin-deficient mice, FGF21 increases, mediated by a methionine-restrictive diet, and insulin sensitivity improves as well (61). In accordance, it could be shown that adiponectin is not required for the acute effects of FGF21 on insulin sensitivity because FGF21 signals to brown adipocytes to mediate its acute glucose-lowering effects (62). Adiponectin is thus dispensable for the metabolic effects of FGF21, a liver-derived hormone that will be discussed in more detail in the next chapter.

LIVER

The liver's ability to respond to changes in the nutritional status is enormous. By sensing the availability of nutrients and adjusting energy and metabolite production for the entire organism, it is the main regulator of energy homeostasis. Unsurprisingly, poorer health status of the liver leads to the onset and development of metabolic diseases.

After a meal, the liver responds to the release of insulin with an increased uptake of glucose, which will be converted into glycogen or used for *de novo* lipogenesis. In response to fasting, on the other hand, the liver is a fuel supplier for the muscle and brain; via glycogenolysis and gluconeogenesis it increases its glucose production, or it produces ketone bodies from the oxidation of lipids. However, besides the storage or conversion of energy, the liver is also able to exert widespread effects through the secretion of multiple hormones (hepatokines) which regulate insulin's action in the brain; whereas brain insulin is capable of regulating hepatic glucose metabolism (63).

REGULATION OF FIBROBLAST GROWTH FACTOR 21

The liver-derived FGF21 is an endocrine hormone that acts physiologically to maintain energy and nutrient homeostasis, induces a macronutrient-specific preference for protein, and has beneficial effects on glucose homeostasis (64, 65). FGF21 improves glucose homeostasis and insulin sensitivity while altering lipid metabolism and energy expenditure (66–69), and it is linked to the adaptive response to starvation (70–73). These data have led to significant interest in FGF21 with respect to the treatment of diabetes, obesity, and metabolic syndrome.

FGF21 is a 19 kDa protein and belongs to the endocrine subfamily of 22 mammalian FGFs, which are considered intracrine, autocrine, or paracrine. Unlike the majority of FGFs which bind heparan sulfate glycosaminoglycans (HSGAGs) and become predominantly captured within the extracellular matrix, preventing their distribution from their cell of origin (74), FGF21 enters the circulation upon secretion due to its reduced HSGAGs binding. Circulating FGF21 is primarily derived from the liver, but is also expressed in the thymus, gut, brain, white adipose tissue (WAT), brown adipose tissue (BAT), muscle, heart, and pancreas (75–77). FGF21 signals through a cell-surface receptor complex comprised of both the ubiquitously expressed fibroblast growth factor receptor (FGFR), FGFR1c, and the FGF co-receptor β-klotho (78). The extracellular signal-regulated kinases 1/2 (ERK1/2) and other downstream kinases are activated afterward. In dependency on the β-klotho expression, FGF21 targets a variety of different organs, such as the liver, pancreas, WAT, BAT, hypothalamus, and brainstem (75, 79–81). Exogenously administered FGF21 acts directly on adipose tissue to decrease blood glucose levels and improve hepatic and possibly peripheral insulin sensitivity, reduce body weight and serum lipids (82, 83). In adipocytes, FGF21 induces glucose transporter-1 (GLUT1) expression and glucose uptake (84). FGF21 treatment causes browning of WAT, increases thermogenesis in BAT and WAT (83), and reduces hepatic lipogenesis (66). The direct administration of FGF21 into the brain has been shown to induce activation of BAT and to increase energy expenditure and insulin sensitivity in the liver (85, 86). Furthermore, FGF21 increases insulin sensitivity through specific expansion of subcutaneous fat (87). Because circulating FGF21 levels positively correlate with the subcutaneous adipose tissue in insulin-sensitive obese individuals, it might be concluded that elevated endogenous FGF21 in obesity serves as a defense mechanism against systemic insulin resistance (87). FGF21's antidiabetic properties can be attributed to renal glucose reabsorption as well, given that FGF21 enhances hyperglycemia in part via reducing renal glucose reabsorption through the PPARδ-mediated SGLT2 pathway (88). Studies investigating whether FGF21 modulates insulin sensitivity in the brain are still missing. So far, FGF21-stimulated adiponectin secretion might partly explain the effects on insulin sensitivity (59). The improvements in hepatic and muscle insulin resistance - via chronic FGF21 infusion - were abolished in adiponectin knockout mice (59). However, the possible role of FGF21 and its insulin actions in the brain in regulating glucose metabolism remains unknown.

The function and induction of FGF21 are complex. After the discovery of the *Fgf21* gene in 2000 (76), FGF21 was proposed to be a potential anti-diabetic drug due to its positive regulation of glucose uptake in adipocytes (67). Over the following years, FGF21 has been widely considered a fasting-induced hormone with the peroxisome proliferator-activated receptor-alpha (PPARα), which becomes activated by fatty acids (89), as the underlying mechanism. Discovered by multiple groups, FGF21 mediates the adaptive fasting response to induce lipolysis, lipid β-oxidation, ketogenesis, and gluconeogenesis in animal models (70, 71, 90). Consistent with observations in rodents, in humans, circulating levels of FGF21 are also induced during seven days of fasting (91, 92). However, in *Fgf21*-KO mice, no impaired response to fasting could be observed, suggesting that FGF21 is not necessary for the physiological response to caloric restriction (93).

Although FGF21 is elevated in low-caloric environments, increased FGF21 is also observed in seemingly opposite nutritional contexts such as overfeeding, ketogenic diets, and high-carbohydrate diets (94). Carbohydrates, particularly fructose, stimulate the FGF21 production by activating the transcription factor carbohydrate-response element-binding protein (ChREBP) through the intermediate xylulose-5-phosphate (95–97). ChREBP knockout mice do not show elevated FGF21 levels in response to simple sugar intake (98, 99). Just like excess fructose consumption, FGF21 is robustly induced by ethanol in both rodents and humans (100). Interestingly, the overexpression of

FGF21 (101) or exogenous FGF21 treatment (102) is reducing the consumption of alcohol by acting on the brain. Similarly, FGF21 administration reduces the carbohydrate appetite (99) and the overexpression of FGF21 decreases the sucrose preference as well as the desire for noncaloric sweeteners (99, 101). The fact that central insulin via intranasal insulin administration also reduces the food intake of chocolate chip cookies in sated women (103) demonstrates that FGF21 and insulin can exert similar effects on carbohydrate intake.

The seemingly paradoxical gene expression of *Fgf21* caused by fasting (free fatty acids) and feeding signals (fructose, alcohol) suggests that FGF21 is elevated under a nutritional crisis, which includes energy restriction and overfeeding with liver-toxic fructose and alcohol. Since caloric restriction automatically entails a reduction of protein intake, which may be designated as a nutritional crisis as well, it was verified that circulating FGF21 concentrations are robustly induced by dietary protein restriction in rats, mice, and humans – without caloric restriction – which indicates FGF21 as an endocrine signal of dietary protein restriction (104). The general amino acid control pathway is induced by protein restriction and is required for activation of FGF21. The serine/threonine kinase general control nonderepressible 2 (GCN2) is a cellular kinase that phosphorylates the eukaryotic initiation factor 2 alpha (eIF2α) in response to depletion of cellular amino acids (105), with this mechanism coupling amino acid detection to metabolism (106–113). The phosphorylation of eIF2α results in the induction of the activating transcription factor 4 (ATF4), which binds to the FGF21 promoter that contains amino acid response elements (AARE) (114–117).

GLUCOSE TOLERANCE IS IMPROVED BY A DIETARY PROTEIN RESTRICTION IN RESPONSE TO A HIGH-FAT DIET

A consistent, powerful effect of dietary protein restriction is the limitation of weight gain (104, 117). Importantly, this effect is primarily driven by the reduction in fat gain, even in mice consuming a diet with relatively low-fat content. In addition, dietary protein restriction protects against HFD-induced obesity as well (118). This anti-obesity effect is associated with enhanced glucose tolerance and reduced fasting insulin. In contrast, these beneficial effects were absent in *Fgf21*-KO mice, which show no changes in glucose tolerance or insulin levels when fed a high-fat / low-protein diet (118, 119). As a consequence, protein restriction requires FGF21 to protect against HFD-induced glucose intolerance.

The fact that adipose tissue expresses the FGF21 co-receptor β-klotho and responds to FGF21 gives rise to the question of whether low protein-induced FGF21 regulates body weight gain via the adipose tissue. Interestingly, the FGF21 signaling in the adipose tissue is not required to reduce body weight when mice were protein-restricted (118).

FGF21 exerts its multifaceted metabolic effects through its actions on various targets, including the brain. It is therefore of great interest to investigate the brain's contribution to low protein-induced FGF21 actions. In contrast to adipose-specific *Klb*-KO mice, brain-specific *Klb*-KO mice do not show a blunted body weight and fat mass gain on a low-protein diet in the context of a normal-fat (22 kcal%) or high-fat (59 kcal%) diet (118). As stated above, protein restriction improves glucose tolerance via an FGF21-dependent mechanism; the question arises whether that protective effect also requires FGF21 signaling in the brain. Both glucose tolerance and insulin sensitivity were improved on high-fat/low-protein diet in control mice, whereas these improvements in glucose homeostasis were lost in brain-specific *Klb*-KO mice (118). Collectively, these data demonstrate that FGF21 signaling via *Klb* within the brain was required for protein restriction sensing and essential to improve glucose metabolism in high-fat / protein-restricted mice. Thus, these data suggest that liver-derived FGF21 is, in fact, a 'protein sensing signal' which acts on the brain to influence metabolism in the protein-dependent state.

DIETARY AMINO ACID RESTRICTION PROTECTS FROM TYPE 2 DIABETES BY FGF21 SECRETION

Due to its beneficial role in the regulation of insulin sensitivity and glucose homeostasis, the endocrine hormone FGF21 is acknowledged as a promising therapeutic candidate for the treatment of diabetes. FGF21 was shown to improve β-cell function and survival (120), suppress islet glucagon secretion (67), and prevent pancreatic inflammation (121). Whereas FGF21 stimulated insulin secretion in *ex vivo* islets isolated from diabetic rodents (120), it did not affect insulin secretion in islets isolated from healthy mice (122). However, it has also been proposed that obesity results in an FGF21-resistance (123, 124), indicating that FGF21 exposure in obese models did not create any benefit. In addition, both obese and diabetic models showed elevated levels of circulating FGF21 (66, 67, 124). Furthermore, in patients with obesity and type 2 diabetes, FGF21 levels were up to 2-fold higher compared with lean healthy subjects (125, 126). Obese rhesus macaque monkeys displayed high levels of FGF21 and reduced levels of the co-receptor β-klotho in white adipose tissue, whereas monkeys that maintain normal levels of β-klotho were protected against the development of dyslipidemia and hyperglycemia (127). Interestingly, the downregulation of β-klotho expression is not the primary mechanism contributing to weakened FGF21 signaling in white adipose tissue (128). In contrast to a few studies suggesting an FGF21 resistance, various studies (66, 67, 129, 130) were conducted in obese animals which may have impaired FGF21 signaling in response to physiological levels of FGF21, showing a protective metabolic function in regard to pharmacological levels of exogenous FGF21. Treatment of obese, hyperglycemic, insulin-resistant, and leptin-deficient *ob/ob* mice with FGF21 could normalize hyperglycemia despite markedly elevated endogenous FGF21 levels (129), concluding that an apparent FGF21 resistance was not evident in this mouse model for obesity and insulin resistance. Under pathological conditions, the increased plasma FGF21 levels therefore apparently represented an FGF21 resistance, but this was rather a compensatory response that consisted of a glucose homeostasis improvement. To support that, exogenous administration of FGF21 is sufficient to prevent hyperglycemia and improve β-cell function in a model for polygenic obesity and diabetes, the New Zealand Obese (NZO) mouse, which – in contrast to other obese mouse models – develops severe hyperglycemia with β-cell loss (131). Furthermore, in NZO mice dietary protein restriction prevents the onset of hyperglycemia, but only when protein was substituted by fat (132). When protein was substituted by carbohydrates, mice developed hyperglycemia, despite elevated plasma FGF21 levels and increased energy expenditure (132).

As stated above, exogenous FGF21 treatment and dietary protein restriction-induced FGF21 prevented the onset of diabetes in NZO mice (131, 132). Furthermore, restriction of the essential amino acid methionine is known to mimic metabolic effects of energy and protein restriction, with FGF21 as a required endocrine factor (133–136). Dietary methionine restriction has been shown to improve glucose homeostasis as well as insulin sensitivity and to prevent weight gain in rodents and humans (134–138). The potential health benefits of vegan and vegetarian diets, including lower rates of obesity and diabetes (139), may be connected to lower methionine intake due to their naturally low methionine levels (140–142). As a matter of fact, a dietary methionine restriction in the context of obesity and a high-fat diet regimen also protects against hyperglycemia and β-cell loss in NZO mice, presumably via increased plasma adiponectin (143), mediating the beneficial glucose homeostasis response. Nevertheless, adiponectin-KO mice on a methionine restrictive diet increased insulin sensitivity (61), concluding that elevated adiponectin plasma levels under dietary methionine restriction are not the only reason for improved insulin sensitivity. Anyhow, the beneficial effects were mediated by FGF21 as, in the presence of cystine, methionine restriction did not increase FGF21 and did not improve glucose homeostasis (143). Interestingly, in humans circulating FGF21 levels are higher in vegans and vegetarians compared with omnivores, and already a short-term switch to a vegetarian diet increased plasma FGF21 in omnivores (143). This might be of interest because lowering the methionine intake, for instance, with vegetarian or vegan diets, which have naturally low methionine levels, may be linked to health benefits (139). In other words, a methionine-restrictive diet (e.g. vegan) may be a beneficial nutritional strategy in patients at high risk of developing type 2 diabetes.

In conclusion, restriction of dietary protein and methionine elevates plasma FGF21 levels, which protect from hyperglycemia, an effect compromised by high dietary carbohydrates. Patients at high risk of developing type 2 diabetes may benefit from dietary changes in favor of methionine restriction (vegan and vegetarian diets), via increased circulating levels of FGF21, without protein and calorie restriction.

PERSPECTIVE

It is noteworthy to mention that dietary polyphenols such as curcumin (turmeric) and resveratrol (grapes), and the compound betaine (sugar beets) have been shown to alter FGF21 expression and improve glucose homeostasis in mice (144–146).

Driven by a pharmacological interest, various FGF21-class molecules were designed to treat obesity-related diseases. To turn FGF21 into a drug, FGF21 analogs were designed to mimic the activity of FGF21 while increasing the stability (lengthening the circulating half-life; preventing the proteolysis by fibroblast activation protein) and increasing the affinity of binding to β-klotho, the obligate co-receptor of FGF21 (147). Only to mention a recent one, the Fc-FGF21 fusion protein AKR-001 features stabilized N- and C-terminal domains and a half-life of about 3-3.5 days (~1 hour for the native mature form of FGF21) and exhibits positive trends in various metabolic markers (e.g. insulin sensitivity) in phase 1 multiple ascending dose study in individuals with type 2 diabetes (148). To improve FGF21's therapeutic properties, FGF21 was fused with the incretin glucagon-like-peptide 1 (GLP-1) to generate a GLP-1/FGF21 dual agonist (149, 150). This dual agonist can activate both GLP-1 and FGF21 receptors. Treatment of diabetic mice with GLP-1/FGF21 dual agonist results in improved glycemic control, with better efficiencies than marketed GLP-1 analogs or single-targeted FGF21 analogs. Whether these effects can be transferred to humans is worthy of clinical investigation.

INSULIN-LIKE GROWTH FACTORS

Insulin-like growth factors were discovered in 1957 when Salmon and Daughaday began studying the effect of biological substances to stimulate cartilage sulfation *in vitro* (151). IGFs are hormones similar to insulin with regard to molecular structure and functionality and increase insulin sensitivity. IGFs play a pivotal role in growth and development in an autocrine, paracrine and endocrine manner. The two main IGFs, IGF1 and IGF2, are evolutionarily conserved peptide hormones with structural homology to proinsulin (152). The three ligands IGF1, IGF2, and insulin bind with different affinities to four types of receptors; the IGF1-R, the IGF2-R, the hybrid receptor, and the insulin receptor (153). Finally, the binding of the ligand to the receptor elicits a phosphorylation of intracellular adaptor proteins and activation of the MAPK and PI3K/AKT signaling pathways, which leads to GLUT4 translocation from an intracellular compartment to the plasma membrane, enhancing glucose uptake (153, 154). In comparison to insulin, IGFs are about 90% less potent to stimulate glucose transport and induce hypoglycemia *in vitro* and in vivo, respectively, but are about 100 times higher concentrated (155, 156). Furthermore, IGFs are regulated by a family of seven IGF-binding proteins (IGFBP) that transport IGFs, extend their half-life, and regulate their clearance (157). Besides limiting the hypoglycemic effects of the circulating IGF1 and IGF2, IGFBP fulfills IGF-independent functions (158).

IGF1

IGF1 is primarily secreted by the liver – as a response to growth hormone stimulation – and exists almost entirely bound to IGFBPs (152). Important for normal carbohydrate metabolism, IGF1 mimics actions of insulin, e.g. stimulation of glucose and amino acid uptake, and the inhibition of gluconeogenesis (159). Similar to FGF21, central IGF1 can improve insulin sensitivity and glucose tolerance, and stimulate energy expenditure via thermogenesis in mice through intracerebroventricular injection or IGF1 overexpression (160). Importantly, IGF1 can cross the blood-brain barrier (BBB) when bound to binding proteins and is produced by all major cell types in the CNS in the developing brain, whereas IGF1-R is expressed in all neuroepithelial cell types and shows a relatively stable pattern of expression from early development to maturity (161).

In healthy conditions, IGF1 acts on the skeletal muscle and elicits its insulin-sensitizing effects to lower glycemia (159, 162). IGF1 does not act on hepatocytes and in terms of insulin sensitizing on adipocytes (163), and therefore its primary insulin-sensitizing action is supposed to be mediated through the skeletal muscle. As GLP-1 and insulin can partly signal through the same receptor, single IGF1-R or insulin receptor (IR) knockout models do not completely eliminate their action. In mice where both IGF1-R and IR were inactivated in the brain, the glucose tolerance was impaired (164). However, at the beginning of insulin resistance, with blood glucose levels at a normal range, circulating IGF1 bioactivity increases until reaching a plateau (165). Eventually, in a progressing development of type 2 diabetes, patients with increased blood glucose levels show declined circulating bioactivity of IGF1 (165). Furthermore, patients with type 2 diabetes show elevated IGFBP1 levels, indicating an initial compensation of increased IGF1 secretion to counteract the hypoglycemic effect of insulin via inhibition of IGF1 (158, 166). Despite the capacity of IGF1 to significantly lower blood glucose and increase insulin sensitivity in human subjects with type 2 diabetes, IGF1 has become less attractive for therapeutic usage due to some adverse effects (167).

IGF2

Mentioned for the sake of completeness, the hepatokine IGF2 predominantly binds to IGF2-R. In contrast to the activation of IGF1-R by IGF1, which leads to intracellular signaling as described above, IGF2-R will be degraded after internalization (153). IGF2 functions predominantly in fetal development (168). Older studies demonstrate that IGF2 gene variants might affect insulin sensitivity and BMI (153).

Besides IGF1 and FGF21, there are other beneficial hepatokines that improve glucose homeostasis, such as activin-E, ANGPTL6, GDF15, lipocalin 13, and SMOC1. They are not further discussed here but can be studied in a review (166).

GASTROINTESTINAL TRACT AND GLP-1

One promising therapeutic target in the treatment of type 2 diabetes is the glucagon-like-peptide 1 (GLP-1), an incretin hormone of 37-amino acids which is released by the enteroendocrine L cells in the gut during digestion and enhances the glucose-induced insulin secretion in the pancreas (169). Discovered in 1980 in anglerfish, GLP-1 integrates energy metabolism, feeding behavior (GLP-1 is anorexigenic) as well as learning and is implicated in the neuroprotection and memory (170, 171). Besides being released by the gut, it has extra-pancreatic effects and acts as a neuropeptide deriving from preproglucagon-expressing neurons in the nucleus tractus solitarius (NTS) of the brain stem (172, 173). Furthermore, GLP-1 and its analogs are able to pass the BBB (174, 175) and facilitate insulin signaling via the GLP-1 receptor (GLP-1R). However, endogenous GLP-1 derived from the gut is unlikely to cross the BBB due to the rapid cleavage by the dipeptidyl peptidase-4 (DPP-4) which shortens GLP-1's half-life to less than 2 min (176). Therefore, pharmaceutical agents were developed that either mimic the physiological effects of GLP-1 through GLP-1 receptor agonists (GLP-1 agonists, e.g. liraglutide, exenatide) or by inhibiting GLP-1 degradation through DPP-4 inhibitors (gliptins). On a side note, exenatide is a synthetic version of exendin-4, a hormone found in the saliva of the "Gila monster" lizard (177, 178). GLP-1 receptor agonists have a longer half-life of several hours. Owing to their structural alterations, they are DPP-4-resistant. That means that endogenous GLP-1 (due to inhibited degradation by gliptins) or GLP-1 receptor agonists signal to the brain not only through neural pathways (vagal afferent fibers) but also through the humoral pathway, meaning passage of the BBB or binding to GLP-1 receptors expressed in the circumventricular organ, which has a leaky BBB (179–181).

The G-protein-coupled GLP-1 receptors (GLP-1R) are expressed in peripheral tissues and in the brain, e.g. in the hypothalamus, thalamus, brainstem nucleus, cortex, and hippocampus (182). Therefore, the effects of GLP-1 on energy metabolism are mediated by both the peripheral nervous system and the central nervous system (183). Because impaired glucose utilization in the brain is characteristic of Alzheimer's disease (AD), an accumulation of aggregated amyloid-β protein (extracellular) and hyper-phosphorylated tau protein (intracellular), AD and type 2 diabetes are pathophysiologically related (184, 185). In AD patients, the hippocampus and cerebral cortex, which are brain areas that are related to cognitive functions, display a decreased glucose utilization (186). The exact connection between type 2 diabetes and Alzheimer's disease is still in debate. Nevertheless, in the brains of AD patients with a cognitive decline, not only insulin resistance but also IGF-1 resistance can be observed (187). According to rodent studies, GLP-1 receptor-deficient mice show impaired memory retention (184), and the intraperitoneal treatment of rats with the GLP-1 agonist exendin-4 leads to an improvement in memory performance (188). Besides a diminished accumulation of endogenous amyloid-β protein and amyloid plaques in the AD brain as a response to a GLP-1 receptor agonist, the glucose metabolism is stimulated and cognitive function is improved (184). Lastly, central insulin resistance leading to impaired cognitive function can be improved by GLP-1.

CONCLUDING REMARKS

As stated above, functional insulin action in the brain is mandatory for the maintenance of systemic glucose homeostasis, whereas impaired central insulin signaling can be observed in obesity, type 2 diabetes, and Alzheimer's disease.

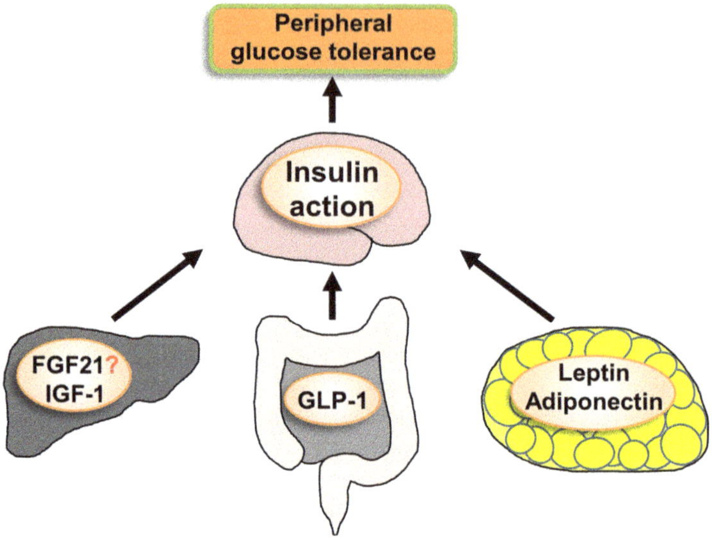

Figure 9.1 Several endocrines that cross the blood-brain barrier are able to influence insulin signaling in the brain. Therefore, the communication between the organs partly relies on insulin action in the brain to ensure the maintenance of systemic homeostasis. Whether FGF21 modulates central insulin sensitivity after passing the blood-brain barrier is unknown. (Abbreviations: FGF21, fibroblast growth factor 21; GLP-1, glucagon-like peptide-1; IGF-1, insulin-like growth factor 1.)

Interestingly, several endocrines that cross the blood-brain barrier are able to influence insulin signaling in the brain, thus extending their range of action by exerting their effects not only in the periphery. These findings support the idea of restoring insulin sensitivity in the brain using these circulating endocrines (Figure 9.1).

ACKNOWLEDGMENTS

Linguistic refinements of the text by Nicole Kühn are gratefully acknowledged. Thomas Laeger was supported by a research fellowship from the Deutsche Forschungsgemeinschaft (DFG) LA 3042/-2-1 and by grants LA 3042/3-1 and LA 3042/4-1.

REFERENCES

1. C. Muller et al., Effect of chronic intracerebroventricular infusion of insulin on brown adipose tissue activity in fed and fasted rats. *Int J Obes Relat Metab Disord* **21**, 562–566 (1997).
2. Y. Zhang et al., Positional cloning of the mouse obese gene and its human homologue. *Nature* **372**, 425–432 (1994).
3. D. L. Coleman, A historical perspective on leptin. *Nat Med* **16**, 1097–1099 (2010).
4. D. L. Coleman, Effects of parabiosis of obese with diabetes and normal mice. *Diabetologia* **9**, 294–298 (1973).
5. E. L. Air, S. C. Benoit, D. J. Clegg, R. J. Seeley, S. C. Woods, Insulin and leptin combine additively to reduce food intake and body weight in rats. *Endocrinology* **143**, 2449–2452 (2002).
6. P. Cohen et al., Selective deletion of leptin receptor in neurons leads to obesity. *J Clin Invest* **108**, 1113–1121 (2001).
7. R. C. Frederich et al., Leptin levels reflect body lipid content in mice: evidence for diet-induced resistance to leptin action. *Nat Med* **1**, 1311–1314 (1995).
8. J. S. Flier, E. Maratos-Flier, Leptin's physiologic role: does the emperor of energy balance have no clothes? *Cell Metab* **26**, 24–26 (2017).
9. D. J. Dyck, G. J. Heigenhauser, C. R. Bruce, The role of adipokines as regulators of skeletal muscle fatty acid metabolism and insulin sensitivity. *Acta Physiol (Oxf)* **186**, 5–16 (2006).
10. I. S. Farooqi et al., Effects of recombinant leptin therapy in a child with congenital leptin deficiency. *N Engl J Med* **341**, 879–884 (1999).
11. M. S. Westerterp-Plantenga, W. H. Saris, C. J. Hukshorn, L. A. Campfield, Effects of weekly administration of pegylated recombinant

human OB protein on appetite profile and energy metabolism in obese men. *Am J Clin Nutr* **74**, 426–434 (2001).

12. H. Munzberg, C. D. Morrison, Structure, production and signaling of leptin. *Metabolism* **64**, 13–23 (2015).
13. A. Boucsein, K. Kamstra, A. Tups, Central signalling cross-talk between insulin and leptin in glucose and energy homeostasis. *J Neuroendocrinol*, **33**, e12944 (2021).
14. M. G. Myers, M. A. Cowley, H. Munzberg, Mechanisms of leptin action and leptin resistance. *Annu Rev Physiol* **70**, 537–556 (2008).
15. J. German et al., Hypothalamic leptin signaling regulates hepatic insulin sensitivity via a neurocircuit involving the vagus nerve. *Endocrinology* **150**, 4502–4511 (2009).
16. G. J. Morton et al., Leptin regulates insulin sensitivity via phosphatidylinositol-3-OH kinase signaling in mediobasal hypothalamic neurons. *Cell Metab* **2**, 411–420 (2005).
17. X. Yu, B. H. Park, M. Y. Wang, Z. V. Wang, R. H. Unger, Making insulin-deficient type 1 diabetic rodents thrive without insulin. *Proc Natl Acad Sci U S A* **105**, 14070–14075 (2008).
18. C. Y. Lin, D. A. Higginbotham, R. L. Judd, B. D. White, Central leptin increases insulin sensitivity in streptozotocin-induced diabetic rats. *Am J Physiol Endocrinol Metab* **282**, E1084–1091 (2002).
19. S. Hidaka et al., Chronic central leptin infusion restores hyperglycemia independent of food intake and insulin level in streptozotocin-induced diabetic rats. *FASEB J* **16**, 509–518 (2002).
20. S. E. Arnold et al., Brain insulin resistance in type 2 diabetes and Alzheimer disease: concepts and conundrums. *Nat Rev Neurol* **14**, 168–181 (2018).
21. S. M. Son et al., Altered APP processing in insulin-resistant conditions is mediated by autophagosome accumulation via the inhibition of mammalian target of rapamycin pathway. *Diabetes* **61**, 3126–3138 (2012).
22. K. D. Niswender et al., Insulin activation of phosphatidylinositol 3-kinase in the hypothalamic arcuate nucleus: a key mediator of insulin-induced anorexia. *Diabetes* **52**, 227–231 (2003).
23. R. Coppari, C. Bjorbaek, Leptin revisited: its mechanism of action and potential for treating diabetes. *Nat Rev Drug Discov* **11**, 692–708 (2012).
24. D. Pretz, C. Le Foll, M. Z. Rizwan, T. A. Lutz, A. Tups, Hyperleptinemia as a contributing factor for the impairment of glucose intolerance in obesity. *FASEB J* **35**, e21216 (2021).
25. P. E. Scherer, S. Williams, M. Fogliano, G. Baldini, H. F. Lodish, A novel serum protein similar to C1q, produced exclusively in adipocytes. *J Biol Chem* **270**, 26746–26749 (1995).
26. S. F. Martinez-Huenchullan et al., Skeletal muscle adiponectin induction in obesity and exercise. *Metabolism* **102**, 154008 (2020).
27. T. Yamauchi et al., The fat-derived hormone adiponectin reverses insulin resistance associated with both lipoatrophy and obesity. *Nat Med* **7**, 941–946 (2001).
28. A. H. Berg, T. P. Combs, X. Du, M. Brownlee, P. E. Scherer, The adipocyte-secreted protein Acrp30 enhances hepatic insulin action. *Nat Med* **7**, 947–953 (2001).
29. J. Fruebis et al., Proteolytic cleavage product of 30-kDa adipocyte complement-related protein increases fatty acid oxidation in muscle and causes weight loss in mice. *Proc Natl Acad Sci U S A* **98**, 2005–2010 (2001).
30. T. Yamauchi et al., Adiponectin stimulates glucose utilization and fatty-acid oxidation by activating AMP-activated protein kinase. *Nat Med* **8**, 1288–1295 (2002).
31. T. Kadowaki et al., Adiponectin and adiponectin receptors in insulin resistance, diabetes, and the metabolic syndrome. *J Clin Invest* **116**, 1784–1792 (2006).
32. T. Schraw, Z. V. Wang, N. Halberg, M. Hawkins, P. E. Scherer, Plasma adiponectin complexes have distinct biochemical characteristics. *Endocrinology* **149**, 2270–2282 (2008).
33. T. P. Combs et al., Sexual differentiation, pregnancy, calorie restriction, and aging affect the adipocyte-specific secretory protein adiponectin. *Diabetes* **52**, 268–276 (2003).
34. K. Hotta et al., Plasma concentrations of a novel, adipose-specific protein, adiponectin, in type 2 diabetic patients. *Arterioscler Thromb Vasc Biol* **20**, 1595–1599 (2000).
35. P. Flachs et al., Polyunsaturated fatty acids of marine origin induce adiponectin in mice fed a high-fat diet. *Diabetologia* **49**, 394–397 (2006).
36. K. Nagao, N. Inoue, Y. M. Wang, T. Yanagita, Conjugated linoleic acid enhances plasma adiponectin level and alleviates hyperinsulinemia and hypertension in Zucker diabetic fatty (fa/fa)

rats. *Biochem Biophys Res Commun* **310**, 562–566 (2003).
37. A. Nagasawa et al., Effects of soy protein diet on the expression of adipose genes and plasma adiponectin. *Horm Metab Res* **34**, 635–639 (2002).
38. D. Gomez-Arbelaez et al., Aged garlic extract improves adiponectin levels in subjects with metabolic syndrome: a double-blind, placebo-controlled, randomized, crossover study. *Mediators Inflamm* **2013**, 285795 (2013).
39. T. Pischon, C. J. Girman, N. Rifai, G. S. Hotamisligil, E. B. Rimm, Association between dietary factors and plasma adiponectin concentrations in men. *Am J Clin Nutr* **81**, 780–786 (2005).
40. T. Yamauchi et al., Cloning of adiponectin receptors that mediate antidiabetic metabolic effects. *Nature* **423**, 762–769 (2003).
41. C. Hug et al., T-cadherin is a receptor for hexameric and high-molecular-weight forms of Acrp30/adiponectin. *Proc Natl Acad Sci U S A* **101**, 10308–10313 (2004).
42. M. Liu, F. Liu, Regulation of adiponectin multimerization, signaling and function. *Best Pract Res Clin Endocrinol Metab* **28**, 25–31 (2014).
43. M. Kumada et al., Adiponectin specifically increased tissue inhibitor of metalloproteinase-1 through interleukin-10 expression in human macrophages. *Circulation* **109**, 2046–2049 (2004).
44. J. Y. Kim et al., Obesity-associated improvements in metabolic profile through expansion of adipose tissue. *J Clin Invest* **117**, 2621–2637 (2007).
45. N. Kubota et al., Adiponectin stimulates AMP-activated protein kinase in the hypothalamus and increases food intake. *Cell Metab* **6**, 55–68 (2007).
46. A. Coope et al., AdipoR1 mediates the anorexigenic and insulin/leptin-like actions of adiponectin in the hypothalamus. *FEBS Lett* **582**, 1471–1476 (2008).
47. M. Bassi, J. M. do Carmo, J. E. Hall, A. A. da Silva, Chronic effects of centrally administered adiponectin on appetite, metabolism and blood pressure regulation in normotensive and hypertensive rats. *Peptides* **37**, 1–5 (2012).
48. Y. Qi et al., Adiponectin acts in the brain to decrease body weight. *Nat Med* **10**, 524–529 (2004).
49. A. Tsuchida et al., Insulin/Foxo1 pathway regulates expression levels of adiponectin receptors and adiponectin sensitivity. *J Biol Chem* **279**, 30817–30822 (2004).
50. Y. Arita et al., Paradoxical decrease of an adipose-specific protein, adiponectin, in obesity. *Biochem Biophys Res Commun* **257**, 79–83 (1999).
51. H. Waki et al., Impaired multimerization of human adiponectin mutants associated with diabetes. Molecular structure and multimer formation of adiponectin. *J Biol Chem* **278**, 40352–40363 (2003).
52. Y. Seino, H. Hirose, I. Saito, H. Itoh, High molecular weight multimer form of adiponectin as a useful marker to evaluate insulin resistance and metabolic syndrome in Japanese men. *Metabolism* **56**, 1493–1499 (2007).
53. M. Liu et al., Fat-specific DsbA-L overexpression promotes adiponectin multimerization and protects mice from diet-induced obesity and insulin resistance. *Diabetes* **61**, 2776–2786 (2012).
54. T. P. Combs et al., Induction of adipocyte complement-related protein of 30 kilodaltons by PPARgamma agonists: a potential mechanism of insulin sensitization. *Endocrinology* **143**, 998–1007 (2002).
55. S. A. Phillips et al., Modulation of circulating and adipose tissue adiponectin levels by antidiabetic therapy. *Diabetes* **52**, 667–674 (2003).
56. N. Kubota et al., Pioglitazone ameliorates insulin resistance and diabetes by both adiponectin-dependent and -independent pathways. *J Biol Chem* **281**, 8748–8755 (2006).
57. H. Y. Wang et al., Insulin and adipokine signaling and their cross-regulation in postmortem human brain. *Neurobiol Aging* **84**, 119–130 (2019).
58. W. L. Holland et al., An FGF21-adiponectin-ceramide axis controls energy expenditure and insulin action in mice. *Cell Metab* **17**, 790–797 (2013).
59. Z. Lin et al., Adiponectin mediates the metabolic effects of FGF21 on glucose homeostasis and insulin sensitivity in mice. *Cell Metab* **17**, 779–789 (2013).
60. S. Talukdar et al., A long-acting FGF21 molecule, PF-05231023, decreases body weight and improves lipid profile in non-human primates and type 2 diabetic subjects. *Cell Metab* **23**, 427–440 (2016).
61. K. P. Stone et al., Compromised responses to dietary methionine restriction in adipose tissue but not liver of ob/ob mice. *Obesity (Silver Spring)* **23**, 1836–1844 (2015).
62. L. D. BonDurant et al., FGF21 regulates metabolism through adipose-dependent and

-independent mechanisms. *Cell Metab* **25**, 935–944 e934 (2017).

63. H. Inoue et al., Role of hepatic STAT3 in brain-insulin action on hepatic glucose production. *Cell Metab* **3**, 267–275 (2006).
64. C. M. Hill et al., FGF21 and the physiological regulation of macronutrient preference. *Endocrinology* **161**, (2020).
65. K. H. Flippo, M. J. Potthoff, metabolic messengers: FGF21. *Nat Metab* **3**, 309–317 (2021).
66. T. Coskun et al., Fibroblast growth factor 21 corrects obesity in mice. *Endocrinology* **149**, 6018–6027 (2008).
67. A. Kharitonenkov et al., FGF-21 as a novel metabolic regulator. *J Clin Invest* **115**, 1627–1635 (2005).
68. A. Kharitonenkov et al., The metabolic state of diabetic monkeys is regulated by fibroblast growth factor-21. *Endocrinology* **148**, 774–781 (2007).
69. J. Xu et al., Fibroblast growth factor 21 reverses hepatic steatosis, increases energy expenditure, and improves insulin sensitivity in diet-induced obese mice. *Diabetes* **58**, 250–259 (2009).
70. M. K. Badman et al., Hepatic fibroblast growth factor 21 is regulated by PPARalpha and is a key mediator of hepatic lipid metabolism in ketotic states. *Cell Metab* **5**, 426–437 (2007).
71. T. Inagaki et al., Endocrine regulation of the fasting response by PPARalpha-mediated induction of fibroblast growth factor 21. *Cell Metab* **5**, 415–425 (2007).
72. M. J. Potthoff et al., FGF21 induces PGC-1alpha and regulates carbohydrate and fatty acid metabolism during the adaptive starvation response. *Proc Natl Acad Sci U S A* **106**, 10853–10858 (2009).
73. M. J. Potthoff, S. A. Kliewer, D. J. Mangelsdorf, Endocrine fibroblast growth factors 15/19 and 21: from feast to famine. *Genes Dev* **26**, 312–324 (2012).
74. A. Beenken, M. Mohammadi, The FGF family: biology, pathophysiology and therapy. *Nat Rev Drug Discov* **8**, 235–253 (2009).
75. K. Fon Tacer et al., Research resource: comprehensive expression atlas of the fibroblast growth factor system in adult mouse. *Mol Endocrinol* **24**, 2050–2064 (2010).
76. T. Nishimura, Y. Nakatake, M. Konishi, N. Itoh, Identification of a novel FGF, FGF-21, preferentially expressed in the liver. *Biochim Biophys Acta* **1492**, 203–206 (2000).
77. S. Geller et al., Tanycytes regulate lipid homeostasis by sensing free fatty acids and signaling to key hypothalamic neuronal populations via FGF21 secretion. *Cell Metab* **30**, 833–844 e837 (2019).
78. H. Kurosu et al., Tissue-specific expression of betaKlotho and fibroblast growth factor (FGF) receptor isoforms determines metabolic activity of FGF19 and FGF21. *J Biol Chem* **282**, 26687–26695 (2007).
79. A. C. Adams, C. C. Cheng, T. Coskun, A. Kharitonenkov, FGF21 requires betaklotho to act in vivo. *PLoS One* **7**, e49977 (2012).
80. A. L. Bookout et al., FGF21 regulates metabolism and circadian behavior by acting on the nervous system. *Nat Med* **19**, 1147–1152 (2013).
81. K. Hultman et al., The central fibroblast growth factor receptor/beta klotho system: Comprehensive mapping in Mus musculus and comparisons to nonhuman primate and human samples using an automated in situ hybridization platform. *J Comp Neurol* **527**, 2069–2085 (2019).
82. A. C. Adams et al., The breadth of FGF21's metabolic actions are governed by FGFR1 in adipose tissue. *Mol Metab* **2**, 31–37 (2012).
83. R. E. Gimeno, D. E. Moller, FGF21-based pharmacotherapy--potential utility for metabolic disorders. *Trends Endocrinol Metab* **25**, 303–311 (2014).
84. X. Ge et al., Fibroblast growth factor 21 induces glucose transporter-1 expression through activation of the serum response factor/Ets-like protein-1 in adipocytes. *J Biol Chem* **286**, 34533–34541 (2011).
85. D. A. Sarruf et al., Fibroblast growth factor 21 action in the brain increases energy expenditure and insulin sensitivity in obese rats. *Diabetes* **59**, 1817–1824 (2010).
86. A. Kharitonenkov, R. DiMarchi, Fibroblast growth factor 21 night watch: advances and uncertainties in the field. *J Intern Med* **281**, 233–246 (2017).
87. H. Li et al., Fibroblast growth factor 21 increases insulin sensitivity through specific expansion of subcutaneous fat. *Nat Commun* **9**, 272 (2018).
88. S. Li et al., Fibroblast growth factor 21 regulates glucose metabolism in part by reducing renal glucose reabsorption. *Biomed Pharmacother* **108**, 355–366 (2018).

89. S. Kersten et al., Peroxisome proliferator-activated receptor alpha mediates the adaptive response to fasting. *J Clin Invest* **103**, 1489–1498 (1999).
90. Y. Izumiya et al., FGF21 is an Akt-regulated myokine. *FEBS Lett* **582**, 3805–3810 (2008).
91. P. K. Fazeli et al., FGF21 and the late adaptive response to starvation in humans. *J Clin Invest* **125**, 4601–4611 (2015).
92. C. Galman et al., The circulating metabolic regulator FGF21 is induced by prolonged fasting and PPARalpha activation in man. *Cell Metab* **8**, 169–174 (2008).
93. P. J. Antonellis, M. P. Hayes, A. C. Adams, Fibroblast growth factor 21-null mice do not exhibit an impaired response to fasting. *Front Endocrinol (Lausanne)* **7**, 77 (2016).
94. S. M. Solon-Biet et al., Defining the nutritional and metabolic context of FGF21 using the geometric framework. *Cell Metab* **24**, 555–565 (2016).
95. A. Iroz et al., A specific ChREBP and PPARalpha cross-talk is required for the glucose-mediated FGF21 response. *Cell Rep* **21**, 403–416 (2017).
96. K. Iizuka, J. Takeda, Y. Horikawa, Glucose induces FGF21 mRNA expression through ChREBP activation in rat hepatocytes. *FEBS Lett* **583**, 2882–2886 (2009).
97. T. Uebanso et al., Paradoxical regulation of human FGF21 by both fasting and feeding signals: is FGF21 a nutritional adaptation factor? *PLoS One* **6**, e22976 (2011).
98. F. M. Fisher et al., A critical role for ChREBP-mediated FGF21 secretion in hepatic fructose metabolism. *Mol Metab* **6**, 14–21 (2017).
99. S. von Holstein-Rathlou et al., FGF21 mediates endocrine control of simple sugar intake and sweet taste preference by the liver. *Cell Metab* **23**, 335–343 (2016).
100. B. N. Desai et al., Fibroblast growth factor 21 (FGF21) is robustly induced by ethanol and has a protective role in ethanol associated liver injury. *Mol Metab* **6**, 1395–1406 (2017).
101. S. Talukdar et al., FGF21 regulates sweet and alcohol preference. *Cell Metab* **23**, 344–349 (2016).
102. G. Schumann et al., KLB is associated with alcohol drinking, and its gene product beta-Klotho is necessary for FGF21 regulation of alcohol preference. *Proc Natl Acad Sci U S A* **113**, 14372–14377 (2016).
103. M. Hallschmid, S. Higgs, M. Thienel, V. Ott, H. Lehnert, Postprandial administration of intranasal insulin intensifies satiety and reduces intake of palatable snacks in women. *Diabetes* **61**, 782–789 (2012).
104. T. Laeger et al., FGF21 is an endocrine signal of protein restriction. *J Clin Invest* **124**, 3913–3922 (2014).
105. S. A. Wek, S. Zhu, R. C. Wek, The histidyl-tRNA synthetase-related sequence in the eIF-2 alpha protein kinase GCN2 interacts with tRNA and is required for activation in response to starvation for different amino acids. *Mol Cell Biol* **15**, 4497–4506 (1995).
106. T. G. Anthony et al., Preservation of liver protein synthesis during dietary leucine deprivation occurs at the expense of skeletal muscle mass in mice deleted for eIF2 kinase GCN2. *J Biol Chem* **279**, 36553–36561 (2004).
107. B. M. Baker, A. M. Nargund, T. Sun, C. M. Haynes, Protective coupling of mitochondrial function and protein synthesis via the eIF2alpha kinase GCN-2. *PLoS Genet* **8**, e1002760 (2012).
108. S. M. Dudek, C. F. Semenkovich, Essential amino acids regulate fatty acid synthase expression through an uncharged transfer RNA-dependent mechanism. *J Biol Chem* **270**, 29323–29329 (1995).
109. F. Guo, D. R. Cavener, The GCN2 eIF2alpha kinase regulates fatty-acid homeostasis in the liver during deprivation of an essential amino acid. *Cell Metab* **5**, 103–114 (2007).
110. R. B. Hamanaka, B. S. Bennett, S. B. Cullinan, J. A. Diehl, PERK and GCN2 contribute to eIF2alpha phosphorylation and cell cycle arrest after activation of the unfolded protein response pathway. *Mol Biol Cell* **16**, 5493–5501 (2005).
111. Z. Talloczy et al., Regulation of starvation- and virus-induced autophagy by the eIF2alpha kinase signaling pathway. *Proc Natl Acad Sci U S A* **99**, 190–195 (2002).
112. F. Xiao et al., Leucine deprivation increases hepatic insulin sensitivity via GCN2/mTOR/S6K1 and AMPK pathways. *Diabetes* **60**, 746–756 (2011).
113. P. Zhang et al., The GCN2 eIF2alpha kinase is required for adaptation to amino acid deprivation in mice. *Mol Cell Biol* **22**, 6681–6688 (2002).
114. F. G. Schaap, A. E. Kremer, W. H. Lamers, P. L. Jansen, I. C. Gaemers, Fibroblast growth factor 21 is induced by endoplasmic reticulum stress. *Biochimie* **95**, 692–699 (2013).
115. A. L. De Sousa-Coelho, P. F. Marrero, D. Haro, Activating transcription factor 4-dependent induction of FGF21 during amino acid deprivation. *Biochem J* **443**, 165–171 (2012).

116. R. Maruyama, M. Shimizu, J. Li, J. Inoue, R. Sato, Fibroblast growth factor 21 induction by activating transcription factor 4 is regulated through three amino acid response elements in its promoter region. *Biosci Biotechnol Biochem* **80**, 929–934 (2016).
117. T. Laeger et al., Metabolic responses to dietary protein restriction require an increase in FGF21 that is delayed by the absence of GCN2. *Cell Rep* **16**, 707–716 (2016).
118. C. M. Hill et al., FGF21 signals protein status to the brain and adaptively regulates food choice and metabolism. *Cell Rep* **27**, 2934–2947 e2933 (2019).
119. T. Laeger, C. M. Hill, D. C. Albarado, C. D. Morrison, Abstracts, in 35th *Annual Conference of the Society for Adiposity (DAG)*, A. Bosy-Westphal, ed. (Georg Thieme Verlag KG Stuttgart, New York, Kiel, 2019), p. 167.
120. W. Wente et al., Fibroblast growth factor-21 improves pancreatic beta-cell function and survival by activation of extracellular signal-regulated kinase 1/2 and Akt signaling pathways. *Diabetes* **55**, 2470–2478 (2006).
121. G. Singhal et al., Fibroblast growth factor 21 (FGF21) protects against high fat diet induced inflammation and islet hyperplasia in pancreas. *PLoS One* **11**, e0148252 (2016).
122. J. Xu et al., Acute glucose-lowering and insulin-sensitizing action of FGF21 in insulin-resistant mouse models--association with liver and adipose tissue effects. *Am J Physiol Endocrinol Metab* **297**, E1105–1114 (2009).
123. F. M. Fisher et al., Obesity is a fibroblast growth factor 21 (FGF21)-resistant state. *Diabetes* **59**, 2781–2789 (2010).
124. X. Zhang et al., Serum FGF21 levels are increased in obesity and are independently associated with the metabolic syndrome in humans. *Diabetes* **57**, 1246–1253 (2008).
125. W. W. Chen et al., Circulating FGF-21 levels in normal subjects and in newly diagnose patients with Type 2 diabetes mellitus. *Exp Clin Endocrinol Diabetes* **116**, 65–68 (2008).
126. J. Dushay et al., Increased fibroblast growth factor 21 in obesity and nonalcoholic fatty liver disease. *Gastroenterology* **139**, 456–463 (2010).
127. E. B. Nygaard, C. L. Moller, P. Kievit, K. L. Grove, B. Andersen, Increased fibroblast growth factor 21 expression in high-fat diet-sensitive non-human primates (Macaca mulatta). *Int J Obes (Lond)* **38**, 183–191 (2014).
128. K. R. Markan et al., FGF21 resistance is not mediated by downregulation of beta-klotho expression in white adipose tissue. *Mol Metab*, (2017).
129. C. Hale et al., Lack of overt FGF21 resistance in two mouse models of obesity and insulin resistance. *Endocrinology* **153**, 69–80 (2012).
130. A. C. Adams et al., LY2405319, an engineered FGF21 variant, improves the metabolic status of diabetic monkeys. *PLoS One* **8**, e65763 (2013).
131. T. Laeger et al., FGF21 improves glucose homeostasis in an obese diabetes-prone mouse model independent of body fat changes. *Diabetologia* **60**, 2274–2284 (2017).
132. T. Laeger et al., Dietary carbohydrates impair the protective effect of protein restriction against diabetes in NZO mice used as a model of type 2 diabetes. *Diabetologia* **61**, 1459–1469 (2018).
133. N. Orentreich, J. R. Matias, A. DeFelice, J. A. Zimmerman, Low methionine ingestion by rats extends life span. *J Nutr* **123**, 269–274 (1993).
134. R. A. Miller et al., Methionine-deficient diet extends mouse lifespan, slows immune and lens aging, alters glucose, T4, IGF-I and insulin levels, and increases hepatocyte MIF levels and stress resistance. *Aging Cell* **4**, 119–125 (2005).
135. D. Wanders et al., FGF21 mediates the thermogenic and insulin-sensitizing effects of dietary methionine restriction but not its effects on hepatic lipid metabolism. *Diabetes* **66**, 858–867 (2017).
136. E. K. Lees et al., Methionine restriction restores a younger metabolic phenotype in adult mice with alterations in fibroblast growth factor 21. *Aging Cell* **13**, 817–827 (2014).
137. H. M. Brown-Borg, R. Buffenstein, Cutting back on the essentials: can manipulating intake of specific amino acids modulate health and lifespan? *Ageing Res Rev* **39**, 87–95 (2017).
138. K. P. Stone, D. Wanders, M. Orgeron, C. C. Cortez, T. W. Gettys, Mechanisms of increased in vivo insulin sensitivity by dietary methionine restriction in mice. *Diabetes* **63**, 3721–3733 (2014).
139. S. Tonstad, T. Butler, R. Yan, G. E. Fraser, Type of vegetarian diet, body weight, and prevalence of type 2 diabetes. *Diabetes Care* **32**, 791–796 (2009).
140. M. F. McCarty, J. Barroso-Aranda, F. Contreras, The low-methionine content of vegan diets may make methionine restriction feasible as a life extension strategy. *Med Hypotheses* **72**, 125–128 (2009).

141. J. A. Schmidt et al., Plasma concentrations and intakes of amino acids in male meat-eaters, fish-eaters, vegetarians and vegans: a cross-sectional analysis in the EPIC-Oxford cohort. *Eur J Clin Nutr* **70**, 306–312 (2016).
142. G. E. Fraser, Vegetarian diets: what do we know of their effects on common chronic diseases? *Am J Clin Nutr* **89**, 1607S–1612S (2009).
143. T. Castano-Martinez et al., Methionine restriction prevents onset of type 2 diabetes in NZO mice. *FASEB J* **33**, 7092–7102 (2019).
144. K. Zeng et al., Diet polyphenol curcumin stimulates hepatic Fgf21 production and restores its sensitivity in high-fat-diet-fed male mice. *Endocrinology* **158**, 277–292 (2017).
145. A. Ejaz et al., Dietary betaine supplementation increases Fgf21 levels to improve glucose homeostasis and reduce hepatic lipid accumulation in mice. *Diabetes* **65**, 902–912 (2016).
146. Y. Li et al., Hepatic SIRT1 attenuates hepatic steatosis and controls energy balance in mice by inducing fibroblast growth factor 21. *Gastroenterology* **146**, 539–549 e537 (2014).
147. J. Sonoda, M. Z. Chen, A. Baruch, FGF21-receptor agonists: an emerging therapeutic class for obesity-related diseases. *Horm Mol Biol Clin Investig* **30**, (2017).
148. A. Kaufman, L. Abuqayyas, W. S. Denney, E. J. Tillman, T. Rolph, AKR-001, an Fc-FGF21 analog, showed sustained pharmacodynamic effects on insulin sensitivity and lipid metabolism in type 2 diabetes patients. *Cell Rep Med* **1**, 100057 (2020).
149. Q. Pan et al., A novel GLP-1 and FGF21 dual agonist has therapeutic potential for diabetes and non-alcoholic steatohepatitis. *EBioMedicine* **63**, 103202 (2021).
150. C. A. Gilroy et al., Sustained release of a GLP-1 and FGF21 dual agonist from an injectable depot protects mice from obesity and hyperglycemia. *Sci Adv* **6**, eaaz9890 (2020).
151. W. D. Salmon, Jr., W. H. Daughaday, A hormonally controlled serum factor which stimulates sulfate incorporation by cartilage in vitro. *J Lab Clin Med* **49**, 825–836 (1957).
152. J. I. Jones, D. R. Clemmons, Insulin-like growth factors and their binding proteins: biological actions. *Endocr Rev* **16**, 3–34 (1995).
153. N. J. Haywood, T. A. Slater, C. J. Matthews, S. B. Wheatcroft, The insulin like growth factor and binding protein family: Novel therapeutic targets in obesity & diabetes. *Mol Metab* **19**, 86–96 (2019).
154. E. Menu et al., Specific roles for the PI3K and the MEK-ERK pathway in IGF-1-stimulated chemotaxis, VEGF secretion and proliferation of multiple myeloma cells: study in the 5T33MM model. *Br J Cancer* **90**, 1076–1083 (2004).
155. C. Poggi, Y. Le Marchand-Brustel, J. Zapf, E. R. Froesch, P. Freychet, Effects and binding of insulin-like growth factor I in the isolated soleus muscle of lean and obese mice: comparison with insulin. *Endocrinology* **105**, 723–730 (1979).
156. J. Zapf, C. Hauri, M. Waldvogel, E. R. Froesch, Acute metabolic effects and half-lives of intravenously administered insulinlike growth factors I and II in normal and hypophysectomized rats. *J Clin Invest* **77**, 1768–1775 (1986).
157. S. M. Firth, R. C. Baxter, Cellular actions of the insulin-like growth factor binding proteins. *Endocr Rev* **23**, 824–854 (2002).
158. K. Rajkumar, M. Krsek, S. T. Dheen, L. J. Murphy, Impaired glucose homeostasis in insulin-like growth factor binding protein-1 transgenic mice. *J Clin Invest* **98**, 1818–1825 (1996).
159. K. Sjogren et al., Liver-derived IGF-I is of importance for normal carbohydrate and lipid metabolism. *Diabetes* **50**, 1539–1545 (2001).
160. H. Hong et al., Central IGF1 improves glucose tolerance and insulin sensitivity in mice. *Nutr Diabetes* **7**, 2 (2017).
161. S. Wrigley, D. Arafa, D. Tropea, Insulin-like growth factor 1: at the crossroads of brain development and aging. *Front Cell Neurosci* **11**, 14 (2017).
162. A. M. Fernandez et al., Functional inactivation of the IGF-I and insulin receptors in skeletal muscle causes type 2 diabetes. *Genes Dev* **15**, 1926–1934 (2001).
163. N. Kloting et al., Autocrine IGF-1 action in adipocytes controls systemic IGF-1 concentrations and growth. *Diabetes* **57**, 2074–2082 (2008).
164. M. Soto, W. Cai, M. Konishi, C. R. Kahn, Insulin signaling in the hippocampus and amygdala regulates metabolism and neurobehavior. *Proc Natl Acad Sci U S A* **116**, 6379–6384 (2019).
165. M. P. Brugts et al., Igf-I bioactivity in an elderly population: relation to insulin sensitivity, insulin levels, and the metabolic syndrome. *Diabetes* **59**, 505–508 (2010).
166. S. O. Jensen-Cody, M. J. Potthoff, Hepatokines and metabolism: deciphering communication from the liver. *Mol Metab* **44**, 101138 (2021).

167. A. C. Moses, S. C. Young, L. A. Morrow, M. O'Brien, D. R. Clemmons, Recombinant human insulin-like growth factor I increases insulin sensitivity and improves glycemic control in type II diabetes. *Diabetes* **45**, 91–100 (1996).
168. J. Baker, J. P. Liu, E. J. Robertson, A. Efstratiadis, Role of insulin-like growth factors in embryonic and postnatal growth. *Cell* **75**, 73–82 (1993).
169. I. M. Varndell et al., Localization of glucagon-like peptide (GLP) immunoreactants in human gut and pancreas using light and electron microscopic immunocytochemistry. *J Histochem Cytochem* **33**, 1080–1086 (1985).
170. M. P. Mattson, T. Perry, N. H. Greig, Learning from the gut. *Nat Med* **9**, 1113–1115 (2003).
171. P. K. Lund, The discovery of glucagon-like peptide 1. *Regul Pept* **128**, 93–96 (2005).
172. P. J. Larsen, M. Tang-Christensen, J. J. Holst, C. Orskov, Distribution of glucagon-like peptide-1 and other preproglucagon-derived peptides in the rat hypothalamus and brainstem. *Neuroscience* **77**, 257–270 (1997).
173. J. J. Holst, R. Burcelin, E. Nathanson, Neuroprotective properties of GLP-1: theoretical and practical applications. *Curr Med Res Opin* **27**, 547–558 (2011).
174. K. Hunter, C. Holscher, Drugs developed to treat diabetes, liraglutide and lixisenatide, cross the blood brain barrier and enhance neurogenesis. *BMC Neurosci* **13**, 33 (2012).
175. P. L. McClean, V. Parthsarathy, E. Faivre, C. Holscher, The diabetes drug liraglutide prevents degenerative processes in a mouse model of Alzheimer's disease. *J Neurosci* **31**, 6587–6594 (2011).
176. T. J. Kieffer, C. H. McIntosh, R. A. Pederson, Degradation of glucose-dependent insulinotropic polypeptide and truncated glucagon-like peptide 1 in vitro and in vivo by dipeptidyl peptidase IV. *Endocrinology* **136**, 3585–3596 (1995).
177. J. P. Raufman, Bioactive peptides from lizard venoms. *Regul Pept* **61**, 1–18 (1996).
178. J. Eng, W. A. Kleinman, L. Singh, G. Singh, J. P. Raufman, Isolation and characterization of exendin-4, an exendin-3 analogue, from Heloderma suspectum venom. Further evidence for an exendin receptor on dispersed acini from guinea pig pancreas. *J Biol Chem* **267**, 7402–7405 (1992).
179. A. J. Kastin, V. Akerstrom, W. Pan, Interactions of glucagon-like peptide-1 (GLP-1) with the blood-brain barrier. *J Mol Neurosci* **18**, 7–14 (2002).
180. M. Tang-Christensen, N. Vrang, P. J. Larsen, Glucagon-like peptide 1(7–36) amide's central inhibition of feeding and peripheral inhibition of drinking are abolished by neonatal monosodium glutamate treatment. *Diabetes* **47**, 530–537 (1998).
181. C. Orskov, S. S. Poulsen, M. Moller, J. J. Holst, Glucagon-like peptide I receptors in the subfornical organ and the area postrema are accessible to circulating glucagon-like peptide I. *Diabetes* **45**, 832–835 (1996).
182. S. C. Cork et al., Distribution and characterisation of Glucagon-like peptide-1 receptor expressing cells in the mouse brain. *Mol Metab* **4**, 718–731 (2015).
183. J. J. Holst, The physiology of glucagon-like peptide 1. *Physiol Rev* **87**, 1409–1439 (2007).
184. C. S. Bae, J. Song, The role of glucagon-like peptide 1 (GLP1) in type 3 diabetes: GLP-1 controls insulin resistance, neuroinflammation and neurogenesis in the brain. *Int J Mol Sci* **18**, (2017).
185. K. Akter et al., Diabetes mellitus and Alzheimer's disease: shared pathology and treatment? *Br J Clin Pharmacol* **71**, 365–376 (2011).
186. P. M. Doraiswamy et al., Amyloid-beta assessed by florbetapir F 18 PET and 18-month cognitive decline: a multicenter study. *Neurology* **79**, 1636–1644 (2012).
187. K. Talbot et al., Demonstrated brain insulin resistance in Alzheimer's disease patients is associated with IGF-1 resistance, IRS-1 dysregulation, and cognitive decline. *J Clin Invest* **122**, 1316–1338 (2012).
188. R. Isacson et al., The glucagon-like peptide 1 receptor agonist exendin-4 improves reference memory performance and decreases immobility in the forced swim test. *Eur J Pharmacol* 650, 249–255 (2011).

CHAPTER TEN

Insulin Resistance as a Risk Factor for Alzheimer's Disease

Miren Ettcheto
Amanda Cano
Elena Sanchez-Lopez
Carme Auladell
Jaume Folch
Antoni Camins

CONTENTS

DOI: 10.1201/9781003079927-10

INTRODUCTION

Biomedical research throughout all these years has allowed an advance in the therapeutic strategies of several pathologies such as metabolic disorders, cardiovascular diseases and infectious diseases, which have reduced drastically the mortality in the world population (1). This improvement in the human quality of life mostly in developed countries has led to a significant increase in life expectancy which has been closely related to the ageing population. However, the counterpart of this success is the rise in the incidence of illnesses that appear because of ageing processes, like most dementias and neurodegenerative disorders (1–3). The World Health Organization (WHO) demonstrated that these diseases, nowadays incurable, are responsible for 12% of deaths globally. Among the diverse neurodegenerative disorders, Alzheimer's disease (AD) corresponds to the most common form of dementia. It is neuropathologically characterized by the accumulation of extracellular neuritic plaques and fibrils in the brain, mainly consisting of aggregated amyloid-beta (Aβ) peptides, and intracellular neurofibrillary tangles (NFT), composed of hyperphosphorylated Tau (p-Tau) (4). Soluble Aβ peptides through binding to their receptors would be responsible for generating the neurodegenerative process (5). Likewise, synaptic damage is mediated by soluble Aβ oligomers, which have been shown to better correlate with the disease severity than with the accumulation of insoluble Aβ peptides into plaques triggering AD pathophysiology (5). In this way, Aβ oligomers would be the neurotoxin responsible for the cognitive decline and the hippocampal LTP process (6–8).

Furthermore, insoluble NFT alters axonal transport, synapses causing neurodegeneration (4). In addition, microglial infiltration, brain atrophy and generalized synaptic and neuronal loss have been also observed in these patients (9).

According to its etiology and progression, AD has been classified into two forms, the early onset AD (EOAD) and the late onset AD (LOAD). EOAD is the least common form (5% <of cases) and is also known as familial AD (FAD); it may be attributed to inherited genetic causes associated with mutations in the amyloid precursor protein (APP) and presenilin 1 and 2 (PSEN 1 and PSEN 2) genes (1). By contrast, the LOAD forms constitute over 95% of all AD cases, which have a complex etiology, having been associated with many different risk factors. The apolipoprotein E (APOE) gene ε4 allele represents the main gene associated with LOAD, as multiple genetic studies have consistently reported (10). APOE ε4 allele favors the risk of AD through the interaction with Aβ promoting its aggregation and deposition in insoluble fibrillar deposits. It also plays a key role in the modulation of membrane lipids, synaptic integrity and injury repair (10).

In addition, in the last decades, the increase in the prevalence of LOAD suggests that additional factors, other than genetic and age, may be involved in the neurodegenerative process pointing out that AD is a multifactorial pathology in which several factors and associated comorbidities are involved (1, 11).

Obesity-induced metabolic disorders, commonly caused by the excessive consumption of high caloric foods and high-fat diets (HFD), have been considered as important risk factors for AD development (Figure 10.1). These dysregulations include the appearance of insulin resistance (IR) which has been related to cognitive impairment (12, 13). Moreover, peripheral metabolic disorders associated with the diabetic phenotype (i.e., hyperglycemia, hyperinsulinemia, hypercholesterolemia) have also been directly implicated with brain atrophy, cognitive loss, and the pathological features of LOAD, evidencing the association between type 2 diabetes mellitus (T2DM) and LOAD, in which IR is a common contributor to both disorders (14–16). This concept has gained so much attention that some authors defined LOAD as "diabetes of the brain" or "type 3 diabetes (T3DM), prompting research to understand the molecular mechanisms involved in these two chronic age-related disorders (17).

Here we review the preclinical and clinical evidence of the impact of T2DM on pathological changes in the brain that contribute to a cognitive decline and dementia. Furthermore, we discuss the possible underlying mechanisms that link brain insulin resistance (BIR) to neurodegeneration.

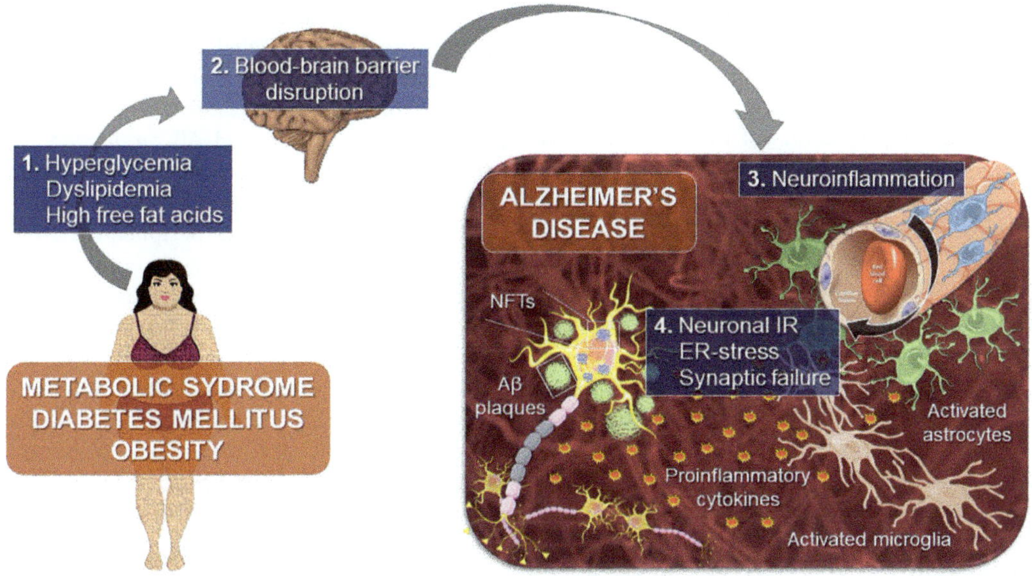

Figure 10.1 Physiopathological mechanisms promoted by metabolic syndromes in the development of AD.

PERIPHERAL INSULIN RESISTANCE AS RISK FACTOR OF COGNITIVE ALTERATION IN LATE ONSET ALZHEIMER'S DISEASE

As mentioned above, several studies suggest that there is a relationship between calorie intake, HFD and consequent obesity which leads to cognitive deficits (18). Within this scenario, clinical studies have found a positive association between IR, not only peripheral but also BIR, and AD. Since cognitive alterations at the hippocampal level have been found in these patients, data suggests that LOAD could be a specific brain form of T2DM (19).

The Rotterdam study was the first to propose the connection between these two pathologies, revealing that diabetes increases the risk of dementia (20, 21). Later, clinical and epidemiological studies have confirmed this association, demonstrating that the alteration of metabolic parameters, such as hyperglycemia and hyperinsulinemia, are positively correlated with the development of LOAD neuropathology. In this line, the Hisayama Study investigates the pathways through which diabetes could influence the pathogenic process of this neurodegenerative disease, evaluating the formation of neuritic plaques and NFTs (12). In this study, the authors examined 2,520 subjects who underwent a metabolic characterization based on blood tests, which included the oral glucose tolerance test (OGTT) profiles of 75 g and fasting insulin levels. Subsequently, the pathology of LOAD was evaluated. Likewise, the authors also examined the combined effects of APOEε4 genotype associated with diabetes-related factors on the risk of the main hallmarks of LOAD, Aβ plaques and NFTs. The study found an association between diabetes plus APOEε4 and Aβ plaques, but not with NFT formation (12).

Moreover, Hokama et al. performed a preclinical and clinical study focused on the analysis of gene expression related to the insulin pathway involved in T2DM development and obesity at the hippocampus level (22). In the preclinical study, APP/PS1 mice, a familial animal model of AD, were evaluated and in the clinical study, 88 autopsies from the Hisayama Study of LOAD patients were carried out. They reported that in both brains of LOAD patients and APP/PSI mice, alterations in the expression profiles of T2DM-related genes were observed, suggesting that peripheral IR exacerbates LOAD pathology which induces the hippocampal alterations of genes related with the insulin pathway (22).

Likewise, Clark et al. evaluated the role of hypertension and obesity in 207 participants enrolled in the Wisconsin Registry for Alzheimer's Prevention study (WRAP). They reported that the presence of hypertension or obesity in midlife may exacerbate the cognitive decline associated with Aβ deposition, through the reduction of the cerebral blood flow and alterations in the blood-brain barrier (18). Thus, it is thought that Aβ clearance could play a key role in the development of LOAD. In line with this study, Starks et al. reported that prediabetic patients with IR and genetic risk of LOAD showed elevated levels of Tau protein in the cerebrospinal fluid (CSF), suggesting that IR may worsen Tau pathology, especially in APOEε4 carriers (23).

Willette et al. reported the first study performed in cognitively healthy middle-aged adults ($N = 186$) from the WRAP, where the association between IR and Aβ brain deposition in LOAD patients was demonstrated, thus confirming that IR is a risk factor in the early stages of LOAD (24, 25). Therefore, the authors highlight the importance of understanding the role of IR, which constitutes one of the modifiable risk factors during the preclinical stage of the pathology and opens a new therapeutic window for the design of new strategies focused on the prevention, early diagnose and treatment of LOAD (24). Likewise, Bruehl et al. showed a negative relationship between IR and cognitive decline that was observed in middle-aged individuals, suggesting that it could be a potential indicator of the risk of dementia in elderly populations (26).

In accordance with this hypothesis, Rasgon and colleagues performed a study focused on the correlation between IR and the risk of LOAD, which was performed on 50 physically healthy and cognitively intact, Caucasian postmenopausal women who had participated in a larger study of brain changes during post menopause. The effects of IR on hippocampal volume and cognitive performance in non-diabetic women as a risk of LOAD (by virtue of carrying apolipoprotein ε-4 or having a family history of LOAD) were evaluated, which also concluded that IR could be a potential LOAD indicator (27).

Moreover, a relationship between peripheral hyperinsulinemia and the increase of inflammatory markers in the brain was reported by Fishel and colleagues (28). In this study, both processes showed that IR was a common hallmark, demonstrating that IR increased the levels of brain inflammatory markers such as interleukine-1beta and 6 (IL-1β, IL-6), tumor necrosis factor alpha (TNF-α), and others, thus, enhancing Aβ deposition. In consequence, Aβ accumulation promotes neuroinflammation through glial activation, which in turn generates the release of more proinflammatory cytokines, causing a continuous vicious circle between inflammation and Aβ production, and contributing to LOAD neurodegeneration (28).

Likewise, Hong and colleagues studied the potential link between the peripheral triglyceride glucose (TyG) index in plasma and dementia using a large-scale population dataset from the National Health Information Database (NHID). They reported that the TyG index was associated with an increased risk of both, LOAD and vascular dementia (29). The authors suggested that treating IR before the LOAD development could lead to an improvement in the assessment of dementia.

PHYSIOLOGICAL ROLES OF INSULIN IN THE BRAIN AND IN LATE ONSET ALZHEIMER'S DISEASE

In 1978, insulin receptors were identified in the brain although the functions of the insulin receptor in the central nervous system were discovered much later (30). Insulin has been shown to play a key role in the regulation of whole-body energy at the peripheral level. It participates in glucose homeostasis in the human body, but also exerts specific effects on the brain through brain insulin receptors and the closely related IGF-1 receptor (IGF-1R) since glucose is the main fuel of the brain to properly perform its functions (14).

As it has been previously mentioned, insulin exerts a decisive regulation on glucose absorption at peripheral level. In consequence, its alterations are closely related to various clinical syndromes such as arterial hypertension and cardiovascular disorders, among others (31). By contrast, brain insulin receptors and IGF-1R, as well as their downstream targets, such as brain insulin receptor substrate 1 and 2 (IRS1 and IRS2), are widely distributed throughout the brain, including the olfactory bulb, cortex, hippocampus, cerebellum and hypothalamus (32, 33). Several studies have demonstrated that insulin plays a key physiological role in the human body metabolism regulation and energy balance acting on the hypothalamus (34). Thus, considering the metabolic disturbances mentioned above, recent studies suggest that the hypothalamus may contribute to the development of LOAD, constituting a future area of interest for further investigation of its role in neurodegeneration (35).

Currently, it has become clear that the physiological actions of insulin are more widespread since it has been demonstrated that the activation of the hippocampal insulin receptor is necessary for the process of neuronal synaptic plasticity, learning and memory (36–38). In line with this, several authors have demonstrated synaptic localization of insulin receptor in the hippocampus which could explain the role of insulin in promoting the formation of synapses, dendritic spines, long-term potentiation (LTP) and neuroprotection (39), suggesting that BIR could be associated and partially responsible of cognitive alterations in LOAD (40, 41).

In this context, Grillo and colleagues performed a preclinical study in a selective IR model of the rat using a lentiviral vector expressing an insulin receptor antisense sequence (42). They demonstrated that the generation of specific IR in the hippocampus affected the synaptic plasticity in this area by regulating the expression and trafficking of the glutamate receptor. Specifically, this study showed that IR decreased the phosphorylation of the serine 845 residue of the glutamate receptor 1 (GluA1) and the expression of glutamate ionotropic receptor N-Methyl D-Aspartate (NMDA) type subunit 2B (GluN2B), which are directly involved in long term potentiation (LTP) (42, 43). In line with these results, it has been reported that insulin administration promotes dendritic spine formation in primary cultures of rat hippocampal neurons and improves synaptic transmission through NMDA receptor (glutamatergic transmission) in the hippocampus, modulating the surface expression of these receptors and stimulating the phosphorylation of GluN2A and GluN2B subunits (36, 44).

Going deeper at the molecular level, it has been demonstrated that insulin not only regulates glutamatergic transmission via NMDA receptor but also induces phosphorylation of the glutamate receptor subunit (GluR2) of the alfa-amino-3-hydroxy-5-methyl-4-isoxazolepropionic acid (AMPA) receptors of hippocampal neurons (45). Moreover, it has also been shown that insulin affects learning and memory processes through acid gamma-aminobutyric acid (GABA) receptors by stimulating their translocation on the post-synaptic and dendritic membranes (36).

Likewise, IRS-2 was identified as a component of the complex system, which mediates synaptic plasticity in the hippocampus. Studies in *Irs2*$^{-/-}$ mice reported an impairment of LTP and an increase of the GSK-3β activity while also exhibiting a role in the regulation of NR1 phosphorylation and in the modulation of NMDA receptor synaptic plasticity (46).

LATE ONSET ALZHEIMER'S DISEASE AND BRAIN INSULIN RESISTANCE

In the 1980s, Dr. Hoyer was one of the first researchers who proposed the concept of BIR associated with LOAD (16, 32, 47–51). The main hypothesis was focused on the idea that hypometabolism of cerebral glucose could be a pathophysiological risk factor of AD. Although the scientific community showed a sceptical attitude at the beginning, several studies have supported this hypothesis as the findings in this field have progressed. In this context, alterations in insulin and insulin-like growth factor (IGF) gene expression and signalling in the brains of LOAD patients have been observed (16, 32). Moreover, Frölich and colleagues also demonstrated reduced activity of insulin signaling molecules in post-mortem regions of the cerebral cortex in patients with LOAD (52, 53). These data were corroborated by Moloney and colleagues which not only showed a significant decrease in the expression of IGF-1 and insulin receptors in LOAD temporal cortex but also in their key downstream signalling proteins IRS1 and IRS2 (54). In agreement, Arvanitakis and colleagues reported alterations in the insulin signalling of post-mortem LOAD subjects. In this study, the authors evaluate the relationship between molecules of the signaling path of insulin and AD. Specifically, they found an increase in brain protein kinase B (PKB or AKT) phosphorylation (specifically pT308AKT), which was associated with LOAD neuropathology and cognitive function decrease (39). The authors suggest the possibility that multiple signaling regulates the AKT and plays a more prominent role in the physiopathology of AD. In this study, they did not find brain changes in the classical markers of IR, such as IRS1 phosphorylation (39). Going further, they also reported that IGF-1 receptors were colocalized with NFTs within neurons in LOAD brain samples, concluding that alterations in the insulin pathway contribute to disease pathogenesis (39). In line with these findings, Kulstad et al. found that insulin disturbances cause a significant increase in the plasmatic Aβ1–42 levels in AD patients, which could be due to a deficiency in the insulin degrading enzyme (IDE) activity (55).

Besides, it has been suggested that oligomeric Aβ may be responsible of trigger BIR since it has been demonstrated that Aβ competes with insulin for binding to insulin receptor, and the inhibition of insulin receptor further accelerates the development of LOAD neuropathology (56, 57). Moreover, this study demonstrated the state of IR in the brain of LOAD patients in the absence of T2DM and the APOEε4, highlighting that BIR appears to be an early and common feature of LOAD even in the absence of T2DM (57).

In addition, de la Monte and colleagues analysed the levels of insulin-related molecules in paired samples of serum and CSF in 19 MCI or LOAD patients in early stages and 21 controls using a multiplex ELISA platform. They demonstrated a significant reduction of insulin-related molecules marker's levels in the CSF and serum in both patients' groups (58–60). Therefore, this data confirms a simultaneous dysregulation in insulin-related endocrine pathways in CSF and serum in an early-stage LOAD subjects and support the hypothesis that AD is a dysregulated metabolic disease affecting the whole organism.

In line with these studies, it has been widely demonstrated that BIR generates an extensive intracellular signalling cascade causing mitochondrial alterations, neuroinflammation and reticulum stress, among others, which lead to cognitive decline due to synapse and cell death, demonstrating that BIR constitutes a risk factor in the neuropathology of LOAD (61, 62).

MOLECULAR PATHWAYS INVOLVED IN CENTRAL INSULIN RESISTANCE ASSOCIATED WITH LATE ONSET ALZHEIMER'S DISEASE NEURODEGENERATION

It has been described that a key feature of BIR in both pathologies, T2DM and LOAD, is the phosphorylation of IRS-1 at serine residues (40). Going deeper at molecular level, insulin receptor is modulated by the phosphorylation and dephosphorylation of IRS1, which in turn regulates several kinases and phosphatases including extracellular signal-regulated kinase (ERK), c-Jun-N-terminal Kinase (JNK) and mammalian target of rapamycin (mTOR). These kinases are known to phosphorylate the serine residues of IRS-1 leading to the inhibition of insulin signal transduction (63, 64). Likewise, IR can be explained by the increase in the tyrosine phosphatase PTP-1B activity, which participates in a very complex molecular pathway (65, 66). In this context, Ser-616 phosphorylation of IRS1 has been demonstrated to be a potential key mediator in the development of AD since it positively correlates with Aβ oligomers promoting memory deficits in MCI and LOAD patients (40). Likewise, its serine hyperphosphorylation has also been observed in previous studies performed with AD transgenic mice and other tauopathies such as Pick's disease (PiD), progressive supranuclear palsy and corticobasal degeneration animal models (67, 68). Furthermore, alterations in insulin signalling contribute to alterations in mitochondrial structure and function due to an energy deficiency, which increases the production of reactive oxygen species (ROS) and neuropathological hallmarks associated with neuroplasticity alterations and cognitive loss in the LOAD (Figure 10.2).

AMYLOID AS A RESPONSIBLE OF BRAIN INSULIN RESISTANCE IN LATE ONSET ALZHEIMER'S DISEASE

The amyloid cascade hypothesis has been one of the most accepted theories focused on explaining the onset of AD. It was based on the deposition of Aβ plaques throughout the brain and its role in the neurodegenerative process (5, 57, 69). However, the research on this initial hypothesis has also focused on the role of Aβ in the deterioration of synapses and dendritic spines (5).

Likewise, it has been described that Aβ oligomers themselves are able to impair the function of insulin receptor, detected at a dendritic level in synapses, inducing insulin receptor loss from the neuronal surface of dendrites (8). In consequence, there is a decrease in glutamatergic neurotransmission. Thus, Aβ oligomers have been considered the responsibility for initiating the process of BIR in AD, independently of obesity T2DM or peripheral metabolic alterations (37, 70).

On the other hand, it has been shown that Aβ oligomers activate c-Jun N-terminal Kinase (JNK), which in turn phosphorylate IRS-1, inactivating insulin receptor and, therefore, causing BIR (41). In line with this, previous studies performed in preclinical models of diabetes have established that

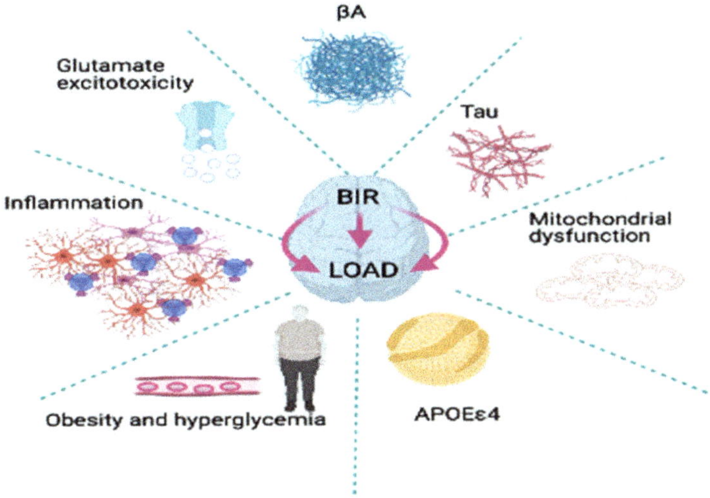

Figure 10.2 Main mediators responsible for the development of LOAD associates to BIR.

the activation of pro-inflammatory TNF-α/JNK signalling constitutes a key process that leads to peripheral IR (71). Thus, following these strategies and extrapolating these results to the CNS, researchers have suggested that there is a molecular similarity between BIR in AD and peripheral IR in T2DM (41).

Likewise, Aβ oligomers aside from acting directly in neurons can activate microglia. The continuous activation results in the release of proinflammatory cytokines (IL-1β, IL-6 and TNFα), chemokines, eicosanoids and nitric oxide (NO). In consequence, their levels are increased in the brain promoting the generation of insoluble Aβ oligomers and plaque formation (5, 8, 72). In consequence, more proinflammatory molecules are released creating a vicious circle. Furthermore, cytokines also bind to their respective receptors, and in a second step, modulate the IRS-1 phosphorylation leading to BIR and finally neuronal damage (41,73). Moreover, cytokines also play a key role in the formation of neuronal Aβ, since they regulate mRNA levels of β-secretase (BACE1) and its enzymatic activity, responsible for APP processing and Aβ formation through the amyloidogenic pathway. Likewise, its transcription is also increased by nuclear factor kappa-light-chain-enhancer of activated B cells (NF-κβ) which is activated by TNF-α. In addition, dsRNA dependent protein kinase (PKR) and eukaryotic translation initiation factor 2alpha (eIF2α-P), two well-characterized downstream mediators of TNFα, have also been associated with cognitive impairments mediated by Aβ oligomers, since the cognitive decline in $PKR^{(-/-)}$ and $TNFR1^{(-/-)}$ mice was prevented by reducing eIF2α-P (74). Therefore, TNF-α represents an initial trigger that orchestrates the activation of multiple stress response pathways that culminate with synapse dysfunction and cognitive loss in LOAD where BIR contributes to this process (74, 75).

Likewise, Aβ oligomers induce endoplasmic reticulum (ER) stress (PKR-mediated phosphorylation of eIF2α-P), which leads to IRS1 inactivation (5, 8). It is highlighted that the Aβ oligomers exert their effects at the CNS level in LOAD. However, this process is much more complex. It has been demonstrated that the administration of Aβ 25–35 or infusion of Aβ oligomers into the hypothalamus promotes peripheral metabolic disturbances and impaired peripheral glucose metabolism linked to LOAD pathology (34, 76). Hypothalamus plays a central role in neuroendocrine interaction between the central system and the periphery. Thus, acting specifically at the hypothalamic level, Aβ itself induces a metabolic alteration. For this reason, a vicious circle modulated by Aβ oligomers can go in both directions, favouring IR throughout the body and turning LOAD into a metabolic disease.

Another important finding is the role of plasma Aβ oligomers in peripheral IR. In line with this concept, it has been demonstrated that glucose overload increases Aβ oligomers levels in the plasma of the AD mouse model, which promotes glucose tolerance and hepatic insulin signalling in AD mice through binding to the insulin receptor, although high concentrations are required (77–82). Moreover, these oligomers also have an indirect inhibitory effect on the insulin signalling pathway through mechanisms mediated by oxidative stress and inflammatory responses, demonstrating that Aβ oligomers modify glucose metabolism in multiple ways. It should consider that, at the brain level, glucose uptake by neurons is largely independent of insulin. For this reason, the effects of Aβ oligomers on insulin signalling could have more important effects on the systemic metabolism of glucose, where insulin is critical for proper glucose homeostasis through effects on the liver, skeletal muscle, and adipose tissue (77–81).

In addition, previous reports demonstrate that brain BACE1 may contribute to metabolic regulation and could be the potential link between T2DM and LOAD. Effectively, using the neuron-specific human BACE1 knock-in mouse model (PLB4), it has been demonstrated that neuronal BACE1 overexpression increases the risk of T2DM development (82). Therefore, these data suggest that AD progression can promote, at least at the preclinical level, the onset peripheral diabetic complications, independently of an obesogenic process. This research study supports the hypothesis of AD as a metabolic disease in which Aβ oligomers production in the brain play a key role in peripheral glucose and insulin alterations.

THE ROLE OF TAU IN THE BRAIN INSULIN RESISTANCE IN THE LATE ONSET ALZHEIMER'S DISEASE

Although it is well known that insulin can modulate the phosphorylation of Tau protein, its role in the regulation of the insulin receptor has been also studied in recent years. Numerous studies have shown that, in addition to synaptic alterations, BIR is also involved in Tau hyperphosphorylation, through the inhibition of glycogen synthase kinase 3β (GSK3β) which exacerbates the development of LOAD. In accordance with this idea, mice that do not express insulin receptors at the neuronal level (NIRKO mice) as well as mice in which IRS-2 has been suppressed showed an increased phosphorylation of Tau through the inhibition of phosphoinositide 3 kinase (PI3K) / AKT signalling pathway (46, 83). This link between insulin and Tau signalling is mainly based on the modulation of downstream signalling pathways involving different kinases such as GSK-3β, JNK and AMP-activated protein kinase (AMPK) and phosphatases including protein phosphatase 1 and 2A (PP1and PP2A, respectively) (84). Surprisingly, Marciniak and colleagues reported that Tau deletion leads to the development of peripheral metabolic abnormalities observed in patients with LOAD, demonstrating a novel function of Tau protein in the control of BIR which contributes to cognitive and metabolic alterations in patients with LOAD and in tauopathies (85). Therefore, chronic BIR promotes the development of Tau pathology by altering the balance between kinases and phosphatases and vice versa.

Moreover, it has been demonstrated in LOAD patients, and in several tauopathies, that Tau hyperphosphorylation leads to an increase in uptake and intraneuronal accumulation of insulin as insoluble oligomeric aggregates (86). Interestingly, this process occurs in non-diabetic patients, suggesting that BIR associated with alterations of insulin signalling pathway is independent of the presence of clinical T2DM.

APOLIPOPROTEIN-E4 AND INSULIN RESISTANCE

It is well known that the ε4 allele of the *APOE* gene is the strongest genetic risk factor for LOAD compared to ε3 allele and the neuroprotective ε2 allele (87, 88). While APOε4 has been shown to impair Aβ clearance and promote its aggregation leading to an enhancement of Aβ pathology, increasing evidence suggests that APOE isoforms differentially regulate brain insulin signalling and glucose metabolism (89–91). Besides, previous studies (The Honolulu-Asia Aging Study and The Cardiovascular Health Study) have shown that the APOEε4 allele associated with T2DM synergistically affects not only its pathogenesis but also increases the risk of suffering LOAD fivefold, showing more severe neuropathology of LOAD (91). Moreover, Zhao and colleagues reported that APOEε4 impairs neuronal insulin signalling in APOE-targeted replacement mice (apoE4-TR). They demonstrated that the aggregating capacity of APOEε4 would be responsible for BIR through the capture of functional insulin receptors and blocking insulin binding (87, 88).

Related to this process, Salameh and colleagues discussed how different risk factors such as APOE, lipids, Aβ or T2DM can interact in the cognitive impairment of AD (13). In a first step, the peripheral inflammatory process could interact with each other decreasing the transport of insulin across the blood-brain barrier and favouring BIR in AD. Likewise, they suggest that peripheral inflammation, elevated triglyceride levels and obesity/T2DM could constitute the initial step of this process, thus, increasing the Aβ levels in the CNS (13). Then, they hypothesized that APOE is also involved in AD through the regulation of brain Aβ clearance. AD patients expressing APOEε4 present a decreased clearance of Aβ from the CNS and an increased Aβ burden compared to APOEε4 non-carriers. Moreover, the peripheral metabolism of Aβ and CNS vicious cycle could be generated by the association and complex interaction of all these risk factors with the consequent process of cognitive loss. Therefore, APOEε4 associated with TD2M are examples of risk factors that all together significantly increase the risk of suffering (1, 2, 10, 92).

PTP-1B AS THE LINK BETWEEN BRAIN INSULIN RESISTANCE ASSOCIATED WITH COGNITIVE LOSS AND LATE ONSET ALZHEIMER'S DISEASE

Protein-tyrosine phosphatase 1B (PTP1B) is an enzyme that has been demonstrated to have a crucial role in the downregulation of insulin signalling (93, 94). The administration of PTP1B inhibitors have been proposed for the treatment of T2DM (95, 96). However, recent studies suggest that they could also have a prominent role in the treatment of neurodegenerative disorders since it has been reported that rodents fed with HFD develop IR and show an increase of PTP1B in the brain (97). Likewise, PTP1B can regulate both the tyrosine phosphorylation of the insulin receptor and p-IRS1, suggesting that this phosphatase could represent a link between T2DM and AD (94).

In this way, PTP1B also regulates leptin signalling. Disturbances in this pathway, especially in the hippocampus, not only have been described in AD brains but also have been associated with Aβ oligomers, Tau phosphorylation in neurons and cognition and memory decline (94). Another important pathway in which PTP1B has been involved is the regulation of the brain derived neurotrophic factor/tropomyosin receptor kinase B (BDNF/TrkB) receptor (94). BDNF/TrkB have been related to synaptic development and evidence indicates that also plays a significant role in the synaptic plasticity (93). In this context, it has been reported that PTP1B overexpression reduces TrkB phosphorylation and, in consequence, the activation of downstream signalling pathway, such as the central BDNF. By contrast, its inhibition has been demonstrated to enhance TrKB signalling and improve memory in mice, thus, supporting the role of PTP1B in neuronal survival, morphogenesis and plasticity (93). Therefore, the inhibition of PTP1B represents a key target involved in the modulation of cognitive process.

In this line, there are many results supporting the role of PTP1B in the control of the synaptic structure and its activity. For instance, high levels of PTP1B have been observed in hippocampal dendritic spines, which were co-localize with synapsin-1, a protein implicated in synaptogenesis and the modulation of neurotransmitter release (98). By contrast, PTP1B has also been identified as crucial for spine maturation and normal synapse formation through modulation of N-cadherin, process in which PTP1B hippocampal disruption has demonstrated to leads to an elongation of dendrites and an improvement in learning and memory processes (98). Therefore, its role in this field needs to be deeply studied.

In addition, PTP1B has also been described as a positive regulator of neuroinflammation (99). This point is because TNFα induces an increase of PTP1B brain levels (100, 101). Likewise, a direct connection between PTP1B and the activation of NF-κB through tyrosine phosphorylation of I-kappa-β-alpha (Iκβ) has been reported (101). All these data indicate the importance of the PTP1B enzyme in the global regulation of BIR. Its inhibition emerges as a potential therapeutic strategy in the treatment of AD due to its ability to increase cognition (94).

NEUROINFLAMMATION AS A LINK BETWEEN BRAIN INSULIN RESISTANCE AND ALZHEIMER'S DISEASE

An important and possible explanation for the vulnerability of the hippocampus to the development of IR may be related to neuroinflammatory processes to which this brain region is subjected. In this regard, previous studies carried out in preclinical and clinical models of AD and obesity have found that the degree of chronic neuroinflammation in both pathologies could be responsible for observed cognitive loss and brain atrophy (9, 102, 103). This process is orchestrated by IKKβ/NF-kβwhich plays a key role as an intermediate in the cellular communication signals from membrane receptors to the nucleus. After activation, NF-kβ translocates into the nucleus, where it participates in the transcription of its target genes such as *TNF-α*, *IL1β*, and *iNOS*, amplifying the inflammatory signal involved in the pathogenesis of BIR (104). Among these cytokines, chronic TNF-α stimulation plays an important role in leading to the phosphorylation of serine residues of the insulin receptor and serine phosphorylation of IRS-1 through JNK activation, thus, disrupting insulin signalling (76, 105). Interestingly, the inhibition of TNFα signalling with specific antibody-mediated targeting has been shown to improve insulin sensitivity in peripheral tissues, helping to restore glucose homeostasis and to improve cognition in LOAD (106). Moreover, apart from JNK and IKKβ, PKR has also been related to inflammatory processes since it coordinates the activity of critical inflammatory kinases to regulate insulin action and metabolism (107–109). Specifically, Nakamura and colleagues demonstrated that PKR responds to HFD as well as to ER stress, leading to inflammatory processes, but also phosphorylates directly serine residue of IRS-1, thereby inactivating it (108, 109). This study has been confirmed by experimental studies reported by Taga and colleagues that showed a correlation between obesity and PKR activation in rodents, pointing out that PKR is a critical component of the inflammatory cascade in response to nutrients (110). By contrast, it has been reported that mice deficient in PKR are protected from obesity and IR induced by HFD through a molecular mechanism associated with JNK and IKKβ inhibition. Collectively, these data suggest the importance of PKR as an insulin signalling modulator (110). ER stress is caused by the disruption of homeostatic mechanisms that cause unfolded proteins, which tend to accumulate and form reactive oxygen species (ROS) (58). Finally, all these processes lead to a progressive neuronal deterioration that may facilitate the transition from a reversible brain injury to chronic progressive LOAD (111).

On the other hand, an additional pathway related to BIR in LOAD is Toll-like receptor 4 (TLR4) signaling. This molecule has been involved in microglial and astrocytes activation, as well as in the generation and release of inflammatory cytokines in LOAD. Microglia and astrocytes are the major brain's resident immune cells that express TLR4 receptors, but TLR4 also is expressed in peripheral tissues, where it acts as a receptor of long-chain saturated fatty acids (112–114). When this binding occurs, IKK /NFκB pathway is activated through the myeloid differentiation of the primary response gene 88 (MyD88), enhancing the inflammatory response in both, periphery and brain tissue. Moreover, it has been reported that TLR4 also is activated in the process of T2DM/obesity (113, 115).

In addition, Aβ oligomers are also able to bind to TLR4 receptor, which activates inflammatory cascade as it has been previously explained. This binding causes a dual effect: in the early stages, it increases Aβ oligomers phagocytosis, conferring neuroprotection; however, as the disease aggravates, the neuroinflammatory responses are exacerbated leading to a chronic long-term activation and, finally, a cognitive decline (114, 115).

Thus, it is possible to hypothesize that TLR4 is a key mediator in the regulation of neuroinflammation, since it is thought that its activation represents a link between T2DM and LOAD. Moreover, apart from Aβ and T2DM/obesity, an additional factor can contribute to the chronic peripheral inflammation process: gut dysbiosis. It is well known that microbiome-generated metabolic products and bacterial components such as lipopolysaccharide (LPS), a glycolipid present in the outer membrane of most Gram-negative bacteria, activate TLR4 enhancing the release of inflammatory cytokines. In consequence, it may have a negative impact in the brain leading to the BIR associated with cognitive decline and neuronal degeneration (116).

In this neuroinflammatory response, the increase of Aβ levels in the brain is also accompanied by a microglial activation, inflammatory responses, oxidative stress, and microglia-specific nucleotide-binding oligomerization domain (NOD)-like receptor protein 3 (NLRP3) inflammasome-mediated chronic neuroinflammation. NLRP3 inflammasome in microglial cells promotes caspase-1 activation, secretion of interleukin (IL)-1β and IL-18 and a chronic brain inflammatory response (102).

Likewise, in the peripheral chronic inflammation associated with T2DM, TLR4 is activated by saturated fatty acids which promote NLRP3 inflammasome activation. In consequence, the inflammation of adipocytes and IR occurs, contributing to LOAD (117, 118). Moreover, mitochondrial ROS in advanced glycation end-products (AGEs) also participate in triggered NLRP3 inflammasome (119). In T2DM, an excess of AGEs is produced in the blood stream, generating oxidative stress and inflammation by binding to receptors on the cell surface and altering the structure and function of surrounding proteins. AGEs have also been shown to be produced in the cytosol of neurons, especially in the hippocampus. Likewise, it should be noted that levels of AGEs and their receptors (119) have been reported to be significantly higher in patients with LOAD and are associated with Aβ plaques and NFT. Thus, NLRP3 inflammasome may also contribute to the development of LOAD (117, 118).

Therefore, it can be collectively suggested that the process of cognitive loss in LOAD could be a perfect storm where Aβ, as well as peripheral and brain inflammatory processes (caused by alterations of the microbiota or by other causes such as obesity/T2DM), would actively participate in this dangerous process.

GLUTAMATE EXCITOTOXICITY AS A RESPONSIBLE FOR BRAIN INSULIN RESISTANCE

Glutamate is the main excitatory neurotransmitter in the brain which exerts a prominent role in the neurodegenerative process in LOAD and other neurological diseases. Memantine, an NMDA antagonist, is currently used for LOAD treatment due to its benefits in terms of improving cognitive decline. However, it is not able to stop LOAD, probably due to the late diagnosis, among other reasons. Interestingly, glutamate exhibits additional physiological roles in peripheral tissues, compared to its brain function (120). Recent preclinical data reported that activation of peripheral NMDA receptors is linked to metabolic disorders induced by HFD in mice (121). Thus, the blockage of NMDA receptors could be a potential strategy for the regulation of glucose and lipid metabolism in the treatment of IR and hepatic diseases (122).

Related to this process, it has been reported that the stimulation of insulin receptors by insulin generates low levels of ROS that act as a second messenger, thus favouring the inhibition of phosphatases, such as PTP1B, and increasing the activity of its own receptor (122). This data indicates the important role of ROS in facilitating insulin action and, at the same time, mitochondria involvement in the regulation of oxidative stress.

However, turning to CNS, it has been demonstrated that glutamate excitotoxicity, in neuronal cell cultures, is associated with an inhibition of neuronal insulin sensitivity (122). This mechanism is associated with a process mediated by the activation of the NMDA receptor, which causes an intracellular increase in calcium levels. In consequence, the membrane potential of mitochondria is decreased, which can affect BIR (123). This data agrees with the studies performed by Zhao and colleagues, which reported that glutamate reduces the tyrosine phosphorylation of insulin receptor in hippocampal cultures when it was added after the prolonged insulin-mediated stimulation (124). Thus, the mitochondrial respiratory chain is involved in insulin-stimulated H_2O_2 production, playing a prominent role in insulin receptor autophosphorylation in neurons and in the regulation of IR at the peripheral level (125, 126).

MITOCHONDRIAL DYSFUNCTION IS RESPONSIBLE FOR BRAIN INSULIN RESISTANCE

It is well known that the normal metabolic process of the cell, mainly in neurons, requires a continuous supply of energy. Mitochondria play a crucial role in this process since it is the main organelle involved in oxidative phosphorylation and ATP synthesis in the cell. In LOAD, it has been described that Aβ oligomers can directly damage the mitochondria, contributing to oxidative damage by enhancing the production of ROS (5). In consequence, the Aβ production is increased which, in turn, is toxic for the mitochondria causing further oxidative damage and creating a vicious circle (57).

In the case of metabolic disorders associated with LOAD, this concept becomes more complex since there is a close relationship between mitochondrial activation and insulin signaling. In this line, previous studies have demonstrated that mitochondrial dysfunction in both, peripheral tissues and the CNS, is involved in impaired insulin signalling (127–130).

Specifically, it has been reported that IR causes a drop in mitochondrial function through a decrease in energy production (131). Likewise, the generation of ROS at the mitochondrial level would favour the activation of JNK, which increases IR, as previously discussed (132). Furthermore, Barone and colleagues have demonstrated that a mitochondrial dysfunction is an early event, confirming its contribution to the onset of BIR (133). Interestingly, it has been observed that oxidative stress appears before AD neuropathology markers including Aβ oligomers, p-Tau and neuroinflammation in 3xTg-AD mice (134).

Going deeper at the molecular level, one of the molecules responsible for the IR state activated by mitochondrial dysfunction is AMP-activated protein kinase (AMPK) (129). AMPK can stimulate glucose uptake by increasing the expression of glucose transporters and/or by translocating transporters to the cell surface (135). For this reason, AMPK signalling is an alternative way to regulate glucose uptake and energy metabolism in neurons. Furthermore, the AMPK signalling pathway would play an important role in neuroprotection through the regulation of its downstream signalling targets, including the - Sirtuin 1(SIRT1), Pparg coactivator 1 alpha (PGC-1α), Forkhead box O (FOXO), and NF-κB proteins regulating the inflammatory process, oxidative stress and mitochondrial biogenesis, among many other issues (84). In this line, previous preclinical studies have shown that drugs that regulate AMPK, such as metformin, confer neuroprotection since it has been reported that they are involved in the regulation of the neuropathology of LOAD by inhibiting the phosphorylation of Tau and the production of Aβ (136).

In summary, it is obvious that mitochondrial alterations, specifically the drop in ATP levels, play an important role in IR and LOAD (123). However, which of these two processes occurs first (if the mitochondrial alteration favours ROS production or *vice versa*) remains unclear.

MAMMALIAN TARGET OF RAPAMYCIN SIGNALLING MODULATES INSULIN RESISTANCE

It has been demonstrated that mTOR signalling plays an important role in maintaining energy homeostasis since nutrient availability is also integrated and transduced through this signalling pathways. In this context, it has been reported that aberrant regulation of mTOR occurs in the brains of individuals with T2DM. These data concur with the study performed by Sun and colleagues, which revealed increased activity of this pathway in the post-mortem brains of LOAD patients (137). Moreover, preclinical studies in 3xTgAD transgenic mice treated with 20% sucrose in the drinking water caused a significant weight gain and IR (138), which favours an increase in Aβ and p-Tau deposition mediated by the upregulation of the mTOR pathway in the brain. In this line, mice treated with an mTOR inhibitor, such as Rapamycin, repressed the effects of sucrose on the pathology of AD, suggesting that alterations in mTOR activation link peripheral IR and brain pathologies (138).

Going more into detail, Tramutola and colleagues reported that in both, MCI and LOAD subjects, the increased phosphorylation of the IRS1 inhibitory serine residue (Ser307) correlated with altered PI3K/AKT/mTOR signaling compared to controls. Thus, BIR in LOAD is maintained by the overactivation of the PI3K / mTOR axis, which inhibits the activity of IRS1, creating a negative feedback mechanism (139). Likewise, something similar happens with mTORC1/S6K1. Its activation feeds back on the pathway to block insulin signalling through inhibitory phosphorylation of insulin receptor substrate 1 IRS1 (Ser307 and Ser636/639) and Grb10 (140, 141). In fact, mice lacking insulin S6K1 are protected against age- and diet-induced obesity and show increased sensitivity to insulin (142).

CERAMIDES REGULATE INSULIN RESISTANCE AND INCREASE THE RISK OF LATE ONSET ALZHEIMER'S DISEASE

Ceramides comprise a family of sphingolipids that are generated from fatty acids and sphingosine. It is well known that ceramides play an important role in the structure of the cell membrane since they participate in the formation of the lipid raft. On the other hand, ceramides exert various physiological and regulatory effects on cell signalling pathways that mediate cellular growth, proliferation and senescence, among other functions (58). Interestingly, ceramides like other neurotoxic lipids can cross the BBB, contributing to the BIR and neurodegeneration. In this line, previous studies have reported that ceramide levels increase in the brain of preclinical mice models of AD and in LOAD patients (143). Likewise, ceramides exert a role in LOAD pathophysiology promoting the regulation of BACE1 and the generation and aggregation of Aβ (144, 145). Besides, ceramides which are located especially in the myelin sheath that surrounds some nerve cell axons, have also been observed in plasma, CSF and brain tissue of LOAD patients, and have also been associated with an increase in Aβ plaques (146, 147).

Related to the insulin pathway, previous studies have reported that ceramides are involved in the IR onset through different processes such as the activation of the NLRP3 sensor which promotes IL-1β-induced inflammation. In consequence, plasma glucose and insulin sensitivity in individuals with T2DM is reduced, inhibiting PI3K/AKT signalling pathway (117). In addition, increased ceramide production can stress the ER contributing to the development of IR and lipolysis, leading to an increased ceramide production and enhancing neuroinflammatory processes. Therefore, the dysregulated lipid metabolism promotes an excess of ceramides generation that exacerbate IR, neuroinflammation, brain damage, and synaptic loss.

CONCLUDING REMARKS

It is widely accepted that LOAD is an extremely complex metabolic disorder hindering the success of a cure. The process of dysregulation of brain insulin receptor signalling in LOAD is a multifunctional process directly associated with an increased Aβ accumulation, Tau hyperphosphorylation, increased levels of glutamate and proinflammatory mediators, as well as mTOR overactivation, among others (148–150). Therefore, if LOAD is really a metabolic progressive disease related to IR in which the brain is also affected, then therapeutic strategies developed for other pathologies related to metabolic alterations and obesity could be effective (151–153). In addition, preventive therapies, such as lifestyle modifications including healthy diet or physical activity could also improve the ability to respond to insulin, avoiding the related development of cognitive decline (154–156).

IR results in a decreased activity of AKT, which reduces the action of insulin on cognition and metabolism. Focusing on the brain, BIR promotes the activation of Tau kinases, thus leading to their hyperphosphorylation, oligomerization and aggregation. Likewise, Tau favours BIR, creating a vicious circle (85, 86). Moreover, Aβ oligomers may also contribute to IR by inducing insulin receptor internalization, but also by promoting the inhibitory phosphorylation of IRS-1 through JNK and others, which lead to the development of LOAD. In addition, neuroinflammatory process is generated triggered by Aβ and due to alterations in the insulin pathway. In line with these processes, mTOR signalling is positioned at the crossroad of BIR and neuroinflammation in LOAD and acts as a vicious positive feed-forward loop to accelerate disease progression. In conclusion, the dangerous combination of peripheral hyperglycemia, obesity, peripheral inflammation and the associated molecular alterations together with brain Aβ, Tau, APOEε4 and BIR could be finally responsible for LOAD.

ACKNOWLEDGEMENTS

This research work was supported by the Spanish Ministry of Economy and Competitiveness under the project SAF2017-84283-R, Biomedical Research Networking Centre in Neurodegenerative Diseases (CIBERNED) under project CB06/05/0024 and the European Regional Development Founds. A. Cano acknowledges the support of the Spanish Ministry of Science, Innovation and Universities under the grant *Juan de la Cierva* (FJC2018-036012-I).

REFERENCES

1. 2021 Alzheimer's disease facts and figures. *Alzheimers Dement* 17, 327–406 (2021).
2. P. Scheltens, et al., Alzheimer's disease. *Lancet* 397(10284), 1577–1590 (2021).
3. G. Deuschl, et al. The burden of neurological diseases in Europe: an analysis for the Global Burden of Disease Study 2017. *Lancet Public Health* 5(10), e551–e567 (2020).
4. J. Avila et al., Multilevel view of the development of Alzheimer's disease. *Neuroscience* 457, 283–293 (2021).
5. K.L. Viola et al., Amyloid beta oligomers in Alzheimer's disease pathogenesis, treatment, and diagnosis. *Acta Neuropathol* 129, 183–206. (2015).
6. P.N. Lacor., Advances on the understanding of the origins of synaptic pathology in AD. *Curr Genomics*. 8, 486–508 (2007).
7. P.N. Lacor, et al., Abeta oligomer-induced aberrations in synapse composition, shape, and density provide a molecular basis for loss of connectivity in Alzheimer's disease. *J Neurosci*. 27, 796–807 (2007).

8. K.L. Viola et al., Why Alzheimer's is a disease of memory: the attack on synapses by A beta oligomers (ADDLs). *J Nutr Health Aging*. 12, 51S–57S (2008).
9. M.T. Heneka, et al., Neuroinflammation in Alzheimer's disease. *Lancet Neurol* 14(4), 388–405 (2015).
10. M.E. Belloy et al., Alzheimer's disease neuroimaging initiative. Association of Klotho-VS Heterozygosity with Risk of Alzheimer Disease in Individuals Who Carry APOE4. *JAMA Neurol* 77, 849–862 (2020).
11. S. Craft. Insulin resistance syndrome and Alzheimer disease: pathophysiologic mechanisms and therapeutic implications. *Alzheimer Dis Assoc Disord* 20, 298–301 (2006).
12. T. Matsuzaki, et al., Insulin resistance is associated with the pathology of Alzheimer disease: the Hisayama study. *Neurology* 75, 764–770 (2010).
13. T.S. Salameh, et al., Insulin resistance, dyslipidemia, and apolipoprotein E interactions as mechanisms in cognitive impairment and Alzheimer's disease. *Exp Biol and Med* 241, 1676–1683 (2016).
14. D. Kellar, et al., Brain insulin resistance in Alzheimer's disease and related disorders: mechanisms and therapeutic approaches. *Lancet Neurol* 19, 758–766 (2020).
15. Hoyer S., Is sporadic Alzheimer disease the brain type of non-insulin dependent diabetes mellitus? A challenging hypothesis. *J Neural Transm*. 105, 415–422 (1998).
16. S. Hoyer., Brain glucose and energy metabolism abnormalities in sporadic Alzheimer disease. Causes and consequences: an update. *Exp Gerontol* 35, 1363–1369 (2020).
17. T.T. Nguyen, et al. Type 3 diabetes and its role implications in Alzheimer's disease. *Int J Mol Sci* 21(9), 3165 (2020).
18. L.R. Clark, et al. Hypertension and obesity moderate the relationship between β-amyloid and cognitive decline in midlife. *Alzheimers Dement*. 15, 418–428 (2019).
19. S.E Arnold, et al. Brain insulin resistance in type 2 diabetes and Alzheimer disease: concepts and conundrums. *Nat Rev Neurol* 14, 168–181 (2018).
20. Ott A, et al., Association of diabetes mellitus and dementia: the Rotterdam Study. *Diabetologia*. 39: 1392–1397 (1996).
21. A Ott, et al., Diabetes mellitus and the risk of dementia: the Rotterdam Study. *Neurology* 1999; 53, 1937–1942 (1999).
22. M. Hokama, et al., Altered expression of diabetes-related genes in Alzheimer's disease brains: the Hisayama study. *Cereb Cortex* 24, 2476–2488 (2014).
23. E.J. Starks, et al, Insulin resistance is associated with higher cerebrospinal fluid tau levels in asymptomatic APOEε4 carriers. *J Alzheimers Dis*. 46, 525–533 (2015).
24. A.A. Willette, et al., Association of insulin resistance with cerebral glucose uptake in late middle-aged adults at risk for Alzheimer disease. *JAMA Neurol* 72, 1013–1020 (2015).
25. A.A. Willette, et al., Insulin resistance, brain atrophy, and cognitive performance in late middle-aged adults. *Diabetes Care* 36, 443–449 (2013).
26. H. Bruehl, et al. Cognitive impairment in non-diabetic middle-aged and older adults is associated with insulin resistance. *J Clin Exp Neuropsychol* 32, 487–493 (2010).
27. N.L. Rasgon, et al., Insulin resistance and hippocampal volume in women at risk for Alzheimer's disease. *Neurobiol Aging* 32, 1942–1948 (2011).
28. M.A. Fishel, et al., Hyperinsulinemia provokes synchronous increases in central inflammation and beta-amyloid in normal adults. *Arch Neurol* 62, 1539–1544 (2005).
29. S. Hong et al., The insulin resistance by triglyceride glucose index and risk for dementia: population-based study. *Alzheimers Res Ther* 13:9 (2021).
30. J. Havrankova, et al., Insulin receptors are widely distributed in the central nervous system of the rat. *Nature* 272, 827–829 (1978).
31. H.E. Lebovitz, Insulin resistance: definition and consequences. *Exp Clin Endocrinol Diabetes* 109, S135–S148 (2001).
32. S. Hoyer., The brain insulin signal transduction system and sporadic (type II) Alzheimer disease: an update. *J Neural Transm*. 109, 341–360 (2002).
33. S. Hoyer et al., Ischemia and the aging brain. Studies on glucose and energy metabolism in rat cerebral cortex. *Neurobiol Aging* 7, 23–29 (1986).
34. I. Arrieta-Cruz et al., Acute exposure of the mediobasal hypothalamus to amyloid-β25–35 perturbs hepatic glucose metabolism. *J Alzheimers Dis* 46, 843–848 (2015).
35. C. Liguori, et al., Hypothalamic dysfunction is related to sleep impairment and CSF biomarkers in Alzheimer's disease. *J Neurol* 264(11), 2215–2223 (2017).
36. M. Spinelli, et al., Brain insulin resistance impairs hippocampal plasticity. *Vitam Horm*. 114: 281–306 (2020).

37. W.Q. Zhao, et al., Amyloid beta oligomers induce impairment of neuronal insulin receptors. *FASEB J* 22, 246–260 (2008).
38. S. Li, et al., A mechanistic hypothesis for the impairment of synaptic plasticity by soluble Abeta oligomers from Alzheimer's brain. *J Neurochem* 154, 583–597 (2020).
39. Z. Arvanitakis et al., Brain insulin signaling, Alzheimer disease pathology, and cognitive function. *Ann Neurol* 88, 513–525 (2020).
40. K. Talbot, et al., Demonstrated brain insulin resistance in Alzheimer's disease patients is associated with IGF-1 resistance, IRS-1 dysregulation, and cognitive decline. *J Clin Invest* 122, 1316–1338. (2012).
41. T.R. Bomfim et al., An anti-diabetes agent protects the mouse brain from defective insulin signaling caused by Alzheimer's disease- associated Aβ oligomers. *J Clin Invest* 122, 1339–1353 (2012).
42. C.A. Grillo et al., Hippocampal insulin resistance impairs spatial learning and synaptic plasticity. *Diabetes* 64, 3927–3936 (2015).
43. C.A. Grillo, et al., Insulin resistance and hippocampal dysfunction: disentangling peripheral and brain causes from consequences. *Exp Neurol* 318, 71–77 (2019).
44. J. Berlanga-Acosta, Insulin resistance at the crossroad of Alzheimer disease pathology: a review. *Front Endocrinol (Lausanne)* 11, 560375 (2020).
45. G. Ahmadian et al., Tyrosine phosphorylation of GluR2 is required for insulin-stimulated AMPA receptor endocytosis and LTD. *EMBO J* 23, 1040–10450 (2004).
46. E.D. Martín et al., IRS-2 Deficiency impairs NMDA receptor-dependent long-term potentiation. *Cereb Cortex*. 22, 1717–27 (2012).
47. S. Hoyer., Senile dementia and Alzheimer's disease. Brain blood flow and metabolism. *Prog Neuropsychopharmacol Biol Psychiatry* 10, 447–478 (1986).
48. S. Hoyer., The biology of the aging brain. Oxidative and related metabolism. *Eur J Gerontol* 1, 157–165 (1992a).
49. S. Hoyer., Oxidative metabolism in Alzheimer brain. Studies in early-onset and late-onset cases. *Mol Chem Neuropathol* 16, 207–224 (1992b).
50. S. Hoyer., Age-related changes in cerebral oxidative metabolism. Implications for drug therapy. *Drugs & Aging* 6, 210–218 (1996).
51. S. Hoyer., Oxidative metabolism deficiencies in brain of patients with Alzheimer's disease. *Acta Neurol Scand [Suppl]* 165, 18–24 (1996).
52. L. Frölich, et al. Brain insulin and insulin receptors in aging and sporadic Alzheimer's disease. *J Neural Transm (Vienna)*. 105, 423–438. (1998).
53. L. Frölich, et al., A disturbance in the neuronal insulin receptor signal transduction in sporadic Alzheimer's disease. *Ann N Y Acad Sci*. 893: 290–293 (1999).
54. A.M. Moloney, et al., Defects in IGF-1 receptor, insulin receptor and IRS-1/2 in Alzheimer's disease indicate possible resistance to IGF-1 and insulin signalling. *Neurobiol Aging* 31, 224–243 (2010).
55. J.J. Kulstad, et al., Differential modulation of plasma beta-amyloid by insulin in patients with Alzheimer disease. *Neurology* 66, 1506–1510 (2006).
56. L. Xie, et al., Alzheimer's beta-amyloid peptides compete for insulin binding to the insulin receptor. *J Neurosci* 22(10), RC221 (2002).
57. E.N. Cline, et al., The amyloid-oligomer hypothesis: beginning of the third decade. *J Alzheimers Dis* 64, S567–S610. (2018).
58. S.M. de la Monte, et al., Dysfunctional proceramide, ER stress, and insulin/IGF signaling networks with progression of Alzheimer's disease. *J Alzheimers Dis* 30 Suppl 2, S217–229. (2012).
59. S.M. de la Monte, et al., Early-stage Alzheimer's disease is associated with simultaneous systemic and central nervous system dysregulation of insulin-linked metabolic pathways. *J Alzheimers Dis* 68, 657–668 (2019).
60. J. Robbins, et al., Dysregulation of insulin-linked metabolic pathways in Alzheimer's disease: co-factor role of apolipoprotein E e4. *J Alzheimers Dis Rep* 4, 479–493 (2020).
61. V.K. Sharma, et al. Insulin resistance and bioenergetic manifestations: targets and approaches in Alzheimer's disease. *Life Sci*. 262:118401 (2020).
62. M.K. Poddar, et al., Metabolic disorder in Alzheimer's disease. *Metab Brain Dis*. 202; 36: 781–813.
63. S. Boura-Halfon et al, Serine kinases of insulin receptor substrate proteins. *Vitam Horm*.80, 313–349 (2009).
64. J.F. Tanti, et al., Cellular mechanisms of insulin resistance: role of stress-regulated serine kinases and insulin receptor substrates (IRS) serine phosphorylation. *Curr Opin Pharmacol* 9,753–62 (2009).
65. F.G. De Felice. Connecting type 2 diabetes to Alzheimer's disease. *Expert Rev Neurother* 13, 1297–1299 (2013).
66. F.G. De Felice, et al., A key role of insulin receptors in memory. *Diabetes* 64, 3653–3655. (2015).

67. C.M. Long-Smith, et al., The diabetes drug liraglutide ameliorates aberrant insulin receptor localisation and signalling in parallel with decreasing both amyloid-β plaque and glial pathology in a mouse model of Alzheimer's disease. *Neuromolecular Med.* 15(1), 102–14 (2013).
68. M. Yarchoan, et al., Abnormal serine phosphorylation of insulin receptor substrate 1 is associated with tau pathology in Alzheimer's disease and tauopathies. *Acta Neuropathol* 128:679–689 (2014).
69. E. McDade, et al., Stop Alzheimer's before it starts. *Nature.* 547, 153–155 (2017).
70. S.T. Ferreira, et al., Soluble amyloid-β oligomers as synaptotoxins leading to cognitive impairment in Alzheimer's disease. *Front Cell Neurosci* 9:191 (2015).
71. M.F. Gregor, et al., Inflammatory mechanisms in obesity. *Annu Rev Immunol* 29, 415–445 (2011).
72. R.B. Maccioni, et al., Inflammation: A Major Target for Compounds to Control Alzheimer's Disease. *J Alzheimers Dis.* 76, 1199–1213 (2020).
73. Q.L. Ma, et al., Beta-amyloid oligomers induce phosphorylation of tau and inactivation of insulin receptor substrate via c-Jun N-terminal kinase signaling: suppression by omega-3 fatty acids and curcumin. *J Neurosci* 29 (28), 9078–9089. (2009).
74. M.V. Lourenco, et al. TNF-α mediates PKR-dependent memory impairment and brain IRS-1 inhibition induced by Alzheimer's β-amyloid oligomers in mice and monkeys. *Cell Metab* 18 (6), 831–843 (2013).
75. G. Sabio, et al., TNF and MAP kinase signalling pathways. *Semin Immunol* 26, 237–245 (2014).
76. J. R. Clarke et al. Alzheimer-associated Aβ oligomers impact the central nervous system to induce peripheral metabolic deregulation. *EMBO Mol Med* 7(2):190–210 (2015).
77. S. Takeda, et al., Elevation of plasma beta-amyloid level by glucose loading in Alzheimer mouse models. *Biochem Biophys Res Commun.* 2009, 385:193–197 (2009).
78. S. Takeda, et al., Oral glucose loading modulates plasma beta-amyloid level in alzheimer's disease patients: potential diagnostic method for Alzheimer's disease. *Dement Geriatr Cogn Disord.* 34, 25–30 (2012).
79. Y. Zhang, et al., Amyloid-β induces hepatic insulin resistance by activating JAK2/STAT3/SOCS-1 signaling pathway. *Diabetes* 61, 1434–1443 (2012).
80. Y. Zhang, et al. Amyloid-β induces hepatic insulin resistance in vivo via JAK2. *Diabetes* 262, 1159–1166 (2013).
81. N. Sato, et al., Plasma Aβ: a possible missing link between Alzheimer disease and diabetes. *Diabetes.* 62, 1005–1006 (2013).
82. K. Plucińska, et al., Neuronal human BACE1 knockin induces systemic diabetes in mice. *Diabetologia.* 59, 1513–1523 (2016).
83. M. Schubert, et al., Role for neuronal insulin resistance in neurodegenerative diseases. *Proc Natl Acad Sci U S A.* 101, 3100–3105 (2004).
84. Salminen, et al., AMP-activated protein kinase: a potential player in Alzheimer's disease. *J Neurochem.* 118, 460–474 (2011).
85. E. Marciniak, et al., Tau deletion promotes brain insulin resistance. *J Exp Med.*214, 2257–2269 (2017).
86. P. Rodriguez-Rodriguez. et al., Tau hyperphosphorylation induces oligomeric insulin accumulation and insulin resistance in neurons. *Brain* 140, 3269–3285 (2017).
87. N. Zhao, et al., Apolipoprotein E, receptors, and modulation of Alzheimer's disease. *Biol Psychiatry* 83, 347–357 (2018).
88. N. Zhao, et al., Apolipoprotein E4 impairs neuronal insulin signaling by trapping insulin receptor in the endosomes. *Neuron* 96, 115–129 (2017).
89. G. W. Small, et al., Apolipoprotein E type 4 allele and cerebral glucose metabolism in relatives at risk for familial Alzheimer disease. *JAMA* 273, 942–947 (1995).
90. E.M. Reiman, et al., Preclinical evidence of Alzheimer's disease in persons homozygous for the epsilon 4 allele for apolipoprotein E. *N Engl J Med.* 334, 752–758 (1995).
91. R. Peila, et al., Type 2 diabetes, APOE gene, and the risk for dementia and related pathologies: the Honolulu-Asia Aging Study. *Diabetes* 5, 1256–1262 (2002).
92. F. Irie, et al., Enhanced risk for Alzheimer disease in persons with type 2 diabetes and APOE epsilon4: the Cardiovascular Health Study Cognition Study. *Arch Neurol* 65, 89–93. (2008).
93. C. Ozek, et al., Protein-tyrosine phosphatase 1B (PTP1B) is a novel regulator of central brain-derived neurotrophic factor and tropomyosin receptor kinase B (TrkB) signaling. *J Biol Chem.* 289, 31682–31692 (2014).
94. M.N. Vieira, et al. Protein Tyrosine Phosphatase 1B (PTP1B): a potential target for Alzheimer's therapy? *Front Aging Neurosci.* 31, 7 (2017).
95. S. Qian, et al., Recent advances in the development of protein tyrosine phosphatase 1B

inhibitors for Type 2 diabetes. *Future Med Chem.* 8, 1239–1258 (2016).
96. H. Hussain, et al., Protein tyrosine phosphatase 1B (PTP1B) inhibitors as potential anti-diabetes agents: patent review (2015–2018). *Expert Opin Ther Pat* 29, 689–702 (2019).
97. M. Ettcheto, et al., Epigallocatechin-3-Gallate (EGCG) improves cognitive deficits aggravated by an obesogenic diet through modulation of unfolded protein response in appswe/ps1de9 mice. *Mol Neurobiol* 57, 1814–1827 (2020).
98. F. Fuentes, et al., Protein tyrosine phosphatase PTP1B is involved in hippocampal synapse formation and learning. *PLoS One* 7, e41536 (2012).
99. G.J. Song, et al. A novel role for protein tyrosine phosphatase 1B as a positive regulator of neuroinflammation. *J Neuroinflammation*. 19, 86 (2016).
100. J.M. Zabolotny et al., Protein-tyrosine phosphatase 1B expression is induced by inflammation in vivo. *J Biol Chem* 283(21), 14230–14241 (2008).
101. Y. Ito, et al., TNFα increases hypothalamic PTP1B activity via the NFκB pathway in rat hypothalamic organotypic cultures. *Regul Pept* 174(1–3), 58–64 (2012).
102. M.T. Heneka, et al,. NLRP3 isactivated in Alzheimer's disease and contributes to pathology in APP/PS1 mice. *Nature* 493, 674–678 (2013).
103. Ardura-Fabregat, et al., Targeting neuroinflammation to treat Alzheimer's disease. *CNS Drugs* 31, 1057–1082 (2017).
104. Guzman-Martinez, et al., Neuroinflammation as a common feature of neurodegenerative disorders. *Front Pharmacol* 10:1008 (2019).
105. G. Sabio, et al., C-Jun NH2-terminal kinase 1 (JNK1): roles in metabolic regulation of insulin resistance. *Trends Biochem Sci* 35, 490–496 (2010).
106. W. Cai., Detection of circulating natural antibodies to inflammatory cytokines in type-2 diabetes and clinical significance. *J Inflamm (Lond)* 14, 24 (2017).
107. L. Liang, et al., Endoplasmic reticulum stress impairs insulin receptor signaling in the brains of obese rats. *PLoS One* 10, e0126384 (2015).
108. T. Nakamura, et al. Double-stranded RNA-dependent protein kinase links pathogen sensing with stress and metabolic homeostasis. *Cell* 140, 338–348 (2010).
109. T. Nakamura, et al. Small-molecule inhibitors of PKR improve glucose homeostasis in obese diabetic mice. *Diabetes* 63, 526–534 (2014).
110. M. Taga, et al., PKR modulates abnormal brain signaling in experimental obesity. *PLoS One*. 13(-5), e0196983 (2018).
111. H. Cheng, et al. The molecular mechanisms underlying mitochondria-associated endoplasmic reticulum membrane-induced insulin resistance. *Front Endocrinol (Lausanne)* 11, 592129 (2020).
112. D. Pal, et al., Fetuin-A acts as an endogenous ligand of TLR4 to promote lipid-induced insulin resistance. *Nat Med* 18, 1279–1285 (2012).
113. N.Q. Huang, et al., TLR4 is a link between diabetes and Alzheimer's disease. *Behav Brain Res.* 316, 234–244 (2017).
114. Y. Zhou, et al., TLR4 Targeting as a Promising Therapeutic Strategy for Alzheimer Disease Treatment. *Front Neurosci.* 14, 602508 (2020).
115. M. Calvo-Rodrigue, et al. Role of toll like receptor 4 in Alzheimer's disease. *Front Immunol* 11, 1588. (2020).
116. S. Hirschberg., Implications of diet and the gut microbiome in neuroinflammatory and neurodegenerative diseases. *Int J Mol Sci* 20 (12), 3109 (2019)
117. B. Vandanmagsar, et al., The NLRP3 inflammasome instigates obesity-induced inflammation and insulin resistance. *Nat Med* 17, 179–188 (2011).
118. Y.S. Feng, et al., The involvement of NLRP3 inflammasome in the treatment of Alzheimer's disease. *Ageing Res Rev* 64, 101192 (2020).
119. A. Halle, et al., The NALP3 infammasome is involved in the innateimmune response to amyloid-beta. *Nat Immunol* 9, 857–865 (2008).
120. A. Welters, et al., Need for better diabetes treatment: the therapeutic potential of NMDA receptor antagonists. *Klin Padiatr* 229, 14–20 (2017)
121. J. Du, et al. Glutamate in peripheral organs: biology and pharmacology. *Eur J Pharmacol*. 784, 42–48 (2016).
122. N.A. Persiyantseva, et al. Mitochondrial H2O2 as an enable signal for triggering autophosphorylation of insulin receptor in neurons. *J Mol Signal* 8, 11 (2013).
123. C. Carvalho, et al. Tortuous paths of insulin signaling and mitochondria in Alzheimer's disease. *Adv Exp Med Biol* 1128, 161–183. (2019).
124. F. Zhao, et al., Insulin modulates excitatory synaptic transmission and synaptic plasticity in the mouse hippocampus. *Neuroscience* 411, 237–254 (2019).
125. T.P. Storozhevykh, et al., Mitochondrial respiratory chain is involved in insulin-stimulated hydrogen peroxide production and plays an

integral role in insulin receptor autophosphorylation in neurons. *BMC Neurosci* 8, 84 (2007).
126. I.A. Pomytkin., H2O2 signalling pathway: a possible bridge between insulin receptor and mitochondria. *Curr Neuropharmacol* 10, 311–320 (2012).
127. M. Schell, et al., Untangling the effect of insulin action on brain mitochondria and metabolism. *J Neuroendocrinol* e12932. (2021).
128. E. Steen, et al., Impaired insulin and insulin-like growth factor expression and signaling mechanisms in Alzheimer's disease–is thistype 3 diabetes? *J Alzheimers Dis* 7, 63–80. (2005).
129. Y. Sun, et al., Metabolism: a novel shared link between diabetes mellitus and Alzheimer's disease. *J Diabetes Res* 2020:4981814. (2020).
130. G.M. Reaven., The insulin resistance syndrome: definition and dietary approaches to treatment. *Annu Rev Nutr* 25: 391–406 (2005).
131. M.A. Schell, et al., Untangling the effect of insulin action on brain mitochondria and metabolism. *J Neuroendocrinol* 33, e12932 (2021).
132. R. Nogueiras, et al., Brain JNK and metabolic disease. *Diabetologia*. 2021;64:265–274.
133. E. Barone, et al., Insulin resistance in Alzheimer disease: is heme oxygenase-1 an Achille's heel? *Neurobiol Dis* 84, 69–77 (2015).
134. E. Barone, et al. Impairment of biliverdin reductase-A promotes brain insulin resistance in Alzheimer disease: a new paradigm. *Free Radic Biol Med* 91, 127–142 (2016).
135. P. Weisová, et al., Regulation of glucose transporter 3 surface expression by the AMP-activated protein kinase mediates tolerance to glutamate excitation in neurons. *J Neurosci* 29, 2997–3008 (2009).
136. M. El Massry, et al., Metformin: a growing journey from glycemic control to the treatment of Alzheimer's disease and depression. *Curr Med Chem* 28(12), 2328–2345 (2020).
137. Y.X. Sun, et al., Differential activation of mTOR complex 1 signaling in human brain with mild to severe Alzheimer's disease. *J Alzheimers Dis* 38, 437–444 (2014).
138. M.E. Orr, et al., Mammalian target of rapamycin hyperactivity mediates the detrimental effects of a high sucrose diet on Alzheimer's disease pathology. *Neurobiol Aging* 35, 1233–1242 (2014).
139. A. Tramutola, et al., Alteration of mTOR signaling occurs early in the progression of Alzheimer disease (AD): analysis of brain from subjects with pre-clinical AD, amnestic mild cognitive impairment and late-stage AD. *J Neurochem* 133, 739–749 (2015).
140. L.S. Harrington, et al., The TSC1–2 tumor suppressor controls insulin-PI3K signaling via regulation of IRS proteins. *J Cell Biol* 166, 213–223 (2004).
141. E.D. Werner, et al., Insulin resistance due to phosphorylation of insulin receptor substrate-1 at serine 302. *J Biol Chem* 279, 35298–35305 (2004).
142. S.H. Um, et al., Absence of S6K1 protects against age- and diet-induced obesity while enhancing insulin sensitivity. *Nature* 431, 200–205 (2004).
143. M.B. Dinkins, et al., Neutral sphingomyelinase-2 deficiency ameliorates Alzheimer's disease pathology and improves cognition in the 5XFAD mouse. *J Neurosci* 36, 8653–8667 (2016).
144. H. Geekiyanage, et al., MicroRNA-137/181c regulates serine palmitoyltransferase and in turn amyloid beta, novel targets in sporadic Alzheimer's disease. *J Neurosci* 31, 14820–14830 (2011).
145. M.B. Dinkins, et al., Neutral Sphingomyelinase-2 deficiency ameliorates Alzheimer's disease pathology and improves cognition in the 5XFAD Mouse. *J Neurosci*. 36, 8653–8667 (2016).
146. M. Kosicek, et al., Elevated cerebrospinal fluid sphingomyelin levels in prodromal Alzheimer's disease. *Neurosci Lett* 516, 302–305 (2012).
147. J.H. Hsiao, et al., Elevation in sphingomyelin synthase activity is associated with increases in amyloid-beta peptide generation. *PLoS One* 8(8), e74016 (2013)
148. T.M. Barber et al. Mechanisms of insulin resistance at the crossroad of obesity with associated metabolic abnormalities and cognitive dysfunction. *Int J Mol Sci* 22, E546 (2021).
149. D.E. Barnes et al., The projected effect of risk factor reduction on Alzheimer's disease prevalence. *Lancet Neurol* 10, 819–28 (2011).
150. G.J. Biessels et al., Risk of dementia in diabetes mellitus: a systematic review. *Lancet Neurol* 5, 64–74 (2006).
151. J. Cummings, et al., Alzheimer's disease drug development pipeline: 2020. *Alzheimers Dement (N Y)* 6, e12050 (2020).
152. J.L. Cummings, et al., Treatment combinations for Alzheimer's disease: current and future

pharmacotherapy options. *J Alzheimers Dis* 67, 779–794 (2019).
153. S.P. Salloway, et al., Advancing combination therapy for Alzheimer's disease. *Alzheimers Dement (N Y)* 6(1), e12073. (2020).
154. E.M. Schrijvers, et al., Insulin metabolism and the risk of Alzheimer disease: the Rotterdam Study. *Neurology* 75, 1982–1987 (2010).
155. M. Schubert, et al., Role for neuronal insulin resistance in neurodegenerative diseases. *Proc Natl Acad Sci U S A* 101, 3100–3105 (2004).
156. M. Gratuze, et al., Mutual Relationship between Tau and central insulin signalling: consequences for AD and tauopathies? *Neuroendocrinology* 107, 181–195 (2018).

CHAPTER ELEVEN

Brain Insulin Action in the Control of Metabolism in Humans

Stephanie Kullmann

CONTENTS

INTRODUCTION

After the revolutionary discovery of insulin around 100 years ago (1), it took another 50 years until insulin receptors were discovered in the brain by Havrankova et al. (2). Since then, numerous studies showed that the brain is an important target for insulin action with tremendous effects on whole-body metabolism, cognition, and mood (3–5). Insulin resistance is a hallmark feature of type 2 diabetes (T2D) and is classically defined as an impaired biological response to insulin stimulation of target tissues, primarily the liver, muscle, and adipose tissue (6). Peripheral insulin resistance appears to impair the transport of insulin via the blood-brain barrier to the brain with a lower transport in case of systematic insulin resistance (7). Decrements in insulin action are not limited to the periphery or metabolic diseases (4). Significant disturbances in central insulin action are described in obesity and T2D but also in brain aging and dementia. It has therefore been proposed that insulin resistance of the brain is a potential link between metabolic and cognitive diseases (8, 9).

DOI: 10.1201/9781003079927-11

Even though the brain takes up glucose in a mostly insulin-independent manner (10), there is an abundance of insulin receptors in the hypothalamus, cerebellum as well as cortical and subcortical regions. Pancreas-derived insulin binds to its receptors after crossing the blood-brain barrier via a saturable, receptor-mediated transport mechanism. Insulin concentrations in the cerebrospinal fluid (CSF) are roughly ten times lower than in serum (7), though, no detailed investigations on changes in CSF insulin concentrations after food intake are available so far. Furthermore, there is a current debate about whether insulin could also be produced locally within the cerebral cortex (11, 12). To experimentally study central insulin action in animals, insulin is often directly infused into the brain. This is done by intracerebroventricular infusion (ICV) (13–17) or direct injection into brain tissue (18, 19). Investigating brain insulin action in humans, on the other hand, is more complex.

INVESTIGATING CENTRAL INSULIN ACTION IN HUMANS

In humans, several methods are established to study peripheral insulin action or insulin secretion in a clinical and experimental setting. These include the oral and the intravenous (iv) glucose tolerance tests, hyperglycemic hyperinsulinemic glucose clamps, as well as continuous iv insulin infusion during euglycemic hyperinsulinemic glucose clamps. The latter is considered the gold standard for measuring whole-body insulin sensitivity or resistance and is primarily used for research purposes due to its limited clinical applicability. Alternatively, it is possible to derive several clinically meaningful surrogates of insulin sensitivity from fasting glucose and insulin or C-peptide levels (e.g. the homeostasis model assessment of insulin resistance (HOMA-IR) (20)) or in their concentrations in response to an oral glucose challenge (6). These surrogate measures of insulin sensitivity are markers of peripheral or whole-body insulin sensitivity. With these methods, insulin effects are not limited to the brain, but occur in most tissues throughout the body.

By the intranasal method, on the other hand, insulin can bypass the body periphery and directly enter the central nervous system (CNS). Meanwhile, we can look back on two decades of research using the intranasal application of insulin (21). Results demonstrate in healthy humans that intranasal insulin exerts significant effects on whole-body metabolism and cognition (3, 21).

INTRANASAL INSULIN

Intranasal insulin application is a non-invasive, effective, and safe method (22–24) to disentangle central from peripheral insulin effects and study the causal role of central insulin action on the periphery. Only very small amounts of the nasally administered insulin are absorbed into the circulation (e.g., around 0.1U after intranasal application of 160U insulin) (25). Hence, systemic exposure is minimized in relation to other administration paradigms (oral glucose tolerance test, intravenous insulin infusion). This facilitates the differentiation between peripheral and central insulin effects. Following intranasal application, insulin enters the nasal mucosa and is transported to the central nervous compartment via olfactory and trigeminal pathways, thereby bypassing the blood-brain barrier (26). Animal research suggests that the olfactory pathway accounts for the majority of intranasal delivery (27). It transports insulin extracellularly through intercellular clefts of the olfactory epithelium leading to the olfactory bulb, and brain stem, and eventually spreads throughout the brain. This insulin transport from the nose to the brain can be followed by using fluorescent tracers in animal models (26). The trigeminal pathway also contributes to the transport following intranasal delivery, since the trigeminal nerve connects the nasal cavity with the brain via the pons by innervating the respiratory and olfactory epithelium.

CSF insulin concentrations in humans after intranasal application (28) show a rise 10 minutes after administration of 40 U insulin. Even 80 minutes after spray delivery, insulin concentrations in the CSF are higher after insulin compared to placebo, indicating a longer half-life of insulin in CSF than what has been reported for the systemic circulation. Comparable findings have been shown in rodents where a long persistence of human insulin in the brain was detected (29). Depending on the nature of the study, different intranasal doses are used. Long-term protocols often apply a lower dose for everyday use (24, 30, 31), whereas studies investigating acute effects, in general, prefer higher doses of up to 160 U (32–35). Nonetheless, since only minor effects on blood glucose levels are usually observed (32), the intranasal application is deemed suitable for the investigation of both short- and long-term insulin effects.

Of note, it is possible to combine intranasal insulin with other insulin stimulation techniques to investigate the causal role of central insulin action on glucose and insulin metabolism in the periphery. For example, in combination with a hyperglycemic clamp technique, it is possible to evaluate the effect of central insulin action on beta-cell insulin secretion from the pancreas (36). In combination with hyperinsulinemic-euglycemic glucose clamp, central-insulin mediated changes in peripheral insulin sensitivity and glucose metabolism (i.e. endogenous glucose production) can be quantified (37, 38).

METHODS TO INVESTIGATE PERIPHERAL AND WHOLE-BODY INSULIN SENSITIVITY

Glucose Challenge

The mixture of different nutrients in a meal triggers several physiological reactions in the periphery and the CNS. Hence, a more standardized way to study insulin action is the oral glucose tolerance test (oGTT), which is strongly implemented in the clinical routine. After oral ingestion of a 75 g glucose solution, blood glucose rises and several endocrine factors, including insulin, are released into the circulation. Peripheral insulin sensitivity can be assessed from this test e.g. by the oGTT-derived Matsuda insulin sensitivity index (39) and insulin secretion can be estimated e.g. by the oral Disposition Index (40). Insulin secretion can also be stimulated by an iv glucose bolus during an iv glucose tolerance test (ivGTT) (41) or by hyperglycemic glucose clamps (42). While the intravenous route of glucose administration is not physiological, it has the advantage of stimulating insulin release without major effects on several other endocrine systems that are activated after oral food or glucose intake, such as incretins, which also act in the brain (43). For both tests, several indices are available to assess peripheral insulin sensitivity and insulin secretion.

In response to glucose ingestion, blood glucose and insulin are significantly elevated in the blood. Blood glucose has its major peak at around 30 minutes and mostly declines thereafter. This provides the possibility to investigate glucose and insulin-mediated effects on the brain. Since glucose is often more rapidly removed from the circulation, predominantly insulin-driven effects are anticipated at later time points after ingestion. However, glucose ingestion triggers a plethora of responses, activating taste receptors and their afferent fibers (44). Neuroimaging findings revealed prominent responses in cortical and subcortical regions, highlighting the role of glucose versus insulin-mediated effects (45, 46). Furthermore, glucose, insulin, and other hormones act directly on peripheral tissues such as skeletal muscle, liver, and adipocytes, making it difficult to differentiate these peripheral effects from central actions.

Hyperinsulinemic-Euglycemic Glucose Clamp

The hyperinsulinemic-euglycemic glucose clamp is considered the "gold standard" for the quantification of peripheral insulin sensitivity/ insulin resistance. During this clamp, a given dose of insulin is continuously infused into the bloodstream to suppress hepatic glucose production and stimulate glucose uptake into insulin-sensitive tissues (e.g. skeletal muscle and adipose tissue). To prevent an insulin-induced decline in blood glucose during this test, a glucose solution is simultaneously infused and frequently adjusted based on frequent blood glucose measurements to keep blood glucose in the euglycemic range (42). The amount of glucose needed to reach a steady state reflects insulin sensitivity. Tracer dilution techniques (d-[6,6-$^2H_2$] glucose) are used to further elaborate possible CNS-based mechanisms responsible for changes in glucose infusion rates. Hence, it is possible to separate the effects of central insulin on endogenous glucose production and glucose disposal to peripheral tissues as the skeletal muscle (37) (Figure 11.1). Like during ivGTT and oGTT, circulating insulin is transported into the CNS via the receptor-mediated pathway. It reaches the brain continuously, while glucose levels are kept constant at normal fasting level. However, insulin effects are not limited to the brain but occur in most tissues throughout the body.

NEUROIMAGING METHODS TO QUANTIFY CENTRAL INSULIN ACTION IN HUMANS

Modern neuroimaging techniques are a valuable tool in the quest for understanding central insulin action in vivo in humans. The most commonly used forms are electroencephalography (EEG), magnetoencephalography (MEG), magnetic resonance

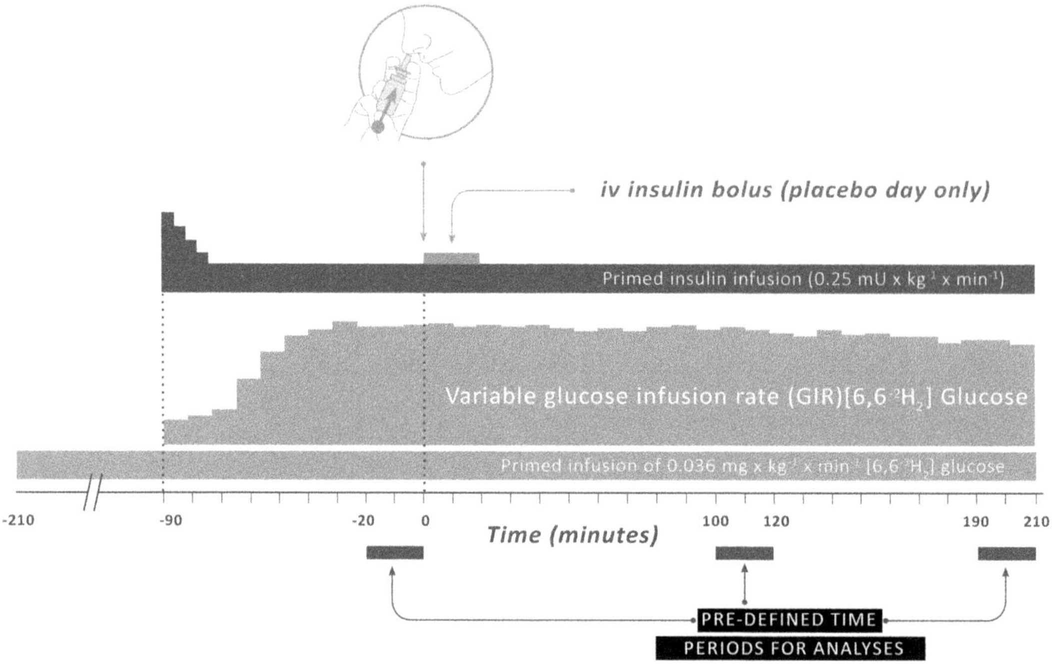

Figure 11.1 Central insulin effects on systemic metabolism. The figure shows intranasal insulin compared to placebo administration during a hyperinsulinemic-euglycemic clamp using d-[6,6-$^2H_2$] glucose. During the experiment, plasma insulin levels are elevated and frequent blood sampling takes place. The glucose infusion rate (GIR) is adjusted to maintain euglycemia (target glucose of 5 mmol/l). It takes approx. 45 minutes for a stable GIR to be reached. At 90 minutes after initiation of the clamp, 160 U of nasal insulin or placebo spray is applied to stimulate insulin-sensitive brain regions. On placebo day, an intravenous bolus of insulin is used to mimic possible spill-over effects of intranasal insulin into the periphery. Endogenous glucose production is determined from the tracer infusion rate of d-[6,6-$^2H_2$] glucose and its enrichment to the hydrogens bound to carbon-6 in relationship to the mean percent enrichment of plasma d-[6,6-$^2H_2$] glucose (37). Insulin-stimulated glucose rate of disappearance can be calculated to evaluate glucose uptake into peripheral tissues (47).

imaging (MRI), and position emission tomography (PET). Generally speaking, all neuroimaging methods detect intrinsic processes within the brain.

Electroencephalography/ Magnetoencephalography

EEG and MEG are non-invasive methods for directly measuring the electric activity of several tens of thousands of simultaneously activated neurons with high temporal resolution (range of milliseconds). While EEG measures the electric signal using electrodes attached to the head, MEG measures the magnetic field generated by the electric activity by superconducting sensors distributed in a helmet over the whole head. The major advantage of EEG and MEG is that they operate on a timescale at which neuronal activity takes place (i.e. milliseconds). In both cases, the sensor activity can be analyzed in the time or frequency domain. Analyses in the time domain are generally related to event-related components. Events include responses to external stimulation by uni- or multimodal stimulation in different sensory systems, e.g. auditory, visual, somatosensory, and others, or reactions such as motor executions. Specific frequency ranges are associated with specific tasks or conditions. In general, adult EEG/MEG activity is divided into the following frequency bands: delta (<4 Hz), theta (4–8 Hz), alpha (8–12 Hz), beta (12–30 Hz), and gamma (30–200 Hz). Specific cognitive domains are associated with certain frequency bands. For example, the theta band has been related to hippocampal memory formation (48, 49) and beta band activity to attentional processing (50). The power in the different frequency bands decreases with increasing frequencies. Due to the strong dependence of the source reconstruction on the signal-to-noise ratio, EEG and MEG have limited sensitivity for processes in deep structures such as the hypothalamus and subcortical regions, and are mainly used to investigate cortical processes.

Nuclear Tomographical Imaging Techniques

In nuclear tomographical imaging, techniques such as PET and single photon emission computed tomography (SPECT) radioactive tracers and gamma-ray detection are applied to probe tissue concentrations of particular molecules of interest. Due to the travel distance of the positron, the temporal (tens of seconds to minutes) and the spatial resolution of these methods (approx. 4 mm) is rather limited, rendering them applicable to specific research questions only. PET typically uses isotopes with a short half-life such as the radionuclides carbon-11, oxygen-15, and fluorine-18. These are either incorporated into compounds normally used by the body as glucose or water, or into molecules that bind to receptors. PET is primarily used to investigate glucose uptake related to tissue metabolism, (^{18}F fluorodeoxyglucose (FDG)), and cerebral blood flow and volume by $H2^{15}O$. PET and SPECT are of great importance in a direct investigation of specific neurotransmitter systems. Obesity research has shown that dopaminergic target regions are particularly vulnerable to increased adiposity (51, 52). Since activation of the striatum could involve both dopamine and opioid modulation, PET is used to directly investigate the synthesis and receptor binding of these neurotransmitters (52). Hence, insulin action on dopamine and/ or opioid modulation can be investigated to detect direct changes on these neurotransmitter systems.^{11}C-raclopride for example is a dopamine D2 receptor ligand widely used to assess synaptic levels of dopamine (53) in health and disease. To increase the spatial resolution, PET was recently combined with MRI to investigate specific neurotransmitters at high resolution.

Functional Magnetic Resonance Imaging (fMRI)

MRI is the most popular non-invasive neuroimaging technique that uses a strong magnetic field (mostly 3 Tesla in humans). No radioactive tracers are required to obtain high spatial resolution (2–3 mm isotropic at 3 Tesla) in vivo measurements of both brain structure and function. Higher spatial resolution can be achieved at ultra-high field, increasing magnetic field strength to 7 or 9.4 Tesla (54, 55). As a proxy for neural activity, fMRI is a valuable tool in measuring the hemodynamic changes related to neuronal activity using blood-oxygen-level-dependent contrast (BOLD) imaging or cerebral blood flow (CBF). BOLD fMRI is currently the most popular functional neuroimaging technique. It relies on the different magnetic properties of oxygenated and deoxygenated hemoglobin. When normal diamagnetic oxyhemoglobin forfeits its oxygen, the resulting deoxyhemoglobin is paramagnetic (56). Due to increased brain activity, the ratio between oxy- and deoxyhemoglobin changes after an enhanced release of oxygen and increased local blood flow. The subsequently decreased concentration of deoxygenated hemoglobin attenuates the local distortion of the magnetic field, resulting in a small increase (0.5%–5%) in the fMRI signal at a magnetic field strength of 3 Tesla. The major limitation of BOLD fMRI is due to the vascular origin of the signal which limits the temporal resolution and quantification (57). BOLD, therefore, does not provide a measure of a single physiological parameter. By virtue of arterial spin labeling, the direct change in CBF provides absolute quantification of the neural signal, thus resulting in a well-characterized physiological parameter in physiological units (ml/100g brain tissue/min). The advantage of using a physiological parameter is that it is easier to interpret, as well as to compare different studies. Hence, arterial spin labeling measurements have been proposed to be ideally suited for pharmacological MRI studies (58). The measurement of perfusion with MRI is based on arterial blood water flowing into the brain after being marked (magnetically 'labeled') by a radiofrequency pulse. The decay of that signal is then measured as a proxy for neural activity.

Thanks to neuroimaging techniques, brain insulin action can be spatially localized by evaluating the brain's response to cognitive demanding tasks such as, for example, the investigation of the role of insulin in memory or reward-related processes. Since the brain is active even in the absence of external cues, spontaneous BOLD fluctuations are located in all brain regions. Brain activity can therefore be evaluated under the so-called "resting-state" or "intrinsic condition". To this end, subjects lie in the MRI scanner with their eyes open or closed without performing any task. Resting-state fMRI thus greatly enhances the translation of fMRI into clinical care (for review, see (59)).

Besides investigating localized brain activity, fMRI can be used to study connected processes. Regions in the brain, with similar functional properties such as, for example, the left and right somatosensory cortices (60), exhibit a consistent pattern of synchronous activity. They are thus considered to be highly functionally connected, reflecting the presence of a functional network (61, 62). Functional connectivity is used for the characterization of brain networks in health and disease, especially under resting-state conditions. Hence, insulin-induced neural activity and changes in insulin-induced functional connectivity can be detected non-invasively using fMRI.

CENTRAL INSULIN ACTION-EVIDENCE OF EEG AND MEG STUDIES

Some of the earliest studies on central insulin action investigated event-related potentials using EEG and MEG providing the first evidence that higher cortical areas are particularly reactive to central acting insulin in healthy humans (Figure 11.2). A reduction in the amplitude of the event-related potential (N100 and P300 component) was found in response to intranasal and hyperinsulinemic clamps (63, 64); and a negative shift in frontal direct current potentials after intravenous insulin injection (65). Concomitantly, MEG studies revealed enhanced amplitudes for auditory and visual evoked responses during a hyperinsulinemic-euglycemic clamp, as well as after intranasal insulin (66, 67). The changes in EEG and MEG findings may indicate that insulin is a rapidly acting feedback signal that enhances cortical signals, which may contribute to the sensation of satiety. More specifically, the first study on central insulin action in overweight and obese humans used MEG. Spontaneous and stimulated cortical activity within the beta and theta frequency band increased during a hyperinsulinemic-euglycemic clamp in normal-weight but not in obese participants (67). Interestingly, in a recent follow-up, this central insulin-induced MEG activity was a significant predictor of body weight trajectory and visceral fat accumulation 9 years after a lifestyle intervention (68). This indicates that persons with high brain insulin sensitivity regain less weight and show a more favorable body fat distribution.

REGIONAL SPECIFIC CENTRAL INSULIN ACTION-EVIDENCE OF PET AND fMRI STUDIES

Evidence of recent modern neuroimaging studies identified regional fMRI and PET activity as well as functional connectivity to be modulated by central insulin action (3) (Figure 11.2, Table 11.1). The hypothalamus, the main area of the brain controlling

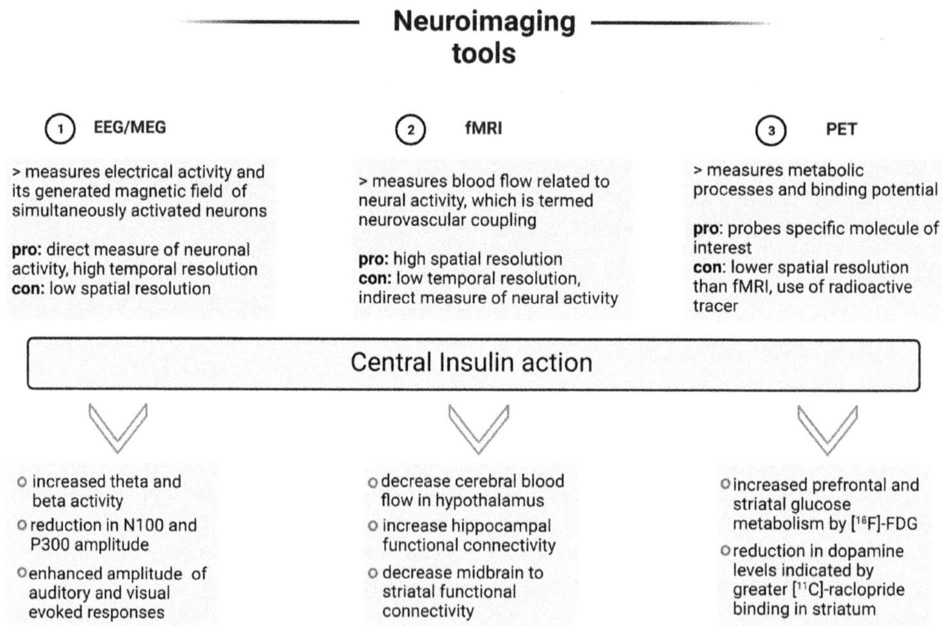

Figure 11.2 Neuroimaging tools investigating central insulin action in humans. The illustration briefly explains the technological background of the methods including advantages and disadvantages; and presents examples of central insulin action in healthy persons using electroencephalography (EEG), magnetoencephalography (MEG), functional magnetic resonance imaging (fMRI) and position emission tomography (PET).

TABLE 11.1

Central Insulin Action in Humans Investigated by fMRI and Intranasal Insulin

Name of Brain Region[a]	Main Functions of Brain Regions as Key Words
Anterior cingulate cortex	Impulse control, attention, decision making
Amygdala	Emotional valence
Anterior medial PFC	Default-mode network, mentalizing, attention
Dorsal lateral PFC	Self-control, decision-making, executive function
Dorsal medial PFC	Default-mode network, processing of self-referenced information, decision-making
Dorsal striatum	Motor function, goal-directed behavior, reward learning (includes parts of the caudate and putamen)
Hippocampus	Memory formation
Hypothalamus	Homeostatic regulation
Insula	Gustatory processing, visceral information processing
Orbitofrontal cortex	Choice valuation, hedonic hotspot, gustatory processing
Putamen	Motor function, learning
Ventral medial PFC	Value signal coding, decision-making
Ventral striatum	Reward, pleasure, motivation (i.e. nucleus accumbens)
Ventral tegmental area	Dopaminergic region, Reward processing

Table includes brain regions and their functions on behavior using key words. Listed regions were identified by functional magnetic resonance imaging (fMRI) in response to intranasal insulin; PFC: prefrontal cortex, [a]listed alphabetically.

energy balance, shows a decrease in fMRI signal in response to intranasal insulin. Conversely, the hypothalamus is responsive to hypoglycemia (69) and ingestion of glucose (70–72). The hypothalamic inhibition of intranasal insulin is weaker in persons with unfavorable fat distribution (68, 73). Hence, individuals with high visceral adipose tissue fail to show a decrease in hypothalamus blood flow after intranasal insulin application, indicating a relationship between brain insulin resistance and metabolically unfavorable abdominal adiposity (68, 73). Likewise, Rebelos and colleagues showed that insulin-induced brain glucose uptake correlates with visceral but not subcutaneous fat mass (74).

Besides the hypothalamus, central insulin affects regional fMRI and PET responses of the mesocorticolimbic circuitry (32, 33, 75–79), which includes in part dopaminergic pathways. These regions encompass the ventral tegmental area (VTA), striatum, prefrontal cortex (PFC), insula, amygdala, and hippocampus (Table 11.1). A recent ^{11}C raclopride-PET study indicates that dopamine plays an important role in the central insulin action of these areas. Intranasal insulin resulted in a reduction in synaptic dopamine levels in striatal areas (80). Similarly, intranasal insulin elicits regional depended increase and decrease in resting-state fMRI activity (32, 33, 75, 76). Intranasal insulin inhibits functional connectivity between the VTA and the striatum (77). Moreover, persons with a strong insulin-induced inhibition of the VTA-striatal connection by intranasal insulin exemplify a more prominent decrease in food palatability (77). Accordingly, in persons with obesity-associated insulin resistance, insulin action of the mesocorticolimbic circuitry is disturbed (32, 73, 78, 81, 82), resulting in a higher preference for palatable food (73, 77). This may further exacerbate insulin resistance favoring weight gain and resulting in a vicious cycle that could lead to the development of T2D.

On the contrary, functional connectivity to parts of the prefrontal cortex seems to be heightened by central insulin (78, 83). Particularly, the hippocampal- medial prefrontal cortex connection is boosted by intranasal insulin in persons of normal weight, overweight as well as patients with type 2 diabetes (83, 84). Besides its prominent role in memory, the hippocampus can detect interceptive signals of hunger and satiety (85, 86) and forms a memory of a meal to inhibit subsequent food intake (86, 87). Concomitantly, the central action of the hippocampus-PFC connection correlates with measures of food craving and hunger (73, 83), which was significantly curbed by central insulin action. The PFC is a key player in human behavior, including the control over food intake and food choice. Moreover, fMRI studies showed that parts of the PFC are particularly sensitive to hormonal changes (32, 33, 45, 78, 88, 89). Intranasal insulin triggered activity changes in the lateral PFC showing a differential activation pattern in normal weight versus persons with obesity, pointing to insulin resistance in the PFC (3). This is further substantiated by the fact that persons with obesity-associated insulin resistance and increased susceptibility to uncontrolled eating and food craving show the most prominent alterations in the PFC in response to central insulin (73, 78, 83). Hence, central insulin action may play an important role in cognitive functions.

CENTRAL EFFECTS OF INSULIN ACTION ON BEHAVIOR

In accordance with animal experiments (90–92), the central effects of insulin show partly opposite effects on energy homeostasis in humans compared to the peripheral impact of the peptide. Central insulin, delivered via the intranasal or intracerebroventricular pathway in humans and animals, respectively, acts in an anorexigenic fashion. In normal-weight male participants, daily administered intranasal insulin for eight weeks, reduced body weight and body fat content (93). Hence, the effect of central insulin action in humans appears to be hypophagic (Figure 11.3). However, central insulin action on behavior seems to trigger sex- and time-dependent effects on caloric intake and food preference (21). In fasted normal weight men, acute doses of intranasal insulin decrease caloric intake, food preference, and appetite ratings (35, 73). In women, however, intranasal insulin curbed the subjective feeling of appetite and hunger as well as snacking only in the postprandial, but not in the fasted state (94). Administrating intranasal insulin before nocturnal sleep decreased food intake during breakfast in men and women (95). Experiments investigating sensory processing and mood also point to sex-specific effects (21). Central insulin action by intranasal insulin acutely improved gustatory sensitivity in healthy men (96). Decreased olfactory sensitivity was observed during hyperinsulinemia (97) and in response to intranasal insulin in men and women (98). However, in a larger sample, the central insulin effect on olfactory sensitivity was restricted to women (99).

Central insulin action also modulates mood and emotions in humans. In response to an 8-week administration of intranasal insulin, normal weight and obese participants reported an improvement in mood (100, 101). Acute administration of intranasal insulin enhanced fear extension, with a stronger impact in women (102). Anxiolytic effects of central insulin have also been reported in animals (5). These findings substantiate that central insulin is a pivotal negative feedback signal in the regulation of the reward-related behaviors (Figure 11.3), possibly in a sex-dependent manner.

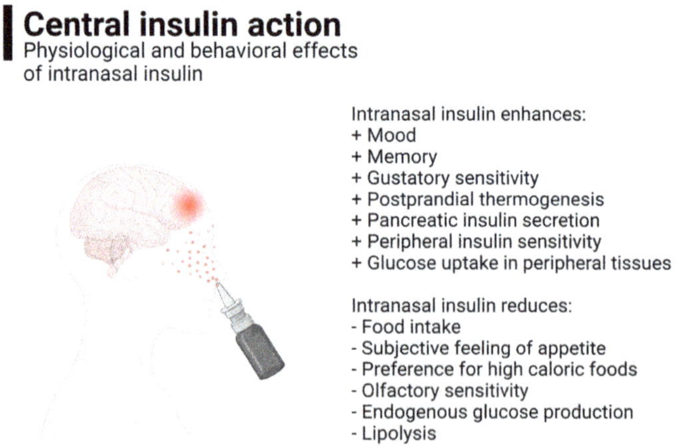

Figure 11.3 Major acute effects of central nervous insulin signaling on behavior and whole-body metabolism in healthy humans investigated by intranasal insulin application (image created using BioRender.com).

CENTRAL EFFECTS OF INSULIN ACTION ON SYSTEMIC METABOLISM

Animal models provide strong evidence that central insulin action has pleiotropic consequences on peripheral metabolism including the liver, skeletal muscle, and adipose tissue (103) (Figure 11.3). First studies in mice clearly showed that neuron-specific knockout of the insulin receptor causes an obese phenotype comprising whole-body insulin resistance and hypertriglyceridemia (104). Further research showed that central insulin action suppresses endogenous glucose production in the liver (105–107) and promotes glucose uptake and storage as glycogen in skeletal muscle (108–110). These mechanisms of action reduce blood glucose levels (111). Furthermore, animal studies showed that central insulin action promotes energy storage in adipocytes by suppressing lipolysis (109, 112, 113).

In humans, central insulin action on peripheral metabolism has only recently become evident showing a significant effect on hepatic glucose metabolism, as well as insulin secretion and peripheral insulin sensitivity. The latter was shown by combining the application of intranasal insulin with clinical surrogate measures of peripheral insulin resistance as HOMA-IR (32, 114) or the hyperinsulinemic-euglycemic glucose clamp procedure - the gold standard for investigating whole-body insulin sensitivity (37, 38, 115). Intranasal insulin administration improved peripheral insulin sensitivity in lean (32, 38), but not obese men (37, 38, 116). Specifically, after insulin nasal spray compared to placebo, higher glucose infusion rate was needed to maintain euglycemia during the hyperinsulinemic-euglycemic glucose clamp (37, 38). Consequently, the clamp-derived whole-body insulin sensitivity index improved after intranasal insulin application (37, 38). This improvement in insulin sensitivity correlated with central insulin-induced hypothalamic activity assessed by fMRI (38). Interestingly, men with obesity did not show these insulin-sensitizing effects (38). In 2017, Heni et al. (37) showed that intranasal insulin triggered higher glucose infusion rates associated with activity changes in the hypothalamus and striatum, which suggests an altered brain-derived modulation of regions in the homeostatic and dopaminergic circuitry (37). Hence, altered brain insulin action may hamper peripheral glucose metabolism in obesity-associated insulin resistance (3).

These results point to the existence of a brain-liver axis. To further elaborate on the involved mechanisms, endogenous glucose production and glucose disappearance were measured by hyperinsulinemic-euglycemic clamps with D-[6,6-$^2H_2$] glucose infusion (37, 115) after nasal insulin compared to placebo. Central insulin suppressed endogenous glucose production and increased glucose disappearance rate; hence increasing glucose uptake into peripheral tissue (37). Thus, central insulin action triggers brain-derived outflow targeting multiple tissues, such as the liver, skeletal muscle, and adipose tissue. Obese men failed to show this central insulin effect on endogenous glucose production suppression (37, 116). Interestingly, the metabolic response in the normal-weight participants was dependent on elevated circulating insulin levels (117). Therefore, the reported finding might be limited to the postprandial state, when insulin levels physiologically rise in the periphery and the brain (117). In line, no suppression of endogenous glucose production was found in response to intranasal insulin in the fasting state (25, 117). Yet, central insulin action enhanced hepatic energy metabolism and reduced liver fat content in healthy persons of normal weight but not in patients with T2D (25). Studies on the central effects of insulin on glucose metabolism were performed in men. Further studies in women are under way pointing to a prominent role of the menstrual cycle on central insulin effects on peripheral metabolism (118).

Another possibility to assess the central effects of insulin action in relation to systemic metabolism is by investigating insulin-stimulated brain glucose metabolism using ^{18}FDG-PET, which quantifies brain glucose uptake at the whole-brain level, most likely driven by astrocytes (119). Hyperinsulinemic clamp increased brain glucose metabolism in persons with obesity (120), but not in lean persons (120), and was related to worse glycemic control (121). Hence, brain insulin resistance, by this method, is defined as insulin-induced increase in brain glucose uptake (74). Weight loss reversed this insulin-driven response (120) and predicted fasting glucose levels 2 years after the intervention (121). Moreover, insulin-stimulated brain glucose metabolism was related to endogenous glucose production in persons with morbid obesity but not normal-weight (121, 122). The mechanisms of insulin-stimulated brain glucose metabolism measured by FDG-PET are currently not known.

Furthermore, research in animals indicates a role of central insulin action in the control of lipolysis in adipose tissue (113). In humans, only a few studies exist with conflicting results (4). Intranasal insulin seems to increase postprandial thermogenesis and decrease prandial free-fatty acids in healthy men (114). In response to exercise, intranasal insulin decreased fat oxidation in normal weight but not overweight participants (123). Furthermore, intranasal insulin caused a reduction in lipolysis measured by the rate of appearance of deuterated glycerol and lowers systemic levels of free-fatty acids concentrations; however, without affecting subcutaneous lipolysis (124). Moreover, other studies could not confirm this finding (25, 115, 125), which is probably related to different study designs. Hence, further research is needed to clarify potential role of the central insulin-adipocyte axis.

Besides peripheral insulin sensitivity, human glucose metabolism is crucially regulated by insulin secretion from pancreatic beta cells. Insulin resistance leads to a compensatory increase in endogenous insulin production, which favors further promotes weight gain. This exacerbates a vicious cycle until the pancreatic beta-cell activity can no longer meet the insulin demand resulting in hyperglycemia (6). Rodent models revealed the role of the CNS in this process by knocking down the insulin receptor of glucose-sensing neurons in the hypothalamus, which impaired insulin secretion and caused glucose intolerance (126). Furthermore, the knockdown of hypothalamic insulin receptors likewise impaired insulin secretion (126). In humans, recent studies using intranasal insulin and fMRI point to a connection between insulin responsiveness of the hypothalamus and pancreatic insulin secretion (127). Intranasal insulin administration during a hyperglycemic glucose clamp (i.e. gold standard for the quantification of pancreatic insulin release) triggered enhanced second phase insulin secretion in persons with strong hypothalamic insulin responsiveness as well as in persons with a lower body mass index, while the other pancreatic hormones glucagon and somatostatin remained unaltered (36). After being released from pancreatic beta cells, insulin drains into the portal vein and reaches the liver at a very high concentration. There it contributes to suppressing hepatic glucose production. Hence, central insulin action may enhance postprandial insulin secretion, which could contribute to the suppression of endogenous glucose production modulating postprandial glucose metabolism. Brain insulin resistance could contribute to altered postprandial energy fluxes and increased visceral fat accumulation (37, 38, 116, 117) (Figure 11.4).

DEFINITION OF BRAIN INSULIN RESISTANCE IN HUMANS

Systemic insulin resistance is well defined based on the impaired response to insulin stimulation of peripheral tissues with several established molecular signaling pathways (128). The definition of obesity-associated brain insulin resistance in humans, on the other hand, is based on a variety of neuroimaging methods. Using fMRI in combination with intranasal insulin, brain insulin resistance is based on the failure to show an insulin-induced inhibition in hypothalamic blood flow and regional specific functional connectivity changes between the striatum, VTA, hippocampus, hypothalamus, and prefrontal cortex (3). Blunted insulin effects on beta and theta activity were observed when MEG together with a hyperinsulinemic clamp was applied (67, 68, 129). FDG-PET studies showed insulin-induced increase in brain glucose uptake as an indication of brain insulin resistance (74, 121). Each of these neuroimaging methods characterizes different brain functions and cell types and has never been used in the same study. More studies are needed to quantify and define brain insulin resistance in humans by combining different approaches to study insulin action in the brain.

CONCLUDING REMARKS

Whole-body insulin resistance precedes the development of T2D by about 10–15 years (6). The developmental trajectory is significantly influenced by central nervous insulin signaling. Impairments in central insulin action (i.e. "brain insulin resistance") promote a metabolically unhealthy phenotype, which exacerbates the vicious cycle of insulin resistance and weight gain further increasing the risk to develop T2D (Figure 11.4).

Central insulin action influences metabolism, eating behavior as well as motivation, reward, and cognition (3–5). Here, we highlighted particularly the role of central insulin on peripheral metabolism and eating behavior and discussed the metabolic consequences of brain insulin resistance. In human studies, the characterization of brain insulin

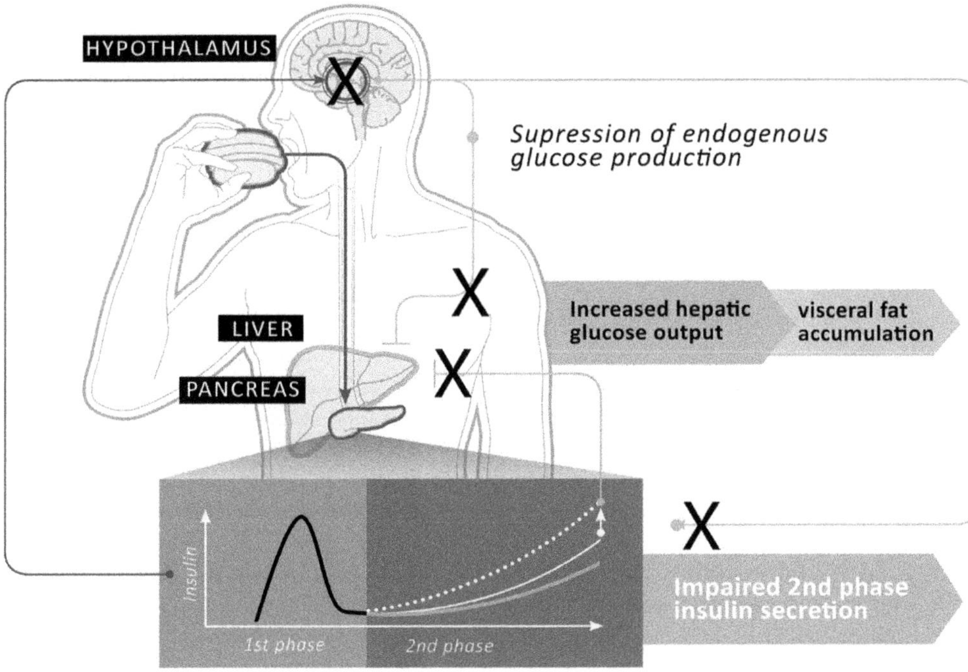

Figure 11.4 Hypothesized effect of decreased insulin sensitivity in the central nervous system. In persons with brain insulin resistance, pancreatic insulin secretion fails to trigger regional changes in activity of insulin-sensitive brain regions-particularly in the hypothalamus (center for energy homeostasis). Signals from the brain to the liver are not sufficient to suppress endogenous glucose production and second-phase insulin secretion in the pancreas. In the insulin-sensitive condition, second-phase insulin secretion would be enhanced and portal insulin would further suppress endogenous glucose production. In the insulin-resistant state, postprandial insulin concentrations are insufficient to suppress endogenous glucose production, which may lead to increased hepatic glucose output and visceral adipose fat accumulation. This can exacerbate the vicious cycle of insulin resistance and weight gain further increasing the risk to develop T2D. (Adapted from Kullmann et al. (3).)

resistance relies on different neuroimaging methods such as PET, MEG, and fMRI in combination with central or peripheral stimulation techniques. Intranasal insulin administration is employed in combination with neuroimaging studies to identify insulin-sensitive brain regions. Findings of these studies attribute central insulin effects mainly to the hypothalamus and the mesocorticolimbic circuitry of the brain with positive implications on whole-body metabolism and eating behavior. Central insulin action curbs food craving increases cognitive control, enhances peripheral insulin sensitivity, and suppresses endogenous glucose production. Moreover, the central insulin response seems to be predictive of the outcome of weight-loss programs. Besides metabolic disturbances, brain insulin resistance may also play a prominent role in psychiatric and neurodegenerative diseases (130). Thus, targeting insulin resistance in the brain may have therapeutic potential to treat and prevent multiple non-communicable diseases.

ACKNOWLEDGEMENTS

SK is supported by grants (01GI0925) from the Federal Ministry of Education and Research (BMBF) to the German Center for Diabetes Research (DZD e.V.).

REFERENCES

1. Banting FG, Best CH. The internal secretion of the pancreas. 1922. *Indian J Med Res.* 2007;125(3): 251–66.
2. Havrankova J, Roth J, Brownstein M. Insulin receptors are widely distributed in the central nervous system of the rat. *Nature.* 1978;272 (5656):827–9.

3. Kullmann S, Kleinridders A, Small DM, Fritsche A, Haring HU, Preissl H, et al. Central nervous pathways of insulin action in the control of metabolism and food intake. *Lancet Diabetes Endocrinol.* 2020;8(6):524–34.
4. Kullmann S, Heni M, Hallschmid M, Fritsche A, Preissl H, Haring HU. Brain Insulin Resistance at the Crossroads of Metabolic and Cognitive Disorders in Humans. *Physiol Rev.* 2016;96(4): 1169–209.
5. Kleinridders A, Pothos EN. Impact of brain insulin signaling on dopamine function, food intake, reward, and emotional behavior. *Curr Nutr Rep.* 2019;8(2):83–91.
6. Freeman AM, Pennings N. Insulin Resistance. StatPearls. *Treasure Island* (FL)2021. https://www.ncbi.nlm.nih.gov/books/NBK507839/
7. Heni M, Schopfer P, Peter A, Sartorius T, Fritsche A, Synofzik M, et al. Evidence for altered transport of insulin across the blood-brain barrier in insulin-resistant humans. *Acta Diabetol.* 2014;51(4):679–81.
8. Cholerton B, Baker LD, Craft S. Insulin, cognition, and dementia. *Eur J Pharmacol.* 2013;719 (1–3):170–9.
9. Verdile G, Fuller SJ, Martins RN. The role of type 2 diabetes in neurodegeneration. *Neurobiol Dis.* 2015; 84:22-38.
10. Seaquist ER, Damberg GS, Tkac I, Gruetter R. The effect of insulin on in vivo cerebral glucose concentrations and rates of glucose transport/metabolism in humans. *Diabetes.* 2001;50(10): 2203–9.
11. Molnar G, Farago N, Kocsis AK, Rozsa M, Lovas S, Boldog E, et al. GABAergic neurogliaform cells represent local sources of insulin in the cerebral cortex. *J Neurosci.* 2014;34(4):1133–7.
12. Steen E, Terry BM, Rivera EJ, Cannon JL, Neely TR, Tavares R, et al. Impaired insulin and insulin-like growth factor expression and signaling mechanisms in Alzheimer's disease--is this type 3 diabetes? *J Alzheimers Dis.* 2005; 7(1):63–80.
13. Woods SC, Lotter EC, McKay LD, Porte D, Jr. Chronic intracerebroventricular infusion of insulin reduces food intake and body weight of baboons. *Nature.* 1979;282(5738):503–5.
14. Brief DJ, Davis JD. Reduction of food intake and body weight by chronic intraventricular insulin infusion. *Brain Res Bull.* 1984;12(5):571–5.
15. Destefano MB, Stern JS, Castonguay TW. Effect of chronic insulin administration on food intake and body weight in rats. *Physiol Behav.* 1991;50(4):801–6.
16. Honda K, Kamisoyama H, Saneyasu T, Sugahara K, Hasegawa S. Central administration of insulin suppresses food intake in chicks. *Neurosci Lett.* 2007;423(2):153–7.
17. Foster LA, Ames NK, Emery RS. Food intake and serum insulin responses to intraventricular infusions of insulin and IGF-I. *Physiol Behav.* 1991;50(4):745–9.
18. McGowan MK, Andrews KM, Grossman SP. Chronic intrahypothalamic infusions of insulin or insulin antibodies alter body weight and food intake in the rat. *Physiol Behav.* 1992; 51(4):753–66.
19. Hatfield JS, Millard WJ, Smith CJ. Short-term influence of intra-ventromedial hypothalamic administration of insulin on feeding in normal and diabetic rats. *Pharmacol Biochem Behav.* 1974;2(2):223–6.
20. Matthews DR, Hosker JP, Rudenski AS, Naylor BA, Treacher DF, Turner RC. Homeostasis model assessment: insulin resistance and beta-cell function from fasting plasma glucose and insulin concentrations in man. *Diabetologia.* 1985;28(7):412–9.
21. Hallschmid M. Intranasal insulin. *J Neuroendocrinol.* 2021:e12934.
22. Schmid V, Kullmann S, Gfrorer W, Hund V, Hallschmid M, Lipp HP, et al. Safety of intranasal human insulin: A review. *Diabetes Obes Metab.* 2018; 20:1563-1577.
23. Gwizdala KL, Ferguson DP, Kovan J, Novak V, Pontifex MB. Placebo controlled phase II clinical trial: Safety and efficacy of combining intranasal insulin & acute exercise. *Metab Brain Dis.* 2021; 36:1289-1303.
24. Craft S, Raman R, Chow TW, Rafii MS, Sun CK, Rissman RA, et al. Safety, efficacy, and feasibility of intranasal insulin for the treatment of mild cognitive impairment and Alzheimer disease dementia: a randomized clinical trial. *JAMA Neurol.* 2020;77(9):1099–109.
25. Gancheva S, Koliaki C, Bierwagen A, Nowotny P, Heni M, Fritsche A, et al. Effects of intranasal insulin on hepatic fat accumulation and energy metabolism in humans. *Diabetes.* 2015;64(6):1966–75.
26. Dhuria SV, Hanson LR, Frey WH, 2nd. Intranasal delivery to the central nervous system: mechanisms and experimental considerations. *J Pharm Sci.* 2010;99(4):1654–73.

27. Lochhead JJ, Thorne RG. Intranasal delivery of biologics to the central nervous system. *Adv Drug Deliv Rev.* 2012;64(7):614–28.
28. Born J, Lange T, Kern W, McGregor GP, Bickel U, Fehm HL. Sniffing neuropeptides: a transnasal approach to the human brain. *Nat Neurosci.* 2002;5(6):514–6.
29. Salameh TS, Bullock KM, Hujoel IA, Niehoff ML, Wolden-Hanson T, Kim J, et al. Central nervous system delivery of intranasal insulin: mechanisms of uptake and effects on cognition. *J Alzheimers Dis.* 2015;47(3):715–28.
30. Claxton A, Baker LD, Wilkinson CW, Trittschuh EH, Chapman D, Watson GS, et al. Sex and ApoE genotype differences in treatment response to two doses of intranasal insulin in adults with mild cognitive impairment or Alzheimer's disease. *J Alzheimers Dis.* 2013;35(4):789–97.
31. Craft S, Baker LD, Montine TJ, Minoshima S, Watson GS, Claxton A, et al. Intranasal insulin therapy for Alzheimer disease and amnestic mild cognitive impairment: a pilot clinical trial. *Arch Neurol.* 2011; 69:29-38.
32. Heni M, Kullmann S, Ketterer C, Guthoff M, Linder K, Wagner R, et al. Nasal insulin changes peripheral insulin sensitivity simultaneously with altered activity in homeostatic and reward-related human brain regions. *Diabetologia.* 2012;55(6):1773–82.
33. Kullmann S, Frank S, Heni M, Ketterer C, Veit R, Haring HU, et al. Intranasal insulin modulates intrinsic reward and prefrontal circuitry of the human brain in lean women. *Neuroendocrinology.* 2013;97(2):176–82.
34. Kullmann S, Heni M, Veit R, Scheffler K, Machann J, Haring HU, et al. Selective insulin resistance in homeostatic and cognitive control brain areas in overweight and obese adults. *Diabetes Care.* 2015.
35. Benedict C, Kern W, Schultes B, Born J, Hallschmid M. Differential sensitivity of men and women to anorexigenic and memory-improving effects of intranasal insulin. *J Clin Endocrinol Metab.* 2008;93(4):1339–44.
36. Heni M, Wagner R, Willmann C, Jaghutriz BA, Vosseler A, Kubler C, et al. Insulin action in the hypothalamus increases second phase insulin secretion in humans. *Neuroendocrinology.* 2019.
37. Heni M, Wagner R, Kullmann S, Gancheva S, Roden M, Peter A, et al. Hypothalamic and striatal insulin action suppresses endogenous glucose production and may stimulate glucose uptake during hyperinsulinemia in lean but not in overweight men. *Diabetes.* 2017;66(7): 1797–806.
38. Heni M, Wagner R, Kullmann S, Veit R, Mat Husin H, Linder K, et al. Central insulin administration improves whole-body insulin sensitivity via hypothalamus and parasympathetic outputs in men. *Diabetes.* 2014;63(12): 4083–8.
39. Matsuda M, DeFronzo RA. Insulin sensitivity indices obtained from oral glucose tolerance testing: comparison with the euglycemic insulin clamp. *Diabetes Care.* 1999;22(9):1462–70.
40. Utzschneider KM, Prigeon RL, Faulenbach MV, Tong J, Carr DB, Boyko EJ, et al. Oral disposition index predicts the development of future diabetes above and beyond fasting and 2-h glucose levels. *Diabetes Care.* 2009;32(2):335–41.
41. Bergman RN, Prager R, Volund A, Olefsky JM. Equivalence of the insulin sensitivity index in man derived by the minimal model method and the euglycemic glucose clamp. *J Clin Invest.* 1987;79(3):790–800.
42. DeFronzo RA, Tobin JD, Andres R. Glucose clamp technique: a method for quantifying insulin secretion and resistance. *Am J Physiol.* 1979;237(3): E214–23.
43. Gallwitz B. GLP-1 agonists and dipeptidyl-peptidase IV inhibitors. *Handbook Exp Pharmacol.* 2011(203): 53–74.
44. Marty N, Dallaporta M, Thorens B. Brain glucose sensing, counterregulation, and energy homeostasis. *Physiology (Bethesda).* 2007;22: 241–51.
45. Heni M, Kullmann S, Ketterer C, Guthoff M, Bayer M, Staiger H, et al. Differential effect of glucose ingestion on the neural processing of food stimuli in lean and overweight adults. *Hum Brain Mapp.* 2013;35(3):918–28.
46. Kroemer NB, Krebs L, Kobiella A, Grimm O, Vollstadt-Klein S, Wolfensteller U, et al. (Still) longing for food: insulin reactivity modulates response to food pictures. *Hum Brain Mapp.* 2013;34(10):2367–80.
47. Steele R. Influences of glucose loading and of injected insulin on hepatic glucose output. *Ann N Y Acad Sci.* 1959;82:420–30.
48. Buzsaki G. Theta oscillations in the hippocampus. *Neuron.* 2002;33(3):325–40.
49. Buzsaki G, Moser EI. Memory, navigation and theta rhythm in the hippocampal-entorhinal system. *Nat Neurosci.* 2013;16(2):130–8.

50. Engel AK, Fries P. Beta-band oscillations--signalling the status quo? *Curr Opin Neurobiol.* 2010;20(2):156–65.
51. Wang GJ, Volkow ND, Logan J, Pappas NR, Wong CT, Zhu W, et al. Brain dopamine and obesity. *Lancet.* 2001;357(9253):354–7.
52. Karlsson HK, Tuominen L, Helin S, Salminen P, Nuutila P, Nummenmaa L. Mesolimbic opioid-dopamine interaction is disrupted in obesity but recovered by weight loss following bariatric surgery. *Translational Psychiatry.* 2021;11(1):259.
53. Volkow ND, Wang GJ, Fowler JS, Logan J, Schlyer D, Hitzemann R, et al. Imaging endogenous dopamine competition with [11C]raclopride in the human brain. *Synapse.* 1994;16(4):255–62.
54. Rauschenberg J, Nagel AM, Ladd SC, Theysohn JM, Ladd ME, Moller HE, et al. Multicenter study of subjective acceptance during magnetic resonance imaging at 7 and 9.4 T. *Invest Radiol.* 2014;49(5):249–59.
55. Ehses P, Bause J, Shajan G, Scheffler K. Efficient generation of T2*-weighted contrast by interslice echo-shifting for human functional and anatomical imaging at 9.4 Tesla. *Magn Reson Med.* 2014; 74:1698–704.
56. Ogawa S, Lee TM, Kay AR, Tank DW. Brain magnetic resonance imaging with contrast dependent on blood oxygenation. *Proc Natl Acad Sci U S A.* 1990;87(24):9868–72.
57. Buxton RB, Uludag K, Dubowitz DJ, Liu TT. Modeling the hemodynamic response to brain activation. *Neuroimage.* 2004;23 Suppl 1:S220–33.
58. Wang DJ, Chen Y, Fernandez-Seara MA, Detre JA. Potentials and challenges for arterial spin labeling in pharmacological magnetic resonance imaging. *J Pharmacol Exp Ther.* 2011;337(2):359–66.
59. Fox MD, Greicius M. Clinical applications of resting state functional connectivity. *Front Syst Neurosci.* 2010;4:19.
60. Biswal B, Yetkin FZ, Haughton VM, Hyde JS. Functional connectivity in the motor cortex of resting human brain using echo-planar MRI. *Magn Reson Med.* 1995;34(4):537–41.
61. Biswal BB, Mennes M, Zuo XN, Gohel S, Kelly C, Smith SM, et al. Toward discovery science of human brain function. *Proc Natl Acad Sci U S A.* 2010;107(10):4734–9.
62. Buckner RL, Andrews-Hanna JR, Schacter DL. The brain's default network: anatomy, function, and relevance to disease. *Ann N Y Acad Sci.* 2008;1124:1–38.
63. Kern W, Born J, Schreiber H, Fehm HL. Central nervous system effects of intranasally administered insulin during euglycemia in men. *Diabetes.* 1999;48(3):557–63.
64. Benedict L, Nelson CA, Schunk E, Sullwold K, Seaquist ER. Effect of insulin on the brain activity obtained during visual and memory tasks in healthy human subjects. *Neuroendocrinology.* 2006;83(1):20–6.
65. Hallschmid M, Schultes B, Marshall L, Molle M, Kern W, Bredthauer J, et al. Transcortical direct current potential shift reflects immediate signaling of systemic insulin to the human brain. *Diabetes.* 2004;53(9):2202–8.
66. Guthoff M, Stingl KT, Tschritter O, Rogic M, Heni M, Stingl K, et al. The insulin-mediated modulation of visually evoked magnetic fields is reduced in obese subjects. *PLoS One.* 2011;6(5):e19482.
67. Tschritter O, Preissl H, Hennige AM, Stumvoll M, Porubska K, Frost R, et al. The cerebrocortical response to hyperinsulinemia is reduced in overweight humans: a magnetoencephalographic study. *Proc Natl Acad Sci U S A.* 2006;103(32):12103–8.
68. Kullmann S, Valenta V, Wagner R, Tschritter O, Machann J, Haring HU, et al. Brain insulin sensitivity is linked to adiposity and body fat distribution. *Nat Commun.* 2020;11(1):1841.
69. Arbelaez AM, Su Y, Thomas JB, Hauch AC, Hershey T, Ances BM. Comparison of regional cerebral blood flow responses to hypoglycemia using pulsed arterial spin labeling and positron emission tomography. *PLoS One.* 2013;8(3):e60085.
70. Page KA, Arora J, Qiu M, Relwani R, Constable RT, Sherwin RS. Small decrements in systemic glucose provoke increases in hypothalamic blood flow prior to the release of counterregulatory hormones. *Diabetes.* 2009; 58(2):448–52.
71. Smeets PA, Vidarsdottir S, de Graaf C, Stafleu A, van Osch MJ, Viergever MA, et al. Oral glucose intake inhibits hypothalamic neuronal activity more effectively than glucose infusion. *Am J Physiol Endocrinol Metab.* 2007;293(3):E754–8.
72. van Opstal AM, van den Berg-Huysmans AA, Hoeksma M, Blonk C, Pijl H, Rombouts S, et al. The effect of consumption temperature on the homeostatic and hedonic responses to glucose ingestion in the hypothalamus and the reward system. *Am J Clin Nutr.* 2018;107(1):20–5.

73. Kullmann S, Heni M, Veit R, Scheffler K, Machann J, Haring HU, et al. Selective insulin resistance in homeostatic and cognitive control brain areas in overweight and obese adults. *Diabetes Care*. 2015;38(6):1044–50.
74. Rebelos E, Nummenmaa L, Dadson P, Latva-Rasku A, Nuutila P. Brain insulin sensitivity is linked to body fat distribution-the positron emission tomography perspective. *Eur J Nucl Med Mol Imaging*. 2021;48(4):966–8.
75. Heni M, Kullmann S, Wagner R, Machicao F, Staiger H, Bruning JC, et al. Evidence for Interaction between the Obesity-Risk Gene FTO and Dopamine D2 Receptor Gene ANKK1/TaqIA. *Diabetes Supplement* 2015. A42–A.
76. Kullmann S, Veit R, Peter A, Pohmann R, Scheffler K, Haring HU, et al. Dose-dependent effects of intranasal insulin on resting-state brain activity. *J Clin Endocrinol Metab*. 2018;103(1):253–62.
77. Tiedemann LJ, Schmid SM, Hettel J, Giesen K, Francke P, Buchel C, et al. Central insulin modulates food valuation via mesolimbic pathways. *Nat Commun*. 2017;8:16052.
78. Edwin Thanarajah S, Iglesias S, Kuzmanovic B, Rigoux L, Stephan KE, Bruning JC, et al. Modulation of midbrain neurocircuitry by intranasal insulin. *Neuroimage*. 2019;194:120–7.
79. Schilling TM, Ferreira de Sa DS, Westerhausen R, Strelzyk F, Larra MF, Hallschmid M, et al. Intranasal insulin increases regional cerebral blood flow in the insular cortex in men independently of cortisol manipulation. *Hum Brain Mapp*. 2013; 35:1944-56.
80. Kullmann S, Blum D, Jaghutriz B, Bender B, Haering H-U, Fougère CL, et al. 1798-P: insulin modulates dopaminergic tone in the human brain—a combined PET/MRI study. *Diabetes*. 2019;68(Supplement 1):1798–P.
81. Heni M, Kullmann S, Ahlqvist E, Wagner R, Machicao F, Staiger H, et al. Interaction between the obesity-risk gene FTO and the dopamine D2 receptor gene ANKK1/TaqIA on insulin sensitivity. *Diabetologia*. 2016;59(12):2622–31.
82. Anthony K, Reed LJ, Dunn JT, Bingham E, Hopkins D, Marsden PK, et al. Attenuation of insulin-evoked responses in brain networks controlling appetite and reward in insulin resistance: the cerebral basis for impaired control of food intake in metabolic syndrome? *Diabetes*. 2006;55(11):2986–92.
83. Kullmann S, Heni M, Veit R, Scheffler K, Machann J, Haring HU, et al. Intranasal insulin enhances brain functional connectivity mediating the relationship between adiposity and subjective feeling of hunger. *Sci Rep*. 2017;7(1):1627.
84. Zhang H, Hao Y, Manor B, Novak P, Milberg W, Zhang J, et al. Intranasal insulin enhanced resting-state functional connectivity of hippocampal regions in type 2 diabetes. *Diabetes*. 2015;64(3):1025–34.
85. Kanoski SE, Grill HJ. Hippocampus contributions to food intake control: mnemonic, neuroanatomical, and endocrine mechanisms. *Biol Psychiatry*. 2015; 81:748-756.
86. Coppin G. The anterior medial temporal lobes: Their role in food intake and body weight regulation. *Physiol Behav*. 2016.
87. Higgs S. Cognitive processing of food rewards. *Appetite*. 2016;104:10–7.
88. Kroemer NB, Krebs L, Kobiella A, Grimm O, Vollstadt-Klein S, Wolfensteller U, et al. (Still) longing for food: Insulin reactivity modulates response to food pictures. *Hum Brain Mapp*. 2012; 34:2367-80.
89. Page KA, Chan O, Arora J, Belfort-Deaguiar R, Dzuira J, Roehmholdt B, et al. Effects of fructose vs glucose on regional cerebral blood flow in brain regions involved with appetite and reward pathways. *JAMA*. 2013;309(1):63–70.
90. Lin HV, Plum L, Ono H, Gutierrez-Juarez R, Shanabrough M, Borok E, et al. Divergent regulation of energy expenditure and hepatic glucose production by insulin receptor in agouti-related protein and POMC neurons. *Diabetes*. 2010;59(2):337–46.
91. Morton GJ, Cummings DE, Baskin DG, Barsh GS, Schwartz MW. Central nervous system control of food intake and body weight. *Nature*. 2006;443(7109):289–95.
92. Porte D, Jr., Baskin DG, Schwartz MW. Insulin signaling in the central nervous system: a critical role in metabolic homeostasis and disease from C. *elegans* to humans. *Diabetes*. 2005;54(5):1264–76.
93. Hallschmid M, Benedict C, Schultes B, Fehm HL, Born J, Kern W. Intranasal insulin reduces body fat in men but not in women. *Diabetes*. 2004;53(11):3024–9.
94. Hallschmid M, Higgs S, Thienel M, Ott V, Lehnert H. Postprandial administration of intranasal insulin intensifies satiety and reduces intake of palatable snacks in women. *Diabetes*. 2012.

95. Santiago JC, Hallschmid M. Central nervous insulin administration before nocturnal sleep decreases breakfast intake in healthy young and elderly subjects. *Front Neurosci.* 2017;11:54.
96. Rodriguez-Raecke R, Yang H, Brunner YF, Freiherr J. Intranasal insulin boosts gustatory sensitivity. *J Neuroendocrinol.* 2016; 29.
97. Ketterer C, Heni M, Thamer C, Herzberg-Schafer SA, Haring HU, Fritsche A. Acute, short-term hyperinsulinemia increases olfactory threshold in healthy subjects. *Int J Obes (Lond).* 2011;35(8):1135–8.
98. Brunner YF, Benedict C, Freiherr J. Intranasal insulin reduces olfactory sensitivity in normosmic humans. *J Clin Endocrinol Metab.* 2013;98(10): E1626–30.
99. Rodriguez-Raecke R, Brunner YF, Kofoet A, Mutic S, Benedict C, Freiherr J. Odor sensitivity after intranasal insulin application is modulated by gender. *Front Endocrinol.* 2018;9:580.
100. Benedict C, Hallschmid M, Hatke A, Schultes B, Fehm HL, Born J, et al. Intranasal insulin improves memory in humans. *Psychoneuroendocrinology.* 2004;29(10):1326–34.
101. Hallschmid M, Benedict C, Schultes B, Born J, Kern W. Obese men respond to cognitive but not to catabolic brain insulin signaling. *Int J Obes (Lond).* 2008;32(2):275–82.
102. Ferreira GM, Albertella L, Moreira-de-Oliveira ME, Piquet-Pessoa M, Yucel M, Lee RSC, et al. Fear driven compulsive behaviour is associated with worse long-term outcome in obsessive-compulsive disorder. *Hum Psychopharmacol.* 2020;35(1):e2720.
103. Kleinridders A, Ferris HA, Cai W, Kahn CR. Insulin action in brain regulates systemic metabolism and brain function. *Diabetes.* 2014;63(7):2232–43.
104. Bruning JC, Gautam D, Burks DJ, Gillette J, Schubert M, Orban PC, et al. Role of brain insulin receptor in control of body weight and reproduction. *Science.* 2000;289(5487):2122–5.
105. Filippi BM, Yang CS, Tang C, Lam TK. Insulin activates Erk1/2 signaling in the dorsal vagal complex to inhibit glucose production. *Cell Metab.* 2012;16(4):500–10.
106. Obici S, Zhang BB, Karkanias G, Rossetti L. Hypothalamic insulin signaling is required for inhibition of glucose production. *Nat Med.* 2002;8(12):1376–82.
107. Pocai A, Lam TK, Gutierrez-Juarez R, Obici S, Schwartz GJ, Bryan J, et al. Hypothalamic K(ATP) channels control hepatic glucose production. *Nature.* 2005;434(7036):1026–31.
108. Coomans CP, Biermasz NR, Geerling JJ, Guigas B, Rensen PC, Havekes LM, et al. Stimulatory effect of insulin on glucose uptake by muscle involves the central nervous system in insulin-sensitive mice. *Diabetes.* 2011;60(12):3132–40.
109. Koch L, Wunderlich FT, Seibler J, Konner AC, Hampel B, Irlenbusch S, et al. Central insulin action regulates peripheral glucose and fat metabolism in mice. *J Clin Invest.* 2008;118(6):2132–47.
110. Perrin C, Knauf C, Burcelin R. Intracerebroventricular infusion of glucose, insulin, and the adenosine monophosphate-activated kinase activator, 5-aminoimidazole-4-carboxamide-1-beta-D-ribofuranoside, controls muscle glycogen synthesis. *Endocrinology.* 2004;145(9):4025–33.
111. Heni M, Kullmann S, Preissl H, Fritsche A, Haring HU. Impaired insulin action in the human brain: causes and metabolic consequences. *Nat Rev Endocrinol.* 2015;11(12):701–11.
112. Coomans CP, Geerling JJ, Guigas B, van den Hoek AM, Parlevliet ET, Ouwens DM, et al. Circulating insulin stimulates fatty acid retention in white adipose tissue via KATP channel activation in the central nervous system only in insulin-sensitive mice. *J Lipid Res.* 2011;52(9):1712–22.
113. Scherer T, O'Hare J, Diggs-Andrews K, Schweiger M, Cheng B, Lindtner C, et al. Brain insulin controls adipose tissue lipolysis and lipogenesis. *Cell Metab.* 2011;13(2):183–94.
114. Benedict C, Brede S, Schioth HB, Lehnert H, Schultes B, Born J, et al. Intranasal insulin enhances postprandial thermogenesis and lowers postprandial serum insulin levels in healthy men. *Diabetes.* 2011;60(1):114–8.
115. Dash S, Xiao C, Morgantini C, Koulajian K, Lewis GF. Intranasal insulin suppresses endogenous glucose production in humans compared with placebo in the presence of similar venous insulin concentrations. *Diabetes.* 2015;64(3):766–74.
116. Xiao C, Dash S, Stahel P, Lewis GF. Effects of intranasal insulin on endogenous glucose production in insulin-resistant men. *Diabetes Obes Metab.* 2018;20(7):1751–4.
117. Plomgaard P, Hansen JS, Ingerslev B, Clemmesen JO, Secher NH, van Hall G, et al. Nasal insulin administration does not affect hepatic glucose

production at systemic fasting insulin levels. *Diabetes Obes Metab*. 2018.

118. Benkendorff CFC, Hummel J, Vosseler A, Kullmann S, Fritsche L, Birkenfeld AL, et al. 150-OR: brain insulin sensitivity is modulated by menstrual cycle. *Diabetes*. 2020;69(Supplement 1):150-OR.
119. Zimmer ER, Parent MJ, Souza DG, Leuzy A, Lecrux C, Kim HI, et al. [(18)F]FDG PET signal is driven by astroglial glutamate transport. *Nat Neurosci*. 2017;20(3):393–5.
120. Tuulari JJ, Karlsson HK, Hirvonen J, Hannukainen JC, Bucci M, Helmio M, et al. Weight loss after bariatric surgery reverses insulin-induced increases in brain glucose metabolism of the morbidly obese. *Diabetes*. 2013;62(8):2747–51.
121. Rebelos E, Immonen H, Bucci M, Hannukainen JC, Nummenmaa L, Honka MJ, et al. Brain glucose uptake is associated with endogenous glucose production in obese patients before and after bariatric surgery and predicts metabolic outcome at follow-up. *Diabetes Obes Metab*. 2019;21(2):218–26.
122. Hirvonen J, Virtanen KA, Nummenmaa L, Hannukainen JC, Honka MJ, Bucci M, et al. Effects of insulin on brain glucose metabolism in impaired glucose tolerance. *Diabetes*. 2011;60(2):443–7.
123. Yokoyama H, Takeda R, Kawai E, Ota A, Morita E, Imai D, et al. Inhibitory effects of intranasal administration of insulin on fat oxidation during exercise are diminished in young overweight individuals. *J Clin Med*. 2018;7(10).
124. Iwen KA, Scherer T, Heni M, Sayk F, Wellnitz T, Machleidt F, et al. Intranasal insulin suppresses systemic but not subcutaneous lipolysis in healthy humans. *J Clin Endocrinol Metab*. 2014;99(2):E246–51.
125. Heni M, Wagner R, Kullmann S, Preissl H, Fritsche A. Response to Comment on Heni et al. Central insulin administration improves whole-body insulin sensitivity via hypothalamus and parasympathetic outputs in men. Diabetes 2014;63:4083–4088. *Diabetes*. 2015; 64(6):e8–9.
126. Paranjape SA, Chan O, Zhu W, Horblitt AM, Grillo CA, Wilson S, et al. Chronic reduction of insulin receptors in the ventromedial hypothalamus produces glucose intolerance and islet dysfunction in the absence of weight gain. *Am J Physiol Endocrinol Metab*. 2011;301(5):E978–83.
127. Kullmann S, Fritsche A, Wagner R, Schwab S, Haring HU, Preissl H, et al. Hypothalamic insulin responsiveness is associated with pancreatic insulin secretion in humans. *Physiol Behav*. 2017;176:134–8.
128. Batista TM, Haider N, Kahn CR. Defining the underlying defect in insulin action in type 2 diabetes. *Diabetologia*. 2021;64(5):994–1006.
129. Tschritter O, Preissl H, Hennige AM, Sartorius T, Stingl KT, Heni M, et al. High cerebral insulin sensitivity is associated with loss of body fat during lifestyle intervention. *Diabetologia*. 2011;55(1):175–82.
130. Ghasemi R, Dargahi L, Haeri A, Moosavi M, Mohamed Z, Ahmadiani A. Brain insulin dysregulation: implication for neurological and neuropsychiatric disorders. *Mol Neurobiol*. 2013;47(3): 1045–65.

CHAPTER TWELVE

Impact of Dietary and Exercise Interventions on Brain Insulin Action and Brain Function

Stefan Kabisch

CONTENTS

THE DIET-INSULIN-BRAIN AXIS IN HUMANS

In the past 100 years, the knowledge about insulin has rapidly expanded from the time of its sole discovery to its description as highly versatile hormone, which affects way more than glucose metabolism. It is involved in other aspects of nutrient disposal, distribution, and recruitment, but also appears to contribute to neural processes, which are tightly connected but not limited to cognitive functions such as reward behavior, control of food intake, and mood. Insulin is a key hormone in the course of diabetes mellitus development that affects and defines the disorder itself, partially represents its treatment and has an impact on neural activity and some neural disorders.

Prediabetic dysglycemia and diabetes mellitus are characterized by an impairment of insulin signaling, which is either caused by reduced insulin secretion rate, lower insulin sensitivity or a combination of both. Most patients with dysglycemia in the context of the metabolic syndrome (i.e. associated with obesity, hypertension, dyslipidemia, but also hyperuricemia and NAFLD) and patients with type 2 diabetes (T2DM) show insulin resistance. The major fraction of type 1 diabetes patients (T1DM), but also some T2DM patients can be diagnosed with insulin deficiency, which is related to autoimmune destruction of the beta-cells (T1DM) or – in the case of T2DM – to a mixture of genetic factors, aging, glucolipotoxicity and hyperinsulinemic exhaustion of beta cells. As far as insulin secretion is intact, diabetes mellitus and its impairment of insulin signaling are reversible. Lifestyle interventions, pharmaceutical approaches, and surgical "last resorts" aim for a preservation of beta-cell function by ameliorating insulin resistance and/or reducing the amount of circulating substrates.

DOI: 10.1201/9781003079927-12

Obesity, the common precursor of insulin-resistant T2DM, is primarily a behavioral and neurological disorder. Its epidemic appearance paralleled the increasing global availability of cheap, high-energy food products. Adiposity can be partially explained by genetics, with many of the relevant single nucleotide polymorphisms (SNPs) residing in genes, which control brain circuits. Only a smaller fraction of genetic loci are primarily related to (nutrigenetic) metabolism.

Insulin-resistant T2DM and obesity are bidirectionally associated with depression, dementia, and some eating disorders (1). Insulin resistance, mirrored by hypoadiponectinemia, seems to reflect global cognitive decline in patients with Alzheimer's disease (AD) or predementia (2, 3). Poor glycemic control is also associated with depressive symptoms and quality of life in diabetes patients (4). Several aspects of lifestyle, which are predisposing to obesity, insulin resistance, and T2DM, are also associated with depression, dementia, and eating disorders.

The connection between obesigenic and/or diabetogenic lifestyle factors and cognitive or emotional outcomes in humans is difficult to investigate. Some epidemiological studies and intervention trials cover hard outcomes (frequency of neurological disorders) or symptom load. Functional magnetic resonance imaging (fMRI) studies have investigated the impact of exercise, but not dietary changes. Brain-derived neurotrophic factor (BDNF) is the only biomarker related to severity and frequency of obesity, eating disorders, depression, and dementia (5–8). BDNF SNPs are risk factors for these disorders (9–14). BDNF levels are reduced by hyperglycemia, independently of insulin levels (15).

ASSOCIATIONS BETWEEN GENERAL DIETARY FACTORS AND NEUROLOGICAL OUTCOMES

A variety of essential amino acids (e.g. tryptophane), fatty acids (omega-3 fatty acids from plant and marine sources), minerals (e.g. iron, magnesium, potassium), vitamins (e.g. A, B1, B2, B6, B9, B12, D) and a sufficient amount of secondary antioxidants (e.g. carotenes, flavonoids, anthocyanins, and other polyphenols) are required to develop and sustain healthy brain functions from childhood to aging. Lack of these factors is linked to lower socioeconomic status (SES), specific medical conditions or a self-chosen, unhealthy diet. Toxins such as nicotine, alcohol, and illicit drugs are more common among subjects with low SES and unhealthy lifestyle. Apart from these aspects of malnutrition or SES-associated harmful behavior, several other dietary factors with rather metabolic linkage have an impact on mood and cognition when investigated in epidemiological studies.

EPIDEMIOLOGICAL IMPACT OF LIFESTYLE FACTORS ON OBESITY, INSULIN RESISTANCE AND T2DM

Specific dietary factors, which are consistently negatively associated with T2DM risk, are overnutrition/obesity, whole grain products and insoluble fiber (16, 17), coffee (with or without caffeine) (18), legumes, and moderate amounts of alcohol. Sugars and other high-glycemic index (GI) foods, artificial sweeteners, and red meat are linked to a higher T2DM risk (19–21). Fish, tea, and dairy have not been consistently shown to affect diabetes risk (22, 23). Some genes within the T2DM risk profile show a nutrigenetic association with either T2DM incidence or continuous parameters, e.g. ACE, FABP2, TCF7L2 (24–26). There are also still inconclusive nutrigenomic associations between fiber intake, diabetes risk or glycemic outcomes, and genes, related to incretin secretion and NAFLD (27–30). High animal protein intake is considered a diabetes risk factor, while moderate intake of plant protein is seen as protective (31). Saturated, cis-monosaturated, and cis-polyunsaturated fatty acids show no epidemiological association with T2DM. Non-ruminant trans fats are classified as diabetogenic, while animal-based trans-fatty acids are not (32).

Obesity itself shares some of these nutritional risk factors, but primarily results from an imbalance between energy intake and energy expenditure, almost irrespective of the consumed nutrients. However, high GI foods are more likely to be overconsumed as they provide only short-term satiety, as the massive postprandial insulinemia is followed by a reflectory hypoglycemia, which induces hunger and food craving despite recent food intake. Protein-rich meals lack such an effect, as "ad libitum open-buffet" experiments with different dietary compositions demonstrated. Risk genes for obesity rarely show a nutrigenetic association; such effects are – for example – described for TCF7L2, FTO, or DRD2 (33–38).

The absence of obesity and even weight loss are not unanimous indicators of good health and long-life expectancy. The "obesity paradox" addresses a multifactorial problem of normal- to underweight elderly adults: increased mortality, even and especially when suffering from T2DM, CVD or cancer (39–41). Maintaining desirable depots of adipose tissue (femoral vs. abdominal fat; subcutaneous vs. visceral fat) might be crucial in providing physiological storage for nutrients, assuring sufficient production of estrogen, adiponectin, and other beneficial hormones and protecting mechanically relevant structures in the context of aging, sarcopenia, and osteoporosis.

Lack of physical activity is linked to both obesity and T2DM in a non-linear fashion: moderate-to-vigorous activity seems to provide the strongest protective effect, higher activity provides almost no additional benefit (42, 43). The impact of low levels of physical activity (PA) is not mediated by major diabetes risk genes (44) (Table 12.1).

LIFESTYLE AND DEMENTIA

Obesity at younger age, a major driver of insulin resistance, seems to promote dementia, while in elderly subjects a pre-obese BMI seems to be protective (45). Weight loss in older age seems to contribute to onset of dementia, irrespective of baseline BMI and the impact of dementia-protective physical activity (46).

High intake of high glycemic index (GI) foods and fruits was associated with poorer school marks in pupils, possibly reflecting low SES, impaired supply of micronutrients and an undesirable glycemic profile (47). High GI foods also strongly predict different types of dementia and their structural correlates (48, 49). Adherence to the Mediterranean diet is associated with better memory, verbal and visuospatial ability, while highly processed foods were linked to poorer cognitive abilities. As far as such associations are strongly explained by childhood abilities, diet and brain structure are not associated (50). Advanced glycation endproducts (AGE) might be

TABLE 12.1
Diabetes Classification

Diabetes Type	Major Pathophysiological and Clinical Features
Type 1 Diabetes	Rapid Autoimmune Destruction of All Beta-Cells Resulting in Complete Insulin Deficiency; Age of Onset Typically in Childhood or Young Adults
Late-onset autoimmune diabetes of the adult (LADA)	Delayed autoimmune destruction of all beta-cells resulting in insulin deficiency; age of onset in adults of all ages
Type 2 Diabetes (T2DM)	Diabetes mellitus with insulin resistance and/or insulin deficiency of variable magnitude; age of onset typically in middle and late adulthood
Severe insulin-deficient diabetes (SIDD)	T2DM with predominant beta-cell failure; average age of onset in middle adulthood; weak contribution of the Metabolic Syndrome; high risk for neuropathy and retinopathy
Severe insulin-resistant diabetes (SIRD)	T2DM with massive insulin resistance and NAFLD; age of onset in late adults; strong connection to Metabolic Syndrome; high risk for nephropathy
Mild obesity-related diabetes (MOD)	T2DM with moderate insulin resistance and beta-cell dysfunction; average age of onset in early adulthood; strong association with obesity
Mild age-related diabetes (MARD)	T2DM with moderate insulin resistance and beta-cell dysfunction; average age of onset in late adulthood; low risk for long-term complications
Type 3 Diabetes	Heterogeneous groups of monogenetic, syndromal, and secondary diabetes
Type 3A = Monogenetic diabetes of the young (MODY)	Monogenetic autosomal (mostly dominant) mutations, > 15 subtypes (most of them leading to insulin deficiency) with variable prognosis
Type 3B-H = various other specific causes of diabetes	Diabetes due to pancreas infections, tumor, injuries or surgery, endocrine disorders, medication, toxins, rare autoimmune conditions or specific genetic syndromes; prognosis and treatment depending on subtype
Type 4 Diabetes = Gestational diabetes (GDM)	Mild diabetes mellitus which is diagnosed during pregnancy; insulin resistance due to hormonal changes; high risk for later T2DM onset

molecular modulators for the risk of dementia (51). Polyamines, abundant in whole grain, dairy, nuts and legumes and thus the Mediterranean diet, are both micronutrients and dementia biomarkers (52, 53).

Alcohol intake is associated with cognitive decline in dose-dependent manner with high intake increasing the risk for dementia (54).

Coffee intake and moderate alcohol consumption are negatively associated with risk for both type 2 diabetes and dementia (55). Cognitive alterations are also linked to coffee consumption in a non-dementia cohort as well (56). Other studies found no effect for coffee (57). The relevant compound (in contrast to associations with T2DM risk) seems to be caffeine, as consumption of regular but not decaffeinated coffee was associated with better cognitive performance (58). Fish, green tea and vegetables appear to reduce the risk for dementia, even though lacking an impact on diabetes risk (45, 59, 60). Dementia mortality seems to be reduced in subjects with higher intake of poultry, eggs and plant protein, but lower intake of processed red meat, contradicting T2DM associations (61). A protective effect is also described for dairy products (62). All food groups with high protein content (including legumes) show a nutrigenetic association with obesity and a BDNF SNP (8). High-fat diets seem to elicit a nutrigenetic effect on cognition, which is mediated by VEGF and GLUT1 (63). The Mediterranean diet seems to protect from cognitive in cohort studies (64).

Physical activity is deemed to protect from cognitive decline and Alzheimer's dementia based on cohort data, showing a dose-response relationship (59, 65–67). Possibly, physical fitness is a more precise predictor for actual functional abilities (68).

LIFESTYLE AND DEPRESSION

High GI and ultraprocessed foods (69, 70), sugary sweets, low intake of legumes, fruits and vegetables, and lack of fiber (71–74) are also associated with depression, including pregnancy-related depression (75). In a Korean cohort study, only fibers from mushrooms and seaweed (but not cereals, fruits and vegetables) were identified as protective types of fiber when evaluating the risk for depression (76). With respect to obesity, both sugar-sweetened beverages (SSB) and non-caloric sweetened beverages (NSB) are considered risk factors (20, 77). SSBs also seem to increase the risk for depression, while NSB consumption is negatively associated (78).

Protein seems to have a differential impact on depression and mood. While high protein intake is associated with reduced risk for depression in men, women show an increased risk for depression with increasing intake of protein (79). Sex differences in protein quality might explain this discrepancy. An investigation of Chinese women showed, that both a mainly plant-based (carbohydrate-rich) dietary pattern as well as a high load of animal protein were connected to a low risk of gestational depression. In the same cohort, women with high-fat diet or a high intake of plant protein (beans) experienced a higher chance of mood disorder (80).

Processed eggs and meat were found to increase the risk for depression, while dairy and fresh eggs seemed to be protective food products (81). As for T2DM risk, especially red meat and processed meat are also risk factors for depression (82). In an Asian cohort, vegetables and fruits, snacks, drinks and dairy were identified as protective factors, while meat and fish intake were not associated with depression in cross-sectional analysis (83). Moderate, but not high intake of fatty fish and long-chain omega-3 fatty acids was associated with decreased risk for depression in the PREDIMED-Plus trial (84). However, the omega-3-index does not seem to correlate with depression symptoms in patients with coronary heart disease and plasma PUFA levels are not associated with depressive disorders (85, 86). Other studies disputed the impact of PUFAs on depression risk (87).

Intake of green and black tea is negatively connected to geriatric depression symptoms (88). Other studies found no such clear associations for fruits and vegetables (89), dietary fiber (90, 91), tea (78), or coffee (92). The role of caffeine as a preventive or causal factor of bipolar depression is still unclear (93).

While certain metabolically unhealthy or healthy food components seem to affect the risk for depression, overall healthy European food patterns (including the Mediterranean diet and the DASH diet) do not seem to have a consistent protective effect on depression (94–96). However, the healthy Japanese diet – based on plants and fish, but free of alcoholic beverages – is associated with a lower risk for depression (97). In US citizens of Asian ethnicity, the vegetarian diet shows linkage to reduced odds for depression, while other studies show an increased risk (98, 99). An additional impact on anxiety is described especially in younger persons (100). The causality between food patterns and mood disorders is still disputed (101).

Objectively measured (but not self-reported) physical activity is also negatively associated with the risk for clinical and subclinical depression, showing a dose-response relation and independence of age and regionality (102–105).

METABOLIC IMPACT OF DIFFERENT LIFESTYLE INTERVENTIONS ON OBESITY, GLYCEMIA, AND INSULIN RESISTANCE

Lifestyle changes include the cessation of specific toxins leading to insulin resistance or beta-cell apoptosis (excess alcohol intake, smoking), reduction of energy intake for patients with overweight, an improvement of diet quality and an increase in physical activity. Smoking cessation often leads to massive weight gain and in the end might even increase the risk for diabetes onset (106, 107). All other approaches have been shown to provide more or less dramatic benefits to glycemic state and other aspects of the metabolic syndrome.

Complex lifestyle interventions combining diet and exercise consistently reduce the risk for T2DM onset in patients at elevated risk (108–110), including a legacy effect for diabetes and CVD prevention (111). All these studies used a low-fat diet combined with or compared to increased physical activity. Hypocaloric low-fat approaches also provide the opportunity of T2DM remission, fewer diabetes-related comorbidities and improved quality of life, even though there is yet no evidence for a concomitant benefit in long-term CVD outcomes (112). T2DM remission might be even more successful with low-carb diets, but evidence is limited (113). The most convincing data for both glycemic amelioration and CVD outcome are published for the Mediterranean diet, a variant of the moderate carbohydrate-restricted diet with specific recommendations for certain food products, which improves glycemia, lipid profile and blood pressure in a superior fashion compared to low-fat (114–116). Other types of carbohydrate-restricted diets lead to almost similar benefits in the aforementioned axes, but epidemiological evidence points out, that only plant-based low-carb diets might reduce CVD risks (117). A clear prospective evidence for glycemic harm by saturated fat or glycemic benefit by any kind of unsaturated fat is missing (118).

zEpidemiological health claims for whole grain and dietary fiber, coffee, and alcohol with respect to T2DM prevention and management are not sufficiently based on interventional data. Interventions with whole grain and soluble cereal show short-term effects on glycemia and insulin resistance, but long-term studies are virtually non-existing (126, 127). Sparsely available randomized controlled trials (RCTs) on insoluble cereal fiber (cellulose, hemicellulose) indicate effects on glucose tolerance and NAFLD after one year, possibly limited to susceptible subgroups (128–131). RCTs, actively promoting alcohol intake, suffer the ethical dilemma of introducing a toxic, addictive substance to non-consumers. However, few earlier trials show moderate effects of red wine and beer on lipid levels, inflammation (132–134), but not glycemia or insulinemia (135). Non-alcoholic wine, or extracts of their polyphenols, seem to be mainly ineffective in improving glucose metabolism (136–138). For resveratrol, results are negative with respect to NAFLD (139), but positive for glucose (140), lipid levels, blood pressure, and inflammation (141). Beer polyphenols have been shown to ameliorate insulin resistance, lipemia, and inflammation (142, 143).

Epidemiological warnings on red (processed) meat are not confirmed by intervention studies. The only six RCTs specifically comparing meat types primarily investigated body composition and lipid outcomes rather than glycemia and insulin resistance, but did not find any relevant differences at all (144–149).

Different types of physical exercise have been shown to beneficially affect insulin resistance and other aspects of the Metabolic Syndrome. Meta-analyses report an improvement of insulin sensitivity due to aerobic training (149a), medium-term (but not long-term) resistance training (150), high-intensity interval training (151), and yoga (152).

EFFECTS OF METABOLICALLY BENEFICIAL COMPLEX LIFESTYLE CHANGES ON COGNITION

Complex lifestyle interventions have been shown to improve cognition, depression symptoms, and anxiety mood on several, but not all occasions.

Cognition improved due to lifestyle modification in women with irregular menses (153), but not in old, non-demented subjects (154). Diet and exercise seem to exert partially opposite effects, for example on BDNF (155–157), which may reflect neurological improvements after dietary interventions (158, 159). Intermittent fasting does not seem to change IGF-1 (160) or BDNF levels, contradicting rodent studies (161–163).

Part of the beneficial effects of lifestyle changes on mood and cognition in humans may be attributed to the caring, comforting situation of the lifestyle program itself, irrespective of the actual content of the training sessions (164). Therefore, a closer look at different lifestyle approaches, diet comparisons and single nutrients is warranted.

Simple weight loss as strategy to improve cognition in the elderly is seen with doubt; for once with respect to the obesity paradox, but also, as most weight loss approaches lack a sufficient coverage of micronutrients and vitamins, which are essential to even maintain pre-existing cognitive state. However, some studies indicate a potential cognitive benefit of weight loss in patients with mild cognitive impairment (MCI) (165, 166), which might be limited to the transient phase of weight loss rather the final post-interventional maintenance phase (167). Improved memory was also seen in healthy overweight elderly subjects after three months of caloric restriction, linked to improving insulin levels, subclinical inflammation at baseline and better adherence (168).

Within the Look AHEAD trial, lifestyle intervention based on low-fat diet, partially with meal replacements, and increased physical activity led to weight loss and glycemic improvements. However, not all patients achieved a maintainable weight reduction or an actually lowered glucose level. In this context, only improvement of glycemia was clearly associated with improved cognitive functions (169). Importantly, intensive lifestyle intervention had a less favorable effect on cognition in overweight and obese patients compared to standard care (170). Hyperglycemia, which is resistant to weight loss, might be linked to already impaired and irretrievable beta-cell function, ergo: insulin deficiency with or without insulin resistance. This could imply, that restoration of cognitive functions in the context of T2DM is only realistic if beta-cell function is still sufficiently intact. Patients with severe (primary, non-autoimmune) insulin-deficient diabetes (SIDD) and those with long-term severe insulin-resistant diabetes (SIRD), according to the Ahlqvist classification (171), would not benefit from this low-fat/exercise lifestyle intervention.

Low-carb or ketogenic diets have not yet been largely tested for cognitive or emotional effects, but could be worth investigating due to their potential to reduce glucose levels more strongly, independently of weight loss, and possibly irrespective of insulin resistance (166, 172, 173). Also, ketone bodies themselves, rather than the metabolic amelioration which leads to their increased production, might facilitate beneficial effects on cognitive decline (174, 175). A potential nutrigenetic effect of APOE, a major risk gene for Alzheimer's disease, is described in dietary interventions with low carbohydrate/high protein intake (165, 176–178). This seems partially explainable by the allele-dependent utilization of ketone bodies (175, 179).

INTERVENTIONAL EFFECTS OF SPECIFIC LIFESTYLE COMPONENTS ON GLYCEMIC METABOLISM AND COGNITION

Interventions focusing on satisfying the demand for essential nutrients under the clear condition of malnutrition showed the benefits of supplementation. Restoration of requirements for iron and potassium in women with nutritional deficiency (180) and for vitamin B12 in children with B12-deficit proved to significantly increase cognitive and motor skills compared to baseline and depending on the achieved levels (181).

Supplementation without clearly detected deficiency is an unfortunate common condition. Often, multivitamin formulations are consumed over long time to improve fatigue, neuropathy or depression. Vitamin D supplementation is effective in improving insulin resistance, especially when given in large doses and for a short period of time, and to patients who were non-obese, Middle Eastern or vitamin D deficient (182, 183). In adults with depression, a multivitamin-enhanced formulation of N-acetylcysteine (NAC), compared to NAC alone, induced only potential minor delayed benefits in depression and social functioning after 16 weeks of intervention and additional follow-up (184). Micronutrient supplementation without deficiency does not seem to relevantly affect mood and cognition, even though it might alter insulin secretion and insulin sensitivity (185).

For omega-3 acids, a still debated protective factor for T2DM, data on cognitive outcomes is controversial. In children, fish meals (vs. meat meals or fish oil supplement over 12 weeks) assured a higher procession speed (d2 attention test) (186). Fish oil supplementation over the same period induced a stronger improvement in the Trail Making Test, but no other cognitive tests (187). Other trials completely failed to show benefits in cognition or fine motor skills by enforcing salmon or meat intake over 16 weeks, despite achieving a higher intake of EPA, DHA, vitamin D, and iodine. Benefits of marine long-chain n3-PUFAs might be counteracted by mercury intake (188, 189). In elderly subjects, fish oil supplementation improved executive functions, accommodated by changes in white and gray brain matter structure and BDNF levels (190), while studies on patients with dementia or T2DM and depression failed to replicate such effects (191, 192). Within a Nordic-diet lunch package given to shift workers, the only nutrient being linked to improved mood and performance was water, but no nutrient related to fish (193).

An isocaloric increase of food sources with unsaturated fatty acids over three months did not alter memory function in healthy elderly women (168). Olive oil, mainly containing oleic acid, slightly improved dementia-related cognition during one year of intervention in a setting of the Mediterranean diet (194).

Meat, especially (processed) red meat is seen as an unrecommendable food product with respect to inflammation, gout, obesity, and T2DM. Its content of ferritin/iron, saturated fat, cholesterol, and uric acid is thereby valued as a stronger metabolic disadvantage in comparison to the potential benefits of high protein and thiamine load on muscle mass and cognition (195). Surprisingly, intervention studies on red meat report inconsistent effects of red meat: In elderly adults, a 24-week intervention of 3 days/week resistance exercise with a daily intake of red meat led to higher arm lean mass, gait speed, and muscle density, but lower improvements of working memory and learning when compared to a low-meat comparator group (196). Controversially, a 24-weeks RCT comparing a low-fat diet with a diet containing 2–3 servings of fresh lean pork per week showed a better processing speed performance and emotional role functioning in 35 subjects between 45 and 80 years of age (197). Another 12-weeks study comparing pork and chicken meals in older participants (n=48) did not find relevant differences in cognitive outcomes. The essence of chicken was reported to improve short-term memory and stress resistance in 235 subjects after a 2-week intervention compared to placebo (198). Similar effects were seen in a study in 102 participants with work-related stress and depression (199). Mold-fermented cheese, tested in a 2×3 months cross-over trial in 71 women with MCI, increased levels of BDNF, but did not affect mini-mental state examination (MMSE) scores (200). Based on a recent meta-analysis, interventions based on dairy products do not have an impact on cognition (201).

In an acute setting, low- and high GI foods (different breakfasts, sucrose- vs. isomaltulose-sweetened trifle) did not clearly affect the postprandial cognitive performance in young healthy adults. Glycemia and insulinemia did not seem to have an acute strong influence on attention or memory (202, 203). Similarly, those food groups do not seem to differ in their impact on BDNF levels (204). Long-term studies are controversial: In high school students, a 9-week intervention with mixed-grain products led to a better cognitive performance and higher BDNF levels compared to regular diet (205). A polydextrose supplement was shown to improve attention and cognitive flexibility in healthy women after 4 weeks (206). In contrast, children receiving a galacto-oligosaccharide prebiotic supplement for 12 weeks did not show improvements in cognition, mood, or anxiety (207). Dried apples – compared to high GI muffins – did acutely improve glycemic state, but not cognitive performance in healthy, normal weight adults (208).

Plant polyphenols are widely advertised for several health claims, including their potential impact on cognition. However, the scientific background is highly heterogeneous and still debated, effects on major metabolic outcomes are not reported. Acute cognitive benefits (mainly on attention) were seen by berry juices given to adults (209–212) and primary school children (213, 214). Other studies did not replicate these results (215). Long-term trials reported benefits by blueberry juice (14 days prior to major surgery) (216), by blueberry extract (3 months) (217), Montmorency cherry juice (12 weeks) (218) and pomegranate juice (12 months) (219) in elderly non-demented adults. Bergamot polyphenols improved cognition in subjects with schizophrenia (8 weeks) (220), while the coffee-related chlorogenic acid was mainly ineffective in subjects with MCI (12 weeks) (221). Resveratrol, a red wine polyphenol, induced surprisingly minor improvements in memory after 6 months, which were accompanied by worsened lifestyle, inflammation, lipid and glucose metabolism, as well as dropping levels of BDNF and IGF-1 (222). This contradicts meta-analyses on metabolic effects of resveratrol in other studies (140, 141). Green teas and other teas show inconsistent effects on attention, psychomotor speed, memory, information procession and mood (223–225). Metabolically, green tea provides a small benefit on fasting glucose levels (226).

Beetroot juice, a source of vasodilating nitrate, appeared to acutely improve the stroop test performance in adult subjects (227). A 4-weeks supplementation with taurine increased the MMSE score in women with dementia (228). However, there is no consistent evidence for an impact of taurine on insulin secretion and overall metabolism (229–231), even though taurine levels were reported to be reduced in T2DM patients (232). A supplementation with extracts from Hericium erinaceus, a "medical mushroom", minimally improved MMSE scores, but not visual retention or verbal learning, in elderly, non-demented adults after 12 weeks (233). The impact of spermidine and other polyamines is controversially addressed in only a few RCTs (234, 235).

Physical activity does not seem to provide any consistent benefit in cognitive domains related to dementia progression in subjects without dementia (236). Single trials reported improvements in learning abilities, memory and executive functions (237–240), mirrored by changes in functional and structural integrity of frontal brain areas (241, 242) or the hippocampus (243). Reports on the effects of different types of physical activity on BDNF level are controversial, showing an increase (239, 243–247), a decrease (248–253) or no changes at all, with inconsistencies among sex, type of exercise and clinical outcome (254–259) as well as sampling method (platelet, serum or plasma BDNF) (260, 261). An increase in BDNF was even reported in subjects with T1DM undergoing high-intensity exercise, indicating that endogenous insulin action is not required for this response (262). However, changes in BDNF levels poorly reflect changes in brain function (263). Alternative biomarkers representing an interventional impact of exercise on brain structure and improved cognition might be found in IGF-1 (244, 258, 262, 264, 265) or VEGF (258, 266), but results and implications are controversial as well (267, 268).

EFFECTS OF SPECIFIC DIETARY INTERVENTIONS ON GLYCEMIC METABOLISM AND MOOD

Dietary interventions in general seem to improve depression symptoms (more than anxiety) in subjects with nonclinical depression when compared to placebo or other active comparators. Effects are more pronounced in women, when aiming for low fat intake, low red meat intake and/or weight loss and can be seen even as add-on effects on exercise interventions (269, 270). Benefits are seen for different types of healthy diets (271), including the DASH diet, investigated over 14 weeks in postmenopausal women (272), and the vegetarian diet, tested over 2 weeks (273). However, Mediterranean diets seem to be even superior when affecting mood (274–277) or BDNF levels (278). Supplements, baseline inflammatory conditions or weight reduction seem to have a minor contribution to mood improvements (271). Fiber-rich rye breads, in contrast to white wheat bread, were shown to acutely improve valance, but no other aspect of mood or cognition in elderly subjects, despite an apparent association between insulin sensitivity and working memory performance (279). The same intervention also led to an increase of BDNF, a surrogate marker of neural integrity (280). Probiotic, even more than prebiotic supplements might reduce depression symptoms after 8 weeks (281). Supplementation with zinc showed a significant benefit with respect to both BDNF levels and depression severity (BDI) (282). BDNF levels were unaffected by 12 weeks of DHA supplementation in subjects with PTSD and depression and 10 weeks of treatment with vitamin D in older adults with vitamin D deficiency (159, 283).

Plant polyphenols and other specific bioactive non-nutrient compounds were investigated for their mood-stabilizing potential. Benefits were seen with the Tibetan herbal tea blend Agar-35 (after 7 days) (284), lavender tea and chamomile tea (after 2 weeks) (285, 286) and orange juice (after 8 weeks) (287). Results of supplementation with

L-theanine, a green tea compound, are controversial and partially overlapping with caffeine (288–290). Coffee, more intensively (but not exclusively) with vs. without caffeine, improves attention and alertness, demonstrating that caffeine might not be the only psychoactive substance of this beverage (291). The epidemiologically claimed antidiabetic effect of coffee has not been confirmed in intervention studies (226), and caffeine even seems to acutely reduce insulin sensitivity (292). Different coffee bean types may elicit different effects due to different compositions (293).

Meta-analyses show a consistent impact of physical activity (294), in particular aerobic exercise on depression and anxiety (295), similar or superior to meditation (296, 297). Ball sports, dance or running seem to result in weaker effects compared to Tai Chi or Yoga (298). Resistance, exercise and mind-and-body exercise do not differ significantly in their outcome (299). Relations between exercise, improved mood and circulating neurotrophic factors (BDNF, IGF-1, VEGF) are still debated (255). For BDNF, similar to studies on dementia and cognition, a trend-wise benefit of physical interventions in patients with major depression can be seen, but warrants further, larger studies (300) (Table 12.2).

CONCLUDING REMARKS

The metabolic hormone insulin is primarily involved in nutrient disposal, but also elicits delicate effects on satiety and possibly other segments of human cognition and behavior. As diet and physical activity affect the insulin system, modulating the effects of dietary patterns or single food products on neural activity is plausible. However, ethics and technical issues limit our opportunities to directly assess insulin action in the human brain. The putative benefits of certain lifestyle interventions on insulin sensitivity, insulin secretion, and additional aspects of the metabolic syndrome could possibly extend to lifestyle-associated neurological disorders, such as dementia or depression. Currently, the evidence for major dietary patterns supports the encouragement of a traditional Mediterranean diet, which is rich in

TABLE 12.2

Lifestyle Components and Their Impact on Insulin Resistance, Cognition and Mood (Summary)

Intervention	Impact on Insulin Resistance and T2DM	Impact on Cognitive Decline	Impact on Mood
Sugar	- strong T2DM risk factor - clearly detrimental in iso- and hypercaloric conditions	- risk factor in cohort studies - weak RCT evidence	- risk factor in cohort studies - insufficient RCT data
Glycemic index	- inconsistent T2DM risk factor - high GI is detrimental in iso- and hypercaloric conditions	- risk factor in cohort studies - unclear RCT evidence	- risk factor in cohort studies - inconclusive intervention trials
Protein intake	- no T2DM risk factor - inconclusive RCT results for high- and low-protein diets	- reduced dementia mortality with high-protein - no evidence from RCTs	- sex-dependent association with depression - insufficient RCT evidence
Saturated fat	- inconsistent T2DM risk factor - detrimental in iso- and hypercaloric conditions	- potential risk factor for dementia - poor RCT evidence	- risk factor in cohort studies - insufficient intervention trials
Unsaturated fat	- no T2DM risk factor - inconclusive RCT results	- unclear epidemiological association - inconclusive RCT data	- unclear epidemiological association - inconclusive RCT data
n3-PUFA	- no consistent impact on T2DM risk - inconclusive RCT results	- unclear epidemiological association - inconclusive RCT data	- potentially protective association - inconclusive RCT data
(Red) meat	- weak epidemiological association with T2DM - no consistent intervention data	- weak epidemiological association with dementia - inconsistent RCT data	- risk factor in cohort studies - inconclusive RCT data

(Continued)

TABLE 12.2 (*CONTINUED*)

Lifestyle Components and Their Impact on Insulin Resistance, Cognition and Mood (Summary)

Intervention	Impact on Insulin Resistance and T2DM	Impact on Cognitive Decline	Impact on Mood
Fish	- no impact on T2DM risk - inconclusive RCT results	- protective association with dementia - inconsistent RCT data	- potentially protective association with depression - insufficient RCT data
Dairy	- no consistent impact on T2DM risk - inconclusive RCT results	- protective association in cohort studies - no RCT effects	- potentially protective association with depression - insufficient RCT data
Soluble dietary fiber	- no antidiabetic association in cohort studies - small short-term impact on IR in RCTs	- unclear epidemiological association - no intervention studies	- possible, inconclusive protective association in cohort studies - insufficient intervention studies
Insoluble dietary fiber	- strong antidiabetic association in cohort studies - small long-term impact on IR in RCTs	- unclear epidemiological association - no intervention studies	- possible, inconclusive protective association in cohort studies - no intervention studies
Coffee	- moderate antidiabetic association in cohort studies - acute impairment of IR-uncertain long-term effect	- potentially protective epidemiological association - no intervention studies	- possible protective association in cohorts - insufficient interventional
Green tea	- no risk factor for T2DM - no interventional evidence of benefit	- protective association with dementia - no consistent RCT data	- potentially protective association with depression - no consistent RCT data
Alcohol	- moderate antidiabetic association in cohort studies - unclear interventional long-term effect	- moderate association with harm in cohort studies - no intervention studies	- possible protective association in cohorts - no intervention studies
Overeating and obesity	- strong predictive factor - variable intervention effect depending on muscle mass	- risk factor for dementia - inconclusive RCT effects depending on specific diet	- risk factor for depression - inconclusive RCT effects depending on specific diet
DASH	- moderate improvement in RCTs	- unclear epidemiological association - no intervention studies	- possible protective association in cohorts - no intervention studies
MedD	- moderate-to-strong improvement in RCTs	- potentially protective association - inconclusive intervention studies	- possible protective association in cohorts - no intervention studies
Low-carb	- superior impact on glycemia, but insufficient evidence for effects on IR	- unclear epidemiological association - insufficient interventional data	- unknown association in cohorts - no intervention studies
Low-fat	- inferior impact on glycemia, but good improvement of IR	- unclear epidemiological association - insufficient intervention studies	- possible protective association in cohorts - no intervention studies
Vegetarian diet	- inferior impact on glycemia, but good improvement of IR	- unclear epidemiological association - no intervention studies	- potentially reduced or increased risk for depression - insufficient RCT data

(*Continued*)

TABLE 12.2 (*CONTINUED*)

Lifestyle Components and Their Impact on Insulin Resistance, Cognition and Mood (Summary)

Intervention	Impact on Insulin Resistance and T2DM	Impact on Cognitive Decline	Impact on Mood
Intermittent fasting	- comparable effect on weight loss, but variable-to-inferior impact on IR	- no reported effect on cognition	- insufficient data
Physical activity	- protective factor for T2DM - moderate middle-term effects in RCTs; otherwise insufficient data	- protective association with dementia - inconsistent RCT data	- protective association with depression - beneficial effect in RCTs
Yoga	- sparse data showing a benefit on IR	- unclear epidemiological association - insufficient RCT data	- beneficial impact
Resistance training	- moderate beneficial effect on IR	- unclear epidemiological association - insufficient RCT data	- potentially beneficial impact
Aerobic training	- moderate beneficial effect on IR	- unclear epidemiological association - insufficient RCT data	- beneficial impact

fiber, plant oils, and polyphenols, but low in simple and refined carbohydrates. This pattern improves metabolism and has the best chances to preserve cognitive abilities and to stabilize mood. Up to date, we cannot evaluate, which amount of sugar, red meat, insoluble and soluble fiber, and alcohol can be deemed both safe and healthy for all desired outcomes. The vast amount of micronutrients is hardly understood in their impact on metabolism and brain function. Weight loss by fasting or other approaches may be a common recommendation for our predominantly overweight-to-obese patients, but bears the risk of sarcopenia, metabolic deterioration, and cognitive decline, especially in elderly people. Moderate-to-vigorous physical activity seems to be protective in both cohort and intervention studies with respect to overall metabolism and depression, but less clearly for dementia. Larger, long-term intervention trials in healthy and risk-defined subjects as well as several groups of patients are required to replicate claims on improved insulin action, to evaluate the impact on neural function and to pinpoint the underlying mechanisms.

REFERENCES

1. L. Gallardo-Alfaro et al., Leisure-time physical activity, sedentary behaviour and diet quality are associated with metabolic syndrome severity: the PREDIMED-plus study. *Nutrients* **12**, 1013 (2020).
2. S. A. Hejazi et al., The correlation of serum adiponectin and insulin resistance with the presence and severity of dementia in non-obese Alzheimer's patients. *Clin Nutr ESPEN* **40**, 376–382 (2020).
3. Y. Zhong et al., Hyperinsulinemia, insulin resistance and cognitive decline in older cohort. *Biomed Environ Sci* **25**, 8–14 (2012).
4. C. L. P. Vigen et al., Psychosocial and behavioral correlates of A1C and quality of life among young adults with diabetes. *Diabetes Educ* **44**, 489–500 (2018).
5. A. R. Brunoni, M. Lopes, F. Fregni, A systematic review and meta-analysis of clinical studies on major depression and BDNF levels: implications for the role of neuroplasticity in depression. *Int J Neuropsychopharmacol* **11**, 1169–1180 (2008).
6. B. S. Diniz et al., Brain-derived neurotrophic factor levels in late-life depression and comorbid mild cognitive impairment: a longitudinal study. *J Psychiatr Res* **49**, 96–101 (2014).
7. G. Martinotti et al., Agomelatine increases BDNF serum levels in depressed patients in correlation with the improvement of depressive symptoms. *Int J Neuropsychopharmacol* **19**, 1–6 (2016).

8. J. M. McCaffery et al., Obesity susceptibility loci and dietary intake in the Look AHEAD Trial. *Am J Clin Nutr* **95**, 1477–1486 (2012).
9. M. R. Ceccarini et al., 5-HT2AR and BDNF gene variants in eating disorders susceptibility. *Am J Med Genet B Neuropsychiatr Genet* **183**, 155–163 (2020).
10. I. Driscoll et al., A candidate gene study of risk for dementia in older, postmenopausal women: results from the Women's Health Initiative Memory Study. *Int J Geriatr Psychiatry* **34**, 692–699 (2019).
11. N. Franzmeier et al., The BDNFVal66Met SNP modulates the association between beta-amyloid and hippocampal disconnection in Alzheimer's disease. *Mol Psychiatry* **26**, 614–628 (2021).
12. J. Liu et al., Brain-derived neurotrophic factor Val66Met polymorphism is associated with mild cognitive impairment in elderly patients with type 2 diabetes: a case-controlled study. *Aging Clin Exp Res* **33**, 1659–1666 (2021).
13. I. S. Losenkov et al., Association between BDNF gene variant Rs6265 and the severity of depression in antidepressant treatment-free depressed patients. *Front Psychiatry* **11**, 38 (2020).
14. V. Vidovic et al., Association of the brain-derived neurotrophic factor Val66Met polymorphism with body mass index, fasting glucose levels and lipid status in adolescents. *Balkan J Med Genet* **23**, 77–82 (2020).
15. K. S. Krabbe et al., Brain-derived neurotrophic factor (BDNF) and type 2 diabetes. *Diabetologia* **50**, 431–438 (2007).
16. C. InterAct, Dietary fibre and incidence of type 2 diabetes in eight European countries: the EPIC-InterAct Study and a meta-analysis of prospective studies. *Diabetologia* **58**, 1394–1408 (2015).
17. M. B. Schulze et al., Fiber and magnesium intake and incidence of type 2 diabetes: a prospective study and meta-analysis. *Arch Intern Med* **167**, 956–965 (2007).
18. M. Carlstrom, S. C. Larsson, Coffee consumption and reduced risk of developing type 2 diabetes: a systematic review with meta-analysis. *Nutr Rev* **76**, 395–417 (2018).
19. G. Livesey et al., Dietary glycemic index and load and the risk of type 2 diabetes: assessment of causal relations. *Nutrients* **11**, 1436 (2019).
20. P. Qin et al., Sugar and artificially sweetened beverages and risk of obesity, type 2 diabetes mellitus, hypertension, and all-cause mortality: a dose-response meta-analysis of prospective cohort studies. *Eur J Epidemiol* **35**, 655–671 (2020).
21. D. Zeraatkar et al., Red and processed meat consumption and risk for all-cause mortality and cardiometabolic outcomes: a systematic review and meta-analysis of cohort studies. *Ann Intern Med* **171**, 703–710 (2019).
22. V. Bellou, L. Belbasis, I. Tzoulaki, E. Evangelou, Risk factors for type 2 diabetes mellitus: An exposure-wide umbrella review of meta-analyses. *PLoS One* **13**, e0194127 (2018).
23. R. Huxley et al., Coffee, decaffeinated coffee, and tea consumption in relation to incident type 2 diabetes mellitus: a systematic review with meta-analysis. *Arch Intern Med* **169**, 2053–2063 (2009).
24. A. M. Chamberlain et al., Ala54Thr polymorphism of the fatty acid binding protein 2 gene and saturated fat intake in relation to lipid levels and insulin resistance: the Coronary Artery Risk Development in Young Adults (CARDIA) study. *Metabolism* **58**, 1222–1228 (2009).
25. T. Reinehr et al., Evidence for an influence of TCF7L2 polymorphism rs7903146 on insulin resistance and sensitivity indices in overweight children and adolescents during a lifestyle intervention. *Int J Obes (Lond)* **32**, 1521–1524 (2008).
26. R. Schuler et al., Dietary fat intake modulates effects of a frequent ACE gene variant on glucose tolerance with association to type 2 diabetes. *Sci Rep* **7**, 9234 (2017).
27. G. Hindy, I. G. Mollet, G. Rukh, U. Ericson, M. Orho-Melander, Several type 2 diabetes-associated variants in genes annotated to WNT signaling interact with dietary fiber in relation to incidence of type 2 diabetes. *Genes Nutr* **11**, 6 (2016).
28. G. Hindy et al., Role of TCF7L2 risk variant and dietary fibre intake on incident type 2 diabetes. *Diabetologia* **55**, 2646–2654 (2012).
29. S. X. Li et al., Interaction between genes and macronutrient intake on the risk of developing type 2 diabetes: systematic review and findings from European Prospective Investigation into Cancer (EPIC)-InterAct. *Am J Clin Nutr* **106**, 263–275 (2017).
30. J. A. Nettleton et al., Interactions of dietary whole-grain intake with fasting glucose- and insulin-related genetic loci in individuals of European descent: a meta-analysis of 14 cohort studies. *Diabetes Care* **33**, 2684–2691 (2010).
31. J. Ye et al., Dietary protein intake and subsequent risk of type 2 diabetes: a dose-response meta-analysis of prospective cohort studies. *Acta Diabetol* **56**, 851–870 (2019).

32. R. J. de Souza et al., Intake of saturated and trans unsaturated fatty acids and risk of all cause mortality, cardiovascular disease, and type 2 diabetes: systematic review and meta-analysis of observational studies. *BMJ* **351**, h3978 (2015).
33. J. D. Cameron et al., The TaqIA RFLP is associated with attenuated intervention-induced body weight loss and increased carbohydrate intake in post-menopausal obese women. *Appetite* **60**, 111–116 (2013).
34. K. Grau et al., TCF7L2 rs7903146-macronutrient interaction in obese individuals' responses to a 10-wk randomized hypoenergetic diet. *Am J Clin Nutr* **91**, 472–479 (2010).
35. A. Haupt et al., Gene variants of TCF7L2 influence weight loss and body composition during lifestyle intervention in a population at risk for type 2 diabetes. *Diabetes* **59**, 747–750 (2010).
36. J. Mattei, Q. Qi, F. B. Hu, F. M. Sacks, L. Qi, TCF7L2 genetic variants modulate the effect of dietary fat intake on changes in body composition during a weight-loss intervention. *Am J Clin Nutr* **96**, 1129–1136 (2012).
37. C. L. Roth, A. Hinney, E. A. Schur, C. T. Elfers, T. Reinehr, Association analyses for dopamine receptor gene polymorphisms and weight status in a longitudinal analysis in obese children before and after lifestyle intervention. *BMC Pediatr* **13**, 197 (2013).
38. L. Xiang et al., FTO genotype and weight loss in diet and lifestyle interventions: a systematic review and meta-analysis. *Am J Clin Nutr* **103**, 1162–1170 (2016).
39. F. D. C. De Stefani, P. S. Pietraroia, M. M. Fernandes-Silva, J. Faria-Neto, C. P. Baena, Observational evidence for unintentional weight loss in all-cause mortality and major cardiovascular events: a systematic review and meta-analysis. *Sci Rep* **8**, 15447 (2018).
40. Y. Kwon, H. J. Kim, S. Park, Y. G. Park, K. H. Cho, Body mass index-related mortality in patients with type 2 diabetes and heterogeneity in obesity paradox studies: a dose-response meta-analysis. *PLoS One* **12**, e0168247 (2017).
41. R. Mahajan et al., Complex interaction of obesity, intentional weight loss and heart failure: a systematic review and meta-analysis. *Heart* **106**, 58–68 (2020).
42. U. Ekelund et al., Dose-response associations between accelerometry measured physical activity and sedentary time and all cause mortality: systematic review and harmonised meta-analysis. *BMJ* **366**, l4570 (2019).
43. M. Renninger et al., Associations between accelerometry measured physical activity and sedentary time and the metabolic syndrome: a meta-analysis of more than 6000 children and adolescents. *Pediatr Obes* **15**, e12578 (2020).
44. R. Hjort et al., Physical activity, genetic susceptibility, and the risk of latent autoimmune diabetes in adults and type 2 diabetes. *J Clin Endocrinol Metab* **105**, e4112–e4123 (2020).
45. C. Y. Lee et al., Modest overweight and healthy dietary habits reduce risk of dementia: a nationwide survey in Taiwan. *J Prev Alzheimers Dis* **4**, 37–43 (2017).
46. T. Astell-Burt, M. A. Navakatikyan, X. Feng, Behavioural change, weight loss and risk of dementia: a longitudinal study. *Prev Med* **145**, 106386 (2021).
47. R. Bleiweiss-Sande et al., Associations between food group intake, cognition, and academic achievement in elementary schoolchildren. *Nutrients* **11**, 2722 (2019).
48. H. Miao, K. Chen, X. Yan, F. Chen, Sugar in beverage and the risk of incident dementia, Alzheimer's disease and stroke: a prospective cohort study. *J Prev Alzheimers Dis* **8**, 188–193 (2021).
49. M. K. Taylor et al., A high-glycemic diet is associated with cerebral amyloid burden in cognitively normal older adults. *Am J Clin Nutr* **106**, 1463–1470 (2017).
50. J. Corley et al., Dietary patterns, cognitive function, and structural neuroimaging measures of brain aging. *Exp Gerontol* **142**, 111117 (2020).
51. L. Perrone, W. B. Grant, Observational and ecological studies of dietary advanced glycation end products in national diets and Alzheimer's disease incidence and prevalence. *J Alzheimers Dis* **45**, 965–979 (2015).
52. L. D. Morrison, S. J. Kish, Brain polyamine levels are altered in Alzheimer's disease. *Neurosci Lett* **197**, 5–8 (1995).
53. R. Seidl et al., Polyamines in frontal cortex of patients with Down syndrome and Alzheimer disease. *Neurosci Lett* **206**, 193–195 (1996).
54. Y. Lao et al., Association between alcohol intake, mild cognitive impairment and progression to dementia: a dose-response meta-analysis. *Aging Clin Exp Res* **33**, 1175–1185 (2021).
55. L. S. Ran et al., Alcohol, coffee and tea intake and the risk of cognitive deficits: a dose-response

meta-analysis. *Epidemiol Psychiatr Sci* **30**, e13 (2021).
56. S. Antwerpes et al., Coffee intake and neurocognitive performance in HIV/HCV coinfected patients (ANRS CO13 HEPAVIH). *Nutrients* **12**, 2532 (2020).
57. J. M. Bae, History of coffee consumption and risk of Alzheimer's disease: a meta-epidemiological study of population-based cohort studies. *Dement Neurocogn Disord* **19**, 108–113 (2020).
58. I. Paz-Graniel et al., Association between coffee consumption and total dietary caffeine intake with cognitive functioning: cross-sectional assessment in an elderly Mediterranean population. *Eur J Nutr* **60**, 2381–2396 (2021).
59. Z. Huang et al., Associations of lifestyle factors with cognition in community-dwelling adults aged 50 and older: a longitudinal cohort study. *Front Aging Neurosci* **12**, 601487 (2020).
60. E. J. Okello et al., Tea consumption and measures of attention and psychomotor speed in the very old: the Newcastle 85+ longitudinal study. *BMC Nutr* **6**, 57 (2020).
61. Y. Sun et al., Association of major dietary protein sources with all-cause and cause-specific mortality: prospective cohort study. *J Am Heart Assoc* **10**, e015553 (2021).
62. L. Wu, D. Sun, Meta-analysis of milk consumption and the risk of cognitive disorders. *Nutrients* **8**, 824 (2016).
63. R. Schuler et al., VEGF and GLUT1 are highly heritable, inversely correlated and affected by dietary fat intake: consequences for cognitive function in humans. *Mol Metab* **11**, 129–136 (2018).
64. H. J. Coelho-Junior, A. Trichopoulou, F. Panza, Cross-sectional and longitudinal associations between adherence to Mediterranean diet with physical performance and cognitive function in older adults: a systematic review and meta-analysis. *Ageing Res Rev* **70**, 101395 (2021).
65. N. Feter et al., Physical activity in later life and risk of dementia: findings from a population-based cohort study. *Exp Gerontol* **143**, 111145 (2021).
66. J. Lee, The relationship between physical activity and dementia: a systematic review and meta-analysis of prospective cohort studies. *J Gerontol Nurs* **44**, 22–29 (2018).
67. W. Xu et al., Leisure time physical activity and dementia risk: a dose-response meta-analysis of prospective studies. *BMJ Open* **7**, e014706 (2017).
68. L. Daimiel et al., Physical fitness and physical activity association with cognitive function and quality of life: baseline cross-sectional analysis of the PREDIMED-Plus trial. *Sci Rep* **10**, 3472 (2020).
69. D. E. Hosking, T. Nettelbeck, C. Wilson, V. Danthiir, Retrospective lifetime dietary patterns predict cognitive performance in community-dwelling older Australians. *Br J Nutr* **112**, 228–237 (2014).
70. M. M. Lane et al., Ultraprocessed food and chronic noncommunicable diseases: a systematic review and meta-analysis of 43 observational studies. *Obes Rev* **22**, e13146 (2021).
71. S. Fatahi et al., Association of dietary fiber and depression symptom: a systematic review and meta-analysis of observational studies. *Complement Ther Med* **56**, 102621 (2021).
72. J. E. Gangwisch et al., High glycemic index diet as a risk factor for depression: analyses from the Women's Health Initiative. *Am J Clin Nutr* **102**, 454–463 (2015).
73. B. Gopinath, V. M. Flood, G. Burlutksy, J. C. Louie, P. Mitchell, Association between carbohydrate nutrition and prevalence of depressive symptoms in older adults. *Br J Nutr* **116**, 2109–2114 (2016).
74. G. Grases, M. A. Colom, P. Sanchis, F. Grases, Possible relation between consumption of different food groups and depression. *BMC Psychol* **7**, 14 (2019).
75. P. Huang et al., Maternal dietary patterns and depressive symptoms during pregnancy: the born in Guangzhou Cohort Study. *Clin Nutr* **40**, 3485–3494 (2021).
76. C. S. Kim, S. Byeon, D. M. Shin, Sources of dietary fiber are differently associated with prevalence of depression. *Nutrients* **12**, 2813 (2020).
77. L. Schwingshackl et al., Food groups and risk of type 2 diabetes mellitus: a systematic review and meta-analysis of prospective studies. *Eur J Epidemiol* **32**, 363–375 (2017).
78. M. A. Perez-Ara et al., Associations of non-alcoholic beverages with major depressive disorder history and depressive symptoms clusters in a sample of overweight adults. *Nutrients* **12**, 3202 (2020).
79. A. R. Wolfe et al., Dietary protein and protein-rich food in relation to severely depressed mood: A 10 year follow-up of a national cohort. *Prog Neuropsychopharmacol Biol Psychiatry* **35**, 232–238 (2011).
80. Y. Zhan et al., Dietary patterns in relation to gestational depression and sleep disturbance in Chinese pregnant women. *J Obstet Gynaecol Res*, 2618–2628 (2020).

81. F. Huang et al., [Association between dietary patterns and depressive symptom based on reduced rank regression in people aged 55 and above in 4 provinces of China]. *Wei Sheng Yan Jiu* **50**, 29–36 (2021).
82. D. Nucci, C. Fatigoni, A. Amerio, A. Odone, V. Gianfredi, Red and processed meat consumption and risk of depression: a systematic review and meta-analysis. *Int J Environ Res Public Health* **17**, 6686 (2020).
83. R. Chan, D. Chan, J. Woo, A prospective cohort study to examine the association between dietary patterns and depressive symptoms in older Chinese people in Hong Kong. *PLoS One* **9**, e105760 (2014).
84. A. Sanchez-Villegas et al., Seafood consumption, omega-3 fatty acids intake, and life-time prevalence of depression in the PREDIMED-plus trial. *Nutrients* **10**, 2000 (2018).
85. S. Cai et al., There is no association between the omega-3 index and depressive symptoms in patients with heart disease who are low fish consumers. *Heart Lung Circ* **26**, 276–284 (2017).
86. K. Hamazaki et al., Plasma phospholipid n-3 polyunsaturated fatty acids and major depressive disorder in Japanese elderly: the Japan Public Health Center-based Prospective Study. *Sci Rep* **11**, 4003 (2021).
87. K. H. O. Deane et al., Omega-3 and polyunsaturated fat for prevention of depression and anxiety symptoms: systematic review and meta-analysis of randomised trials. *Br J Psychiatry* **218**, 135–142 (2021).
88. T. P. Ng, Q. Gao, X. Gwee, D. Q. L. Chua, Tea consumption and depression from follow up in the Singapore longitudinal ageing study. *J Nutr Health Aging* **25**, 295–301 (2021).
89. P. N. A. Dharmayani, M. Juergens, M. Allman-Farinelli, S. Mihrshahi, Association between fruit and vegetable consumption and depression symptoms in young people and adults aged 15–45: a systematic review of cohort studies. *Int J Environ Res Public Health* **18**, 780 (2021).
90. L. Gougeon et al., A prospective evaluation of the depression-nutrient intake reverse causality hypothesis in a cohort of community-dwelling older Canadians. *Br J Nutr* **117**, 1032–1041 (2017).
91. O. G. Swann et al., Dietary fibre intake and its associations with depressive symptoms in a prospective adolescent cohort. *Br J Nutr* **125**, 1166–1176 (2021).
92. M. K. Kwok, G. M. Leung, C. M. Schooling, Habitual coffee consumption and risk of type 2 diabetes, ischemic heart disease, depression and Alzheimer's disease: a Mendelian randomization study. *Sci Rep* **6**, 36500 (2016).
93. S. Frigerio, R. Strawbridge, A. H. Young, The impact of caffeine consumption on clinical symptoms in patients with bipolar disorder: a systematic review. *Bipolar Disord* **23**, 241–251 (2021).
94. C. Lassale et al., Healthy dietary indices and risk of depressive outcomes: a systematic review and meta-analysis of observational studies. *Mol Psychiatry* **24**, 965–986 (2019).
95. A. P. Matison, K. A. Mather, V. M. Flood, S. Reppermund, Associations between nutrition and the incidence of depression in middle-aged and older adults: a systematic review and meta-analysis of prospective observational population-based studies. *Ageing Res Rev* **70**, 101403 (2021).
96. M. Nicolaou et al., Association of a priori dietary patterns with depressive symptoms: a harmonised meta-analysis of observational studies. *Psychol Med* **50**, 1872–1883 (2020).
97. K. Konishi, Associations between healthy Japanese dietary patterns and depression in Japanese women. *Public Health Nutr* **24**, 1753–1765 (2021).
98. Y. Jin, N. R. Kandula, A. M. Kanaya, S. A. Talegawkar, Vegetarian diet is inversely associated with prevalence of depression in middle-older aged South Asians in the United States. *Ethn Health* **26**, 504–511 (2021).
99. S. Ocklenburg, J. Borawski, Vegetarian diet and depression scores: a meta-analysis. *J Affect Disord* **294**, 813–815 (2021).
100. I. Iguacel, I. Huybrechts, L. A. Moreno, N. Michels, Vegetarianism and veganism compared with mental health and cognitive outcomes: a systematic review and meta-analysis. *Nutr Rev* **79**, 361–381 (2021).
101. L. B. Martins et al., Nutrition-based interventions for mood disorders. *Expert Rev Neurother* **21**, 303–315 (2021).
102. K. W. Choi et al., Assessment of bidirectional relationships between physical activity and depression among adults: a 2-sample Mendelian randomization study. *JAMA Psychiatry* **76**, 399–408 (2019).
103. R. K. Dishman, C. P. McDowell, M. P. Herring, Customary physical activity and odds of depression: a systematic review and meta-analysis of

111 prospective cohort studies. *Br J Sports Med* **55**, 926–934 (2021).
104. V. Gianfredi et al., Depression and objectively measured physical activity: a systematic review and meta-analysis. *Int J Environ Res Public Health* **17**, 3738 (2020).
105. F. B. Schuch et al., Physical activity and incident depression: a meta-analysis of prospective cohort studies. *Am J Psychiatry* **175**, 631–648 (2018).
106. W. H. Li et al., Brief intervention to promote smoking cessation and improve glycemic control in smokers with type 2 diabetes: a randomized controlled trial. *Sci Rep* **7**, 45902 (2017).
107. J. H. Stein et al., Smoking cessation and the risk of diabetes mellitus and impaired fasting glucose: three-year outcomes after a quit attempt. *PLoS One* **9**, e98278 (2014).
108. W. C. Knowler et al., Reduction in the incidence of type 2 diabetes with lifestyle intervention or metformin. *N Engl J Med* **346**, 393–403 (2002).
109. A. Ramachandran et al., The Indian Diabetes Prevention Programme shows that lifestyle modification and metformin prevent type 2 diabetes in Asian Indian subjects with impaired glucose tolerance (IDPP-1). *Diabetologia* **49**, 289–297 (2006).
110. J. Tuomilehto et al., Prevention of type 2 diabetes mellitus by changes in lifestyle among subjects with impaired glucose tolerance. *N Engl J Med* **344**, 1343–1350 (2001).
111. Q. Gong et al., Morbidity and mortality after lifestyle intervention for people with impaired glucose tolerance: 30-year results of the Da Qing Diabetes Prevention Outcome Study. *Lancet Diabetes Endocrinol* **7**, 452–461 (2019).
112. M. E. Lean et al., Primary care-led weight management for remission of type 2 diabetes (DiRECT): an open-label, cluster-randomised trial. *Lancet* **391**, 541–551 (2018).
113. J. Z. Goldenberg et al., Efficacy and safety of low and very low carbohydrate diets for type 2 diabetes remission: systematic review and meta-analysis of published and unpublished randomized trial data. *BMJ* **372**, m4743 (2021).
114. M. Neuenschwander, G. Hoffmann, L. Schwingshackl, S. Schlesinger, Impact of different dietary approaches on blood lipid control in patients with type 2 diabetes mellitus: a systematic review and network meta-analysis. *Eur J Epidemiol* **34**, 837–852 (2019).
115. L. Schwingshackl, A. Chaimani, G. Hoffmann, C. Schwedhelm, H. Boeing, A network meta-analysis on the comparative efficacy of different dietary approaches on glycaemic control in patients with type 2 diabetes mellitus. *Eur J Epidemiol* **33**, 157–170 (2018).
116. L. Schwingshackl et al., Comparative effects of different dietary approaches on blood pressure in hypertensive and pre-hypertensive patients: a systematic review and network meta-analysis. *Crit Rev Food Sci Nutr* **59**, 2674–2687 (2019).
117. S. B. Seidelmann et al., Dietary carbohydrate intake and mortality: a prospective cohort study and meta-analysis. *Lancet Public Health* **3**, e419–e428 (2018).
118. T. J. Brown et al., Omega-3, omega-6, and total dietary polyunsaturated fat for prevention and treatment of type 2 diabetes mellitus: systematic review and meta-analysis of randomised controlled trials. *BMJ* **366**, l4697 (2019).
119. T. Dong, M. Guo, P. Zhang, G. Sun, B. Chen, The effects of low-carbohydrate diets on cardiovascular risk factors: a meta-analysis. *PLoS One* **15**, e0225348 (2020).
120. G. A. Silverii et al., Low-carbohydrate diets and type 2 diabetes treatment: a meta-analysis of randomized controlled trials. *Acta Diabetol* **57**, 1375–1382 (2020).
121. F. L. Santos, S. S. Esteves, A. da Costa Pereira, W. S. Yancy, Jr., J. P. Nunes, Systematic review and meta-analysis of clinical trials of the effects of low carbohydrate diets on cardiovascular risk factors. *Obes Rev* **13**, 1048–1066 (2012).
122. N. Steckhan et al., Effects of different dietary approaches on inflammatory markers in patients with metabolic syndrome: a systematic review and meta-analysis. *Nutrition* **32**, 338–348 (2016).
123. P. S. T. Suyoto, Effect of low-carbohydrate diet on markers of renal function in patients with type 2 diabetes: a meta-analysis. *Diabetes Metab Res Rev* **34**, e3032 (2018).
124. D. D. Campbell, K. A. Meckling, Effect of the protein:carbohydrate ratio in hypoenergetic diets on metabolic syndrome risk factors in exercising overweight and obese women. *Br J Nutr* **108**, 1658–1671 (2012).
125. D. A. Lowe et al., Effects of time-restricted eating on weight loss and other metabolic parameters in women and men with overweight and obesity: the TREAT randomized clinical trial. *JAMA Intern Med* **180**, 1491–1499 (2020).
126. A. Reynolds et al., Carbohydrate quality and human health: a series of systematic reviews and meta-analyses. *Lancet* **393**, 434–445 (2019).

127. A. N. Reynolds, A. P. Akerman, J. Mann, Dietary fibre and whole grains in diabetes management: systematic review and meta-analyses. *PLoS Med* **17**, e1003053 (2020).
128. C. Honsek et al., Fibre supplementation for the prevention of type 2 diabetes and improvement of glucose metabolism: the randomised controlled Optimal Fibre Trial (OptiFiT). *Diabetologia* **61**, 1295–1305 (2018).
129. S. Kabisch et al., Fasting glucose state determines metabolic response to supplementation with insoluble cereal fibre: a secondary analysis of the optimal fibre trial (OptiFiT). *Nutrients* **11**, 2385 (2019).
130. S. Kabisch et al., Obesity does not modulate the glycometabolic benefit of insoluble cereal fibre in subjects with prediabetes-a stratified post hoc analysis of the optimal fibre trial (OptiFiT). *Nutrients* **11**, 2726 (2019).
131. M. O. Weickert et al., Effects of supplemented isoenergetic diets differing in cereal fiber and protein content on insulin sensitivity in overweight humans. *Am J Clin Nutr* **94**, 459–471 (2011).
132. G. Avellone et al., Effects of moderate Sicilian red wine consumption on inflammatory biomarkers of atherosclerosis. *Eur J Clin Nutr* **60**, 41–47 (2006).
133. G. Chiva-Blanch et al., Effects of alcohol and polyphenols from beer on atherosclerotic biomarkers in high cardiovascular risk men: a randomized feeding trial. *Nutr Metab Cardiovasc Dis* **25**, 36–45 (2015).
134. M. Naissides, J. C. Mamo, A. P. James, S. Pal, The effect of acute red wine polyphenol consumption on postprandial lipaemia in postmenopausal women. *Atherosclerosis* **177**, 401–408 (2004).
135. G. Chiva-Blanch et al., Effects of red wine polyphenols and alcohol on glucose metabolism and the lipid profile: a randomized clinical trial. *Clin Nutr* **32**, 200–206 (2013).
136. I. P. Botden et al., Red wine polyphenols do not lower peripheral or central blood pressure in high normal blood pressure and hypertension. *Am J Hypertens* **25**, 718–723 (2012).
137. M. Naissides, J. C. Mamo, A. P. James, S. Pal, The effect of chronic consumption of red wine on cardiovascular disease risk factors in postmenopausal women. *Atherosclerosis* **185**, 438–445 (2006).
138. J. Woerdeman et al., Red wine polyphenols do not improve obesity-associated insulin resistance: a randomized controlled trial. *Diabetes Obes Metab* **20**, 206–210 (2018).
139. S. Rafiee et al., Efficacy of resveratrol supplementation in patients with nonalcoholic fatty liver disease: a systematic review and meta-analysis of clinical trials. *Complement Ther Clin Pract* **42**, 101281 (2021).
140. S. Asgary, R. Karimi, S. Momtaz, R. Naseri, M. H. Farzaei, Effect of resveratrol on metabolic syndrome components: a systematic review and meta-analysis. *Rev Endocr Metab Disord* **20**, 173–186 (2019).
141. X. F. Guo, J. M. Li, J. Tang, D. Li, Effects of resveratrol supplementation on risk factors of non-communicable diseases: a meta-analysis of randomized controlled trials. *Crit Rev Food Sci Nutr* **58**, 3016–3029 (2018).
142. P. A. Kern et al., Effects of KDT501 on metabolic parameters in insulin-resistant prediabetic humans. *J Endocr Soc* **1**, 650–659 (2017).
143. K. Obara, M. Mizutani, Y. Hitomi, H. Yajima, K. Kondo, Isohumulones, the bitter component of beer, improve hyperglycemia and decrease body fat in Japanese subjects with prediabetes. *Clin Nutr* **28**, 278–284 (2009).
144. N. Bergeron, S. Chiu, P. T. Williams, M. K. S, R. M. Krauss, Effects of red meat, white meat, and nonmeat protein sources on atherogenic lipoprotein measures in the context of low compared with high saturated fat intake: a randomized controlled trial. *Am J Clin Nutr* **110**, 24–33 (2019).
145. K. Charlton et al., Pork and chicken meals similarly impact on cognitive function and strength in community-living older adults: a pilot study. *J Nutr Gerontol Geriatr* **35**, 124–145 (2016).
146. M. H. Davidson, D. Hunninghake, K. C. Maki, P. O. Kwiterovich, Jr., S. Kafonek, Comparison of the effects of lean red meat vs lean white meat on serum lipid levels among free-living persons with hypercholesterolemia: a long-term, randomized clinical trial. *Arch Intern Med* **159**, 1331–1338 (1999).
147. D. B. Hunninghake et al., Incorporation of lean red meat into a National Cholesterol Education Program Step I diet: a long-term, randomized clinical trial in free-living persons with hypercholesterolemia. *J Am Coll Nutr* **19**, 351–360 (2000).
148. K. J. Murphy et al., A comparison of regular consumption of fresh lean pork, beef and chicken on body composition: a randomized cross-over trial. *Nutrients* **6**, 682–696 (2014).

149. K. J. Murphy, R. L. Thomson, A. M. Coates, J. D. Buckley, P. R. Howe, Effects of eating fresh lean pork on cardiometabolic health parameters. *Nutrients* **4**, 711–723 (2012).
149a. Marson EC, Delevatti RS, Prado AK, Netto N, Kruel LF. Effects of aerobic, resistance, and combined exercise training on insulin resistance markers in overweight or obese children and adolescents, A systematic review and meta-analysis. Prev Med. 2016 Dec 93,211-218.
150. R. E. Ashton et al., Effects of short-term, medium-term and long-term resistance exercise training on cardiometabolic health outcomes in adults: systematic review with meta-analysis. *Br J Sports Med* **54**, 341–348 (2020).
151. C. Jelleyman et al., The effects of high-intensity interval training on glucose regulation and insulin resistance: a meta-analysis. *Obes Rev* **16**, 942–961 (2015).
152. H. Cramer et al., Effects of yoga on cardiovascular disease risk factors: a systematic review and meta-analysis. *Int J Cardiol* **173**, 170–183 (2014).
153. Y. J. Park, H. Shin, S. Jeon, I. Cho, H. J. Park, Development and effects of college-based lifestyle modification program for menstrual health of young adult women with irregular menses: a randomized controlled trial. *Int J Environ Res Public Health* **18**, 233 (2020).
154. T. C. Kwok, L. C. Lam, M. M. Sea, W. Goggins, J. Woo, A randomized controlled trial of dietetic interventions to prevent cognitive decline in old age hostel residents. *Eur J Clin Nutr* **66**, 1135–1140 (2012).
155. M. Glud, T. Christiansen, L. H. Larsen, B. Richelsen, J. M. Bruun, Changes in circulating BDNF in relation to sex, diet, and exercise: a 12-week randomized controlled study in overweight and obese participants. *J Obes* **2019**, 4537274 (2019).
156. P. Sedlacek et al., Effect of n-3 fatty acids supplementation during life style modification in women with overweight. *Cent Eur J Public Health* **26**, 265–271 (2018).
157. L. Zlibinaite, R. Solianik, D. Vizbaraite, D. Mickeviciene, A. Skurvydas, The effect of combined aerobic exercise and calorie restriction on mood, cognition, and motor behavior in overweight and obese women. *J Phys Act Health* **17**, 204–210 (2020).
158. D. A. de Luis, E. Romero, O. Izaola, D. Primo, R. Aller, Cardiovascular risk factors and insulin resistance after two hypocaloric diets with different fat distribution in obese subjects: effect of the rs10767664 gene variant in brain-derived neurotrophic factor. *J Nutrigenet Nutrigenomics* **10**, 163–171 (2017).
159. Y. Matsuoka et al., Serum pro-BDNF/BDNF as a treatment biomarker for response to docosahexaenoic acid in traumatized people vulnerable to developing psychological distress: a randomized controlled trial. *Transl Psychiatry* **5**, e596 (2015).
160. C. S. Kessler et al., A nonrandomized controlled clinical pilot trial on 8 wk of intermittent fasting (24 h/wk). *Nutrition* **46**, 143–152 e142 (2018).
161. O. Carlson et al., Impact of reduced meal frequency without caloric restriction on glucose regulation in healthy, normal-weight middle-aged men and women. *Metabolism* **56**, 1729–1734 (2007).
162. A. Cherif et al., Repeated sprints in fasted state impair reaction time performance. *J Am Coll Nutr* **36**, 210–217 (2017).
163. M. P. Mattson, Energy intake, meal frequency, and health: a neurobiological perspective. *Annu Rev Nutr* **25**, 237–260 (2005).
164. M. Uemura et al., Obesity and mental health improvement following nutritional education focusing on gut microbiota composition in Japanese women: a randomised controlled trial. *Eur J Nutr* **58**, 3291–3302 (2019).
165. N. C. Horie et al., Cognitive effects of intentional weight loss in elderly obese individuals with mild cognitive impairment. *J Clin Endocrinol Metab* **101**, 1104–1112 (2016).
166. R. Krikorian et al., Dietary ketosis enhances memory in mild cognitive impairment. *Neurobiol Aging* **33**, 425 e419–427 (2012).
167. K. Prehn et al., Caloric restriction in older adults-differential effects of weight loss and reduced weight on brain structure and function. *Cereb Cortex* **27**, 1765–1778 (2017).
168. A. V. Witte, M. Fobker, R. Gellner, S. Knecht, A. Floel, Caloric restriction improves memory in elderly humans. *Proc Natl Acad Sci U S A* **106**, 1255–1260 (2009).
169. O. T. Carmichael et al., Long-term change in physiological markers and cognitive performance in type 2 diabetes: the look AHEAD study. *J Clin Endocrinol Metab* **105**, e4778– e4791 (2020).
170. M. A. Espeland et al., Long-term impact of weight loss intervention on changes in cognitive function: exploratory analyses from the action for health in diabetes randomized

controlled clinical trial. *J Gerontol A Biol Sci Med Sci* **73**, 484–491 (2018).
171. E. Ahlqvist et al., Novel subgroups of adult-onset diabetes and their association with outcomes: a data-driven cluster analysis of six variables. *Lancet Diabetes Endocrinol* **6**, 361–369 (2018).
172. E. Brietzke et al., Ketogenic diet as a metabolic therapy for mood disorders: Evidence and developments. *Neurosci Biobehav Rev* **94**, 11–16 (2018).
173. M. K. Taylor, D. K. Sullivan, J. D. Mahnken, J. M. Burns, R. H. Swerdlow, Feasibility and efficacy data from a ketogenic diet intervention in Alzheimer's disease. *Alzheimers Dement (N Y)* **4**, 28–36 (2018).
174. S. Pavon et al., Ketogenic diet and cognition in neurological diseases: a systematic review. *Nutr Rev* **79**, 802–813 (2021).
175. M. A. Reger et al., Effects of beta-hydroxybutyrate on cognition in memory-impaired adults. *Neurobiol Aging* **25**, 311–314 (2004).
176. S. L. Gardener et al., Increased carbohydrate intake is associated with poorer performance in verbal memory and attention in an APOE genotype-dependent manner. *J Alzheimers Dis* **58**, 193–201 (2017).
177. S. J. Morrill, K. J. Gibas, Ketogenic diet rescues cognition in ApoE4+ patient with mild Alzheimer's disease: a case study. *Diabetes Metab Syndr* **13**, 1187–1191 (2019).
178. R. Nagpal, B. J. Neth, S. Wang, S. Craft, H. Yadav, Modified Mediterranean-ketogenic diet modulates gut microbiome and short-chain fatty acids in association with Alzheimer's disease markers in subjects with mild cognitive impairment. *EBioMedicine* **47**, 529–542 (2019).
179. S. T. Henderson et al., Study of the ketogenic agent AC-1202 in mild to moderate Alzheimer's disease: a randomized, double-blind, placebo-controlled, multicenter trial. *Nutr Metab (Lond)* **6**, 31 (2009).
180. M. J. Wenger et al., Consumption of a double-fortified salt affects perceptual, attentional, and mnemonic functioning in women in a randomized controlled trial in India. *J Nutr* **147**, 2297–2308 (2017).
181. X. Sheng, J. Wang, F. Li, F. Ouyang, J. Ma, Effects of dietary intervention on vitamin B12 status and cognitive level of 18-month-old toddlers in high-poverty areas: a cluster-randomized controlled trial. *BMC Pediatr* **19**, 334 (2019).
182. X. Li, Y. Liu, Y. Zheng, P. Wang, Y. Zhang, The effect of vitamin D supplementation on glycemic control in type 2 diabetes patients: a systematic review and meta-analysis. *Nutrients* **10**, 375 (2018).
183. N. Mirhosseini, H. Vatanparast, M. Mazidi, S. M. Kimball, The effect of improved serum 25-hydroxyvitamin D status on glycemic control in diabetic patients: a meta-analysis. *J Clin Endocrinol Metab* **102**, 3097–3110 (2017).
184. M. Berk et al., A randomised controlled trial of a mitochondrial therapeutic target for bipolar depression: mitochondrial agents, N-acetylcysteine, and placebo. *BMC Med* **17**, 18 (2019).
185. S. Mayengbam et al., Metabolic consequences of discretionary fortified beverage consumption containing excessive vitamin B levels in adolescents. *PLoS One* **14**, e0209913 (2019).
186. K. Handeland et al., Fatty fish intake and attention performance in 14–15 year old adolescents: FINS-TEENS - a randomized controlled trial. *Nutr J* **16**, 64 (2017).
187. S. S. Al-Ghannami et al., Randomized open-label trial of docosahexaenoic acid-enriched fish oil and fish meal on cognitive and behavioral functioning in Omani children. *Nutrition* **57**, 167–172 (2019).
188. H. Demmelmair et al., The effect of Atlantic salmon consumption on the cognitive performance of preschool children: a randomized controlled trial. *Clin Nutr* **38**, 2558–2568 (2019).
189. I. Kvestad et al., Fatty fish, hair mercury and cognitive function in Norwegian preschool children: Results from the randomized controlled trial FINS-KIDS. *Environ Int* **121**, 1098–1105 (2018).
190. A. V. Witte et al., Long-chain omega-3 fatty acids improve brain function and structure in older adults. *Cereb Cortex* **24**, 3059–3068 (2014).
191. M. Bot et al., Supplementation with eicosapentaenoic omega-3 fatty acid does not influence serum brain-derived neurotrophic factor in diabetes mellitus patients with major depression: a randomized controlled pilot study. *Neuropsychobiology* **63**, 219–223 (2011).
192. M. A. Phillips, C. E. Childs, P. C. Calder, P. J. Rogers, No effect of omega-3 fatty acid supplementation on cognition and mood in individuals with cognitive impairment and probable Alzheimer's disease: a randomised controlled trial. *Int J Mol Sci* **16**, 24600–24613 (2015).

193. E. Leedo, A. M. Beck, A. Astrup, A. D. Lassen, The effectiveness of healthy meals at work on reaction time, mood and dietary intake: a randomised cross-over study in daytime and shift workers at an university hospital. *Br J Nutr* **118**, 121–129 (2017).
194. E. Mazza et al., Effect of the replacement of dietary vegetable oils with a low dose of extravirgin olive oil in the Mediterranean diet on cognitive functions in the elderly. *J Transl Med* **16**, 10 (2018).
195. F. Koh, K. Charlton, K. Walton, A. T. McMahon, Role of dietary protein and thiamine intakes on cognitive function in healthy older people: a systematic review. *Nutrients* **7**, 2415–2439 (2015).
196. M. B. Formica et al., Effect of lean red meat combined with a multicomponent exercise program on muscle and cognitive function in older adults: a 6-month randomized controlled trial. *Am J Clin Nutr* **112**, 113–128 (2020).
197. A. T. Wade et al., A Mediterranean diet with fresh, lean pork improves processing speed and mood: cognitive findings from the MedPork randomised controlled trial. *Nutrients* **11**, (2019).
198. P. Suttiwan, P. Yuktanandana, S. Ngamake, Effectiveness of essence of chicken on cognitive function improvement: a randomized controlled clinical trial. *Nutrients* **10**, (2018).
199. L. Chan et al., Effectiveness of essence of chicken in improving cognitive function in young people under work-related stress: a randomized double-blind trial. *Medicine (Baltimore)* **95**, e3640 (2016).
200. T. Suzuki et al., The effects of mold-fermented cheese on brain-derived neurotrophic factor in community-dwelling older japanese women with mild cognitive impairment: a randomized, controlled, crossover trial. *J Am Med Dir Assoc* **20**, 1509–1514 e1502 (2019).
201. J. Lee, Z. Fu, M. Chung, D. J. Jang, H. J. Lee, Role of milk and dairy intake in cognitive function in older adults: a systematic review and meta-analysis. *Nutr J* **17**, 82 (2018).
202. O. M. Marchand, F. E. Kendall, C. M. Rapsey, J. J. Haszard, B. J. Venn, The effect of postprandial glycaemia on cognitive function: a randomised crossover trial. *Br J Nutr* **123**, 1357–1364 (2020).
203. N. Sanchez-Aguadero et al., Postprandial effects of breakfast glycaemic index on cognitive performance among young, healthy adults: a crossover clinical trial. *Nutr Neurosci* **23**, 1–7 (2020).
204. H. Tsukamoto et al., Plasma brain-derived neurotrophic factor and dynamic cerebral autoregulation in acute response to glycemic control following breakfast in young men. *Am J Physiol Regul Integr Comp Physiol* **320**, R69–R79 (2021).
205. Y. C. Chung et al., Improved cognitive performance following supplementation with a mixed-grain diet in high school students: a randomized controlled trial. *Nutrition* **28**, 165–172 (2012).
206. K. Berding et al., A specific dietary fibre supplementation improves cognitive performance-an exploratory randomised, placebo-controlled, crossover study. *Psychopharmacology (Berl)* **238**, 149–163 (2021).
207. L. P. Capitao et al., Prebiotic supplementation does not affect reading and cognitive performance in children: a randomised placebo-controlled study. *J Psychopharmacol* **34**, 148–152 (2020).
208. K. Sansone, M. Kern, M. Y. Hong, C. Liu, S. Hooshmand, Acute effects of dried apple consumption on metabolic and cognitive responses in healthy individuals. *J Med Food* **21**, 1158–1164 (2018).
209. P. A. Jackson et al., A randomized, crossover study of the acute cognitive and cerebral blood flow effects of phenolic, nitrate and botanical beverages in young, healthy humans. *Nutrients* **12**, 2254 (2020).
210. D. Lomiwes et al., Timed consumption of a New Zealand blackcurrant juice support positive affective responses during a self-motivated moderate walking exercise in healthy sedentary adults. *J Int Soc Sports Nutr* **16**, 33 (2019).
211. A. Nilsson, I. Salo, M. Plaza, I. Bjorck, Effects of a mixed berry beverage on cognitive functions and cardiometabolic risk markers; a randomized cross-over study in healthy older adults. *PLoS One* **12**, e0188173 (2017).
212. A. R. Whyte, N. Cheng, L. T. Butler, D. J. Lamport, C. M. Williams, Flavonoid-rich mixed berries maintain and improve cognitive function over a 6h period in young healthy adults. *Nutrients* **11**, 2685 (2019).
213. K. L. Barfoot et al., The effects of acute wild blueberry supplementation on the cognition of 7–10-year-old schoolchildren. *Eur J Nutr* **58**, 2911–2920 (2019).
214. A. R. Whyte, G. Schafer, C. M. Williams, The effect of cognitive demand on performance of an executive function task following wild blueberry supplementation in 7 to 10 years old children. *Food Funct* **8**, 4129–4138 (2017).
215. E. O. Igwe et al., Anthocyanin-rich plum juice reduces ambulatory blood pressure but not acute cognitive function in younger and older

adults: a pilot crossover dose-timing study. *Nutr Res* **47**, 28–43 (2017).
216. I. Traupe et al., Postoperative cognitive dysfunction and short-term neuroprotection from blueberries: a pilot study. *Minerva Anestesiol* **84**, 1352–1360 (2018).
217. A. R. Whyte, N. Cheng, E. Fromentin, C. M. Williams, A Randomized, Double-blinded, placebo-controlled study to compare the safety and efficacy of low dose enhanced wild blueberry powder and wild blueberry extract (thinkblue) in maintenance of episodic and working memory in older adults. *Nutrients* **10**, 660 (2018).
218. S. C. Chai, J. Jerusik, K. Davis, R. S. Wright, Z. Zhang, Effect of Montmorency tart cherry juice on cognitive performance in older adults: a randomized controlled trial. *Food Funct* **10**, 4423–4431 (2019).
219. P. Siddarth et al., Randomized placebo-controlled study of the memory effects of pomegranate juice in middle-aged and older adults. *Am J Clin Nutr* **111**, 170–177 (2020).
220. A. Bruno et al., Bergamot polyphenolic fraction supplementation improves cognitive functioning in Schizophrenia: data from an 8-week, open-label pilot study. *J Clin Psychopharmacol* **37**, 468–471 (2017).
221. R. Ochiai, K. Saitou, C. Suzukamo, N. Osaki, T. Asada, Effect of chlorogenic acids on cognitive function in mild cognitive impairment: a randomized controlled crossover trial. *J Alzheimers Dis* **72**, 1209–1216 (2019).
222. S. Huhn et al., Effects of resveratrol on memory performance, hippocampus connectivity and microstructure in older adults: a randomized controlled trial. *Neuroimage* **174**, 177–190 (2018).
223. O. T. Carmichael et al., A combination of essential fatty acids, panax ginseng extract, and green tea catechins modifies brain fMRI signals in healthy older adults. *J Nutr Health Aging* **22**, 837–846 (2018).
224. C. Dietz, M. Dekker, B. Piqueras-Fiszman, An intervention study on the effect of matcha tea, in drink and snack bar formats, on mood and cognitive performance. *Food Res Int* **99**, 72–83 (2017).
225. E. L. Wightman et al., The acute and chronic cognitive and cerebral blood flow effects of a sideritis scardica (Greek Mountain Tea) extract: a double blind, randomized, placebo controlled, parallel groups study in healthy humans. *Nutrients* **10**, 955 (2018).
226. Y. Kondo et al., Effects of coffee and tea consumption on glucose metabolism: a systematic review and network meta-analysis. *Nutrients* **11**, 48 (2018).
227. L. Stanaway et al., Acute supplementation with nitrate-rich beetroot juice causes a greater increase in plasma nitrite and reduction in blood pressure of older compared to younger adults. *Nutrients* **11**, 1683 (2019).
228. M. A. Bae et al., The development of taurine supplementary menus for the prevention of dementia and their positive effect on the cognitive function in the elderly with dementia. *Adv Exp Med Biol* **1155**, 335–347 (2019).
229. C. Brons, C. Spohr, H. Storgaard, J. Dyerberg, A. Vaag, Effect of taurine treatment on insulin secretion and action, and on serum lipid levels in overweight men with a genetic predisposition for type II diabetes mellitus. *Eur J Clin Nutr* **58**, 1239–1247 (2004).
230. F. T. Rosa, E. C. Freitas, R. Deminice, A. A. Jordao, J. S. Marchini, Oxidative stress and inflammation in obesity after taurine supplementation: a double-blind, placebo-controlled study. *Eur J Nutr* **53**, 823–830 (2014).
231. Y. Zheng et al., Plasma taurine, diabetes genetic predisposition, and changes of insulin sensitivity in response to weight-loss diets. *J Clin Endocrinol Metab* **101**, 3820–3826 (2016).
232. F. Franconi et al., Plasma and platelet taurine are reduced in subjects with insulin-dependent diabetes mellitus: effects of taurine supplementation. *Am J Clin Nutr* **61**, 1115–1119 (1995).
233. Y. Saitsu, A. Nishide, K. Kikushima, K. Shimizu, K. Ohnuki, Improvement of cognitive functions by oral intake of Hericium erinaceus. *Biomed Res* **40**, 125–131 (2019).
234. T. Pekar et al., The positive effect of spermidine in older adults suffering from dementia : First results of a 3-month trial. *Wien Klin Wochenschr* **133**, 484–491 (2021).
235. M. Wirth et al., The effect of spermidine on memory performance in older adults at risk for dementia: a randomized controlled trial. *Cortex* **109**, 181–188 (2018).
236. J. Young, M. Angevaren, J. Rusted, N. Tabet, Aerobic exercise to improve cognitive function in older people without known cognitive impairment. *Cochrane Database Syst Rev*, CD005381 (2015). https://www.cochranelibrary.com/cdsr/doi/10.1002/14651858.CD005381.pub4/epdf/full

237. C. Anderson-Hanley et al., The interactive physical and cognitive exercise system (iPACES): effects of a 3-month in-home pilot clinical trial for mild cognitive impairment and caregivers. *Clin Interv Aging* **13**, 1565–1577 (2018).
238. L. D. Baker et al., Aerobic exercise improves cognition for older adults with glucose intolerance, a risk factor for Alzheimer's disease. *J Alzheimers Dis* **22**, 569–579 (2010).
239. I. T. Liu et al., Therapeutic effects of exercise training on elderly patients with dementia: a randomized controlled trial. *Arch Phys Med Rehabil* **101**, 762–769 (2020).
240. N. Napoli et al., Effect of weight loss, exercise, or both on cognition and quality of life in obese older adults. *Am J Clin Nutr* **100**, 189–198 (2014).
241. N. Lehmann, A. Villringer, M. Taubert, Colocalized white matter plasticity and increased cerebral blood flow mediate the beneficial effect of cardiovascular exercise on long-term motor learning. *J Neurosci* **40**, 2416–2429 (2020).
242. T. Tarumi et al., Cerebral white matter integrity in amnestic mild cognitive impairment: a 1-year randomized controlled trial of aerobic exercise training. *J Alzheimers Dis* **73**, 489–501 (2020).
243. K. I. Erickson et al., Exercise training increases size of hippocampus and improves memory. *Proc Natl Acad Sci U S A* **108**, 3017–3022 (2011).
244. S. Y. Cho, H. T. Roh, Taekwondo enhances cognitive function as a result of increased neurotrophic growth factors in elderly women. *Int J Environ Res Public Health* **16**, (2019).
245. A. Kerling et al., Exercise increases serum brain-derived neurotrophic factor in patients with major depressive disorder. *J Affect Disord* **215**, 152–155 (2017).
246. E. Murawska-Cialowicz, J. Wojna, J. Zuwala-Jagiello, Crossfit training changes brain-derived neurotrophic factor and irisin levels at rest, after wingate and progressive tests, and improves aerobic capacity and body composition of young physically active men and women. *J Physiol Pharmacol* **66**, 811–821 (2015).
247. S. H. Yeh et al., Effects of music aerobic exercise on depression and brain-derived neurotrophic factor levels in community dwelling women. *Biomed Res Int* **2015**, 135893 (2015).
248. H. J. Coelho-Junior et al., Effects of combined resistance and power training on cognitive function in older women: a randomized controlled trial. *Int J Environ Res Public Health* **17**, (2020).
249. A. Damirchi, B. S. Tehrani, K. A. Alamdari, P. Babaei, Influence of aerobic training and detraining on serum BDNF, insulin resistance, and metabolic risk factors in middle-aged men diagnosed with metabolic syndrome. *Clin J Sport Med* **24**, 513–518 (2014).
250. D. S. Pereira et al., Effects of physical exercise on plasma levels of brain-derived neurotrophic factor and depressive symptoms in elderly women--a randomized clinical trial. *Arch Phys Med Rehabil* **94**, 1443–1450 (2013).
251. J. R. Ruiz et al., Resistance training does not have an effect on cognition or related serum biomarkers in nonagenarians: a randomized controlled trial. *Int J Sports Med* **36**, 54–60 (2015).
252. F. B. Schuch et al., The effects of exercise on oxidative stress (TBARS) and BDNF in severely depressed inpatients. *Eur Arch Psychiatry Clin Neurosci* **264**, 605–613 (2014).
253. M. S. Toups et al., Effects of serum brain derived neurotrophic factor on exercise augmentation treatment of depression. *J Psychiatr Res* **45**, 1301–1306 (2011).
254. G. S. Goldfield et al., Effects of aerobic training, resistance training, or both on brain-derived neurotrophic factor in adolescents with obesity: the hearty randomized controlled trial. *Physiol Behav* **191**, 138–145 (2018).
255. J. Krogh et al., The effect of exercise on hippocampal volume and neurotrophines in patients with major depression--a randomized clinical trial. *J Affect Disord* **165**, 24–30 (2014).
256. I. Levinger et al., BDNF, metabolic risk factors, and resistance training in middle-aged individuals. *Med Sci Sports Exerc* **40**, 535–541 (2008).
257. K. L. Szuhany, M. Bugatti, M. W. Otto, A meta-analytic review of the effects of exercise on brain-derived neurotrophic factor. *J Psychiatr Res* **60**, 56–64 (2015).
258. M. W. Voss et al., Neurobiological markers of exercise-related brain plasticity in older adults. *Brain Behav Immun* **28**, 90–99 (2013).
259. J. J. Walsh et al., Changes in the brain-derived neurotrophic factor are associated with improvements in diabetes risk factors after exercise training in adolescents with obesity: the HEARTY randomized controlled trial. *Neural Plast* **2018**, 7169583 (2018).
260. A. V. Araya, X. Orellana, D. Godoy, L. Soto, J. Fiedler, Effect of exercise on circulating levels

of brain-derived neurotrophic factor (BDNF) in overweight and obese subjects. *Horm Metab Res* **45**, 541–544 (2013).
261. J. Titus et al., The role of physical exercise in modulating peripheral inflammatory and neurotrophic biomarkers in older adults: a systematic review and meta-analysis. *Mech Ageing Dev* **194**, 111431 (2021).
262. C. Tonoli et al., Neurotrophins and cognitive functions in T1D compared with healthy controls: effects of a high-intensity exercise. *Appl Physiol Nutr Metab* **40**, 20–27 (2015).
263. I. Salehi et al., Electroconvulsive therapy and aerobic exercise training increased BDNF and ameliorated depressive symptoms in patients suffering from treatment-resistant major depressive disorder. *J Psychiatr Res* **57**, 117–124 (2014).
264. R. C. Cassilhas, H. K. Antunes, S. Tufik, M. T. de Mello, Mood, anxiety, and serum IGF-1 in elderly men given 24 weeks of high resistance exercise. *Percept Mot Skills* **110**, 265–276 (2010).
265. A. Maass et al., Relationships of peripheral IGF-1, VEGF and BDNF levels to exercise-related changes in memory, hippocampal perfusion and volumes in older adults. *Neuroimage* **131**, 142–154 (2016).
266. S. R. Vega et al., Responses of serum neurotrophic factors to exercise in pregnant and postpartum women. *Psychoneuroendocrinology* **36**, 220–227 (2011).
267. M. Goekint et al., Strength training does not influence serum brain-derived neurotrophic factor. *Eur J Appl Physiol* **110**, 285–293 (2010).
268. C. G. Schipke et al., Definition and quantification of six immune- and neuroregulatory serum proteins in healthy and demented elderly. *Neurodegener Dis Manag* **9**, 193–203 (2019).
269. J. Firth et al., The effects of dietary improvement on symptoms of depression and anxiety: a meta-analysis of randomized controlled trials. *Psychosom Med* **81**, 265–280 (2019).
270. R. S. Opie, A. O'Neil, C. Itsiopoulos, F. N. Jacka, The impact of whole-of-diet interventions on depression and anxiety: a systematic review of randomised controlled trials. *Public Health Nutr* **18**, 2074–2093 (2015).
271. I. Bourdel-Marchasson et al., Quality of life: psychological symptoms-effects of a 2-month healthy diet and nutraceutical intervention; a randomized, open-label intervention trial (RISTOMED). *Nutrients* **12**, (2020).
272. S. J. Torres, C. A. Nowson, A moderate-sodium DASH-type diet improves mood in postmenopausal women. *Nutrition* **28**, 896–900 (2012).
273. B. L. Beezhold, C. S. Johnston, Restriction of meat, fish, and poultry in omnivores improves mood: a pilot randomized controlled trial. *Nutr J* **11**, 9 (2012).
274. H. M. Francis et al., A brief diet intervention can reduce symptoms of depression in young adults - A randomised controlled trial. *PLoS One* **14**, e0222768 (2019).
275. F. N. Jacka et al., A randomised controlled trial of dietary improvement for adults with major depression (the 'SMILES' trial). *BMC Med* **15**, 23 (2017).
276. N. Parletta et al., A Mediterranean-style dietary intervention supplemented with fish oil improves diet quality and mental health in people with depression: a randomized controlled trial (HELFIMED). *Nutr Neurosci* **22**, 474–487 (2019).
277. J. Wardle et al., Randomized trial of the effects of cholesterol-lowering dietary treatment on psychological function. *Am J Med* **108**, 547–553 (2000).
278. A. Sanchez-Villegas et al., The effect of the Mediterranean diet on plasma brain-derived neurotrophic factor (BDNF) levels: the PREDIMED-NAVARRA randomized trial. *Nutr Neurosci* **14**, 195–201 (2011).
279. J. C. Sandberg, I. M. E. Bjorck, A. C. Nilsson, Impact of rye-based evening meals on cognitive functions, mood and cardiometabolic risk factors: a randomized controlled study in healthy middle-aged subjects. *Nutr J* **17**, 102 (2018).
280. J. C. Sandberg, I. M. E. Bjorck, A. C. Nilsson, Increased plasma brain-derived neurotrophic factor 10.5 h after intake of whole grain rye-based products in healthy subjects. *Nutrients* **10**, (2018).
281. A. Kazemi, A. A. Noorbala, K. Azam, M. H. Eskandari, K. Djafarian, Effect of probiotic and prebiotic vs placebo on psychological outcomes in patients with major depressive disorder: a randomized clinical trial. *Clin Nutr* **38**, 522–528 (2019).
282. Z. Solati, S. Jazayeri, M. Tehrani-Doost, S. Mahmoodianfard, M. R. Gohari, Zinc monotherapy increases serum brain-derived neurotrophic factor (BDNF) levels and decreases depressive symptoms in overweight or obese subjects: a double-blind, randomized, placebo-controlled trial. *Nutr Neurosci* **18**, 162–168 (2015).

283. S. Pirotta, D. J. Kidgell, R. M. Daly, Effects of vitamin D supplementation on neuroplasticity in older adults: a double-blinded, placebo-controlled randomised trial. *Osteoporos Int* **26**, 131–140 (2015).
284. T. J. Martin, S. Mist, T. Lektsok, N. L. Trent, Tibetan herbal tea agar-35 reduces negative affect and anxiety: a placebo-controlled pilot study. *Explore* (NY) **13**, 367–370 (2017).
285. M. R. Bazrafshan, M. Jokar, N. Shokrpour, H. Delam, The effect of lavender herbal tea on the anxiety and depression of the elderly: a randomized clinical trial. *Complement Ther Med* **50**, 102393 (2020).
286. S. M. Chang, C. H. Chen, Effects of an intervention with drinking chamomile tea on sleep quality and depression in sleep disturbed postnatal women: a randomized controlled trial. *J Adv Nurs* **72**, 306–315 (2016).
287. M. Park, J. Choi, H. J. Lee, Flavonoid-rich orange juice intake and altered gut microbiome in young adults with depressive symptom: a randomized controlled study. *Nutrients* **12**, 1815 (2020).
288. S. Hidese et al., Effects of chronic l-theanine administration in patients with major depressive disorder: an open-label study. *Acta Neuropsychiatr* **29**, 72–79 (2017).
289. C. N. Kahathuduwa et al., l-Theanine and caffeine improve target-specific attention to visual stimuli by decreasing mind wandering: a human functional magnetic resonance imaging study. *Nutr Res* **49**, 67–78 (2018).
290. J. Sarris et al., L-theanine in the adjunctive treatment of generalized anxiety disorder: a double-blind, randomised, placebo-controlled trial. *J Psychiatr Res* **110**, 31–37 (2019).
291. C. F. Haskell-Ramsay et al., The acute effects of caffeinated black coffee on cognition and mood in healthy young and older adults. *Nutrients* **10**, (2018).
292. X. Shi, W. Xue, S. Liang, J. Zhao, X. Zhang, Acute caffeine ingestion reduces insulin sensitivity in healthy subjects: a systematic review and meta-analysis. *Nutr J* **15**, 103 (2016).
293. W. D. M. Alharbi, A. Azmat, M. Ahmed, Comparative effect of coffee robusta and coffee arabica (Qahwa) on memory and attention. *Metab Brain Dis* **33**, 1203–1210 (2018).
294. A. P. Bailey, S. E. Hetrick, S. Rosenbaum, R. Purcell, A. G. Parker, Treating depression with physical activity in adolescents and young adults: a systematic review and meta-analysis of randomised controlled trials. *Psychol Med* **48**, 1068–1083 (2018).
295. A. L. Rebar et al., A meta-meta-analysis of the effect of physical activity on depression and anxiety in non-clinical adult populations. *Health Psychol Rev* **9**, 366–378 (2015).
296. A. Heissel et al., Effects of exercise on depression and anxiety in persons living with HIV: A meta-analysis. *J Psychosom Res* **126**, 109823 (2019).
297. J. Song, Z. Z. Liu, J. Huang, J. S. Wu, J. Tao, Effects of aerobic exercise, traditional Chinese exercises, and meditation on depressive symptoms of college student: a meta-analysis of randomized controlled trials. *Medicine (Baltimore)* **100**, e23819 (2021).
298. S. Guo, F. Liu, J. Shen, M. Wei, Y. Yang, Comparative efficacy of seven exercise interventions for symptoms of depression in college students: a network of meta-analysis. *Medicine (Baltimore)* **99**, e23058 (2020).
299. K. J. Miller et al., Comparative effectiveness of three exercise types to treat clinical depression in older adults: a systematic review and network meta-analysis of randomised controlled trials. *Ageing Res Rev* **58**, 100999 (2020).
300. A. Dinoff, N. Herrmann, W. Swardfager, D. Gallagher, K. L. Lanctot, The effect of exercise on resting concentrations of peripheral brain-derived neurotrophic factor (BDNF) in major depressive disorder: a meta-analysis. *J Psychiatr Res* **105**, 123–131 (2018).

CHAPTER THIRTEEN

Pharmacological and Surgical Interventions to Improve Brain Insulin Resistance

Linus Haberbosch and Lukas Maurer
Reiner Jumpertz-von Schwartzenberg

CONTENTS

DOI: 10.1201/9781003079927-13

GLP-1 AGONIST TREATMENT AND BRAIN INSULIN RESISTANCE

GLP-1 receptor agonists (GLP-1-RA) have attracted a lot of attention during the last decade as potential treatments for neurodegenerative diseases, especially in the context of brain insulin resistance and Alzheimer's disease (AD) (1). Neurodegenerative diseases such as AD are associated with impairments in brain insulin as well as insulin-like growth factor (IGF) signaling. This gave rise to the hypothetical concept of "type 3" diabetes representing a brain-specific aberration of glucose metabolism as a hallmark of the neurodegenerative process (2). It should be noted that the classification "type 3" diabetes in this context is used differently than the conventional attribution of "type 3" diabetes to be an umbrella term for non-classical diabetes i.e., insulin deficit due to pancreatic damage, underlying other endocrine disease, etc. But focusing on the evidence for impaired insulin action at least associated with neurodegeneration makes it intriguing to investigate antidiabetic drugs in this new context as potential mediators of "disease-modification" or even "neuroprotection".

Especially, the successful application of GLP-1-RA to modulate peripheral insulin secretion and action in patients with type 2 diabetes has attracted great attention within this context (3). GLP-1-RA increases glucose-dependent insulin secretion (incretin effect) of the beta cells. This effect is dependent on adenylate cyclase (ADCY) turn-over of ATP to cAMP which can activate exchange protein activated by cAMP (EPAC) leading to increased insulin granule exocytosis. Besides EPAC activation, insulin secretion can also be mediated by a protein kinase A-dependent pathway (Figure 13.1). However, GLP-1 agonism not only triggers insulin secretion but also induces insulin gene transcription and biosynthesis to replenish insulin stores of the beta cells. Insulin gene transcription and its biosynthesis are mediated by protein kinase A (PKA) dependent and independent signaling

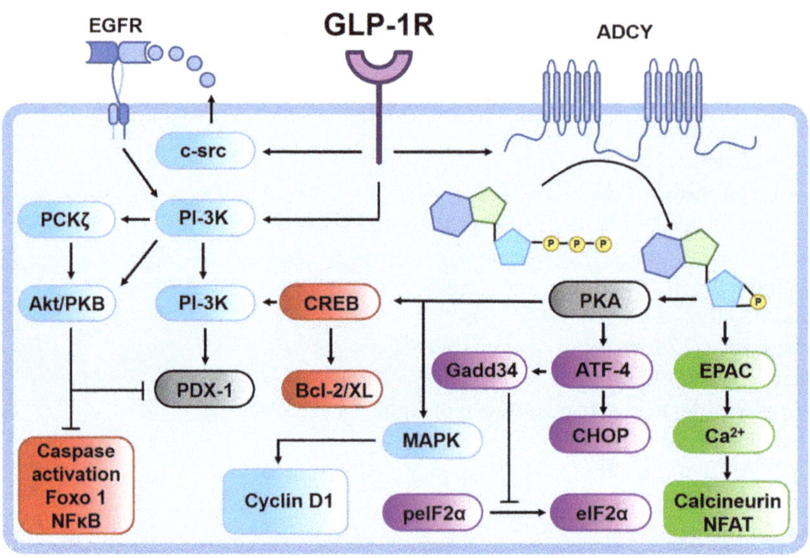

Figure 13.1 Intracellular signaling pathways of GLP-1R in the pancreatic β-cell. Blue: Factors inducing β-cell proliferation; Red: Factors protecting against apoptosis; Green: Factors inducing insulin secretion and biosynthesis. Adenosine triphosphate (ATP) and cyclic adenosine monophosphate (cAMP) depicted as molecules. Abbreviations: EGFR: epidermal growth factor receptor; GLP-1R: glucagon-like peptide-1 receptor; ADCY: Adenylyl cyclase; c-Src: Proto-oncogene tyrosine-protein kinase Src; PKCζ: protein kinase Cζ; PI-3K: phosphatidylinositol-3 kinase; Akt/PKB: Akt-protein kinase B; IRS-2: insulin receptor substrate 2; CREB: cAMP-responsive element binding protein; PKA: Protein kinase A; PDX-1: pancreatic and duodenal homeobox 1; Bcl-2/XL: B-cell lymphoma 2/extra large; GADD34: growth arrest and DNA damage-inducible protein 34; ATF-4: Activating transcription factor 4; EPAC: exchange factor directly activated by cAMP; MAPK: mitogen-activated protein kinase; CHOP: C/EBP homologous protein; Foxo 1: Forkhead-Box-Protein O1; NFκB: Nuclear factor kappa B; (p)eIF2α: (phoyphorylated) eukaryotic initiation factor 2; NFAT: nuclear factor of activated T-cells. This figure was created with BioRender.com. Adapted from: Baggio, L. L., & Drucker, D. J. (2007). Biology of incretins: GLP-1 and GIP. *Gastroenterology*, 132(6), 2131–2157.

cascades, both stimulating the insulin transcription factor PDX-1 (Figure 13.1). The PKA-independent pathway of PDX-1 activation is mediated via phosphoinositide 3-kinase (PI-3K) and insulin receptor substrate 2 (IRS-2) activation (Figure 13.1).

While central and peripheral insulin sensitivity may be two only partially overlapping entities, there is evidence for impaired brain insulin sensitivity accompanying peripheral insulin resistance in patients with metabolic syndrome. Interestingly, brain regions like the ventral striatum and medial prefrontal cortex that are playing a pivotal role in the regulation of reward-related food intake display this attenuated response to insulin in patients with peripheral insulin resistance (4). These results raise the question, if the impaired action of insulin on these regulatory centers constitutes a cross-link to obesity and the metabolic syndrome on the one hand and neurodegeneration on the other.

With regard to GLP-1-RAs, it is evident that the pronounced anorexigenic effect of these drugs is substantially mediated via GLP-1 acting on regulatory centers within the central nervous system, making the brain one of the central target organs of these compounds even in the established therapeutic usage for type 2 diabetes and obesity (5).

While endogenous GLP-1 secreted in the intestinal L-cell is able to activate central GLP-1 receptors, especially in areas belonging to the circumventricular organs like the nucleus of the solitary tract (NTS) or the mediana eminentia (ME) - regions without fully intact blood-brain-barrier - it is hard to quantify their contribution due to the rapid degradation via Dipeptidylpeptidase IV (DPP-4) once released into the blood stream. In parallel endogenous GLP-1 might exert its effect on the central nervous system mainly via afferent neuronal terminals of the vagus nerve (6). In addition, GLP-1 is synthesized within the brain stem itself from preproglucagon-neurons of the NTS. Brain-derived GLP-1 then functions as an endogenous neuropeptide within an extensive network of GLP-1 receptor expressing neurons spanning throughout the central nervous system (7).

GLP-1 analogues, classically used for the treatment of diabetes are able to cross the blood-brain barrier and to interact with complex network brain GLP-1 receptor-positive neurons (8). The pattern of GLP-1 receptor expression inside the brain, while being widely scattered, shows particularly high densities in the circumventricular organs, amygdala, and hypothalamus (9).

Taking into consideration the anatomical overlap between insulin and GLP-1 responsive areas in the brain and their highly intermingled role in peripheral physiology, make it intriguing to elaborate the potential of modulating brain insulin action via GLP-1 receptor activation.

Of note, GLP-1 has been shown to ameliorate brain insulin resistance by modulating the impaired brain IRS-1 signaling in an AD mouse model (10, 11). In combination with numerous preclinical lines of evidence implicating a potential neuroprotective capacity of GLP-1, this has sparked an extensive exploration of GLP-1 treatment in central insulin resistance aiming at disease modification in neurodegenerative diseases (12).

Naturally, this line of investigation has been preformed in specific neuronal culture experiments and rodent models of the different neurodegenerative diseases.

PRECLINICAL EVIDENCE

Alzheimer's Disease Rodent Models

A large amount of data on central GLP-1 effects in this context stems from transgenic protein precursor AD mouse models. Application of GLP-1-RAs like Liraglutide improved memory function and protected against degeneration of hippocampal neurons (13). A reduced neuronal inflammatory response and prevention of synaptic deterioration are also mediated via GLP-1 (14). These protective in vivo effects seem to be based on the GLP-1-dependent prevention of loss of insulin receptors and synapses in AD (15).

Alzheimer's Disease Cell Models

On a cellular level GLP-1 is able to reduce the AD-typical amyloid-β peptide accumulation (16). GLP-1-R activation is also able to reduce oxidative stress (reactive oxygen species formation) and to increase insulin sensitivity in neuronal cells (16).

While AD especially due to the potentially tight link with central insulin resistance constitutes the largest research focus to date, other neurodegenerative and cerebrovascular pathologies have increasingly come into consideration.

Parkinson's Disease (PD) Rodent Models

GLP-1 application in different PD animals has shown a potential benefit. Protection of neurons against metabolic and oxidative insults in several mouse models has been demonstrated for GLP-1-RA and lately for new dual GLP-1/GIP agonists (17, 18). Similar to AD, brain insulin resistance and a potential amelioration through GLP-1 application are considered important features in the neurodegenerative process of PD (19).

Numerous investigations targeting a potential link between central insulin resistance and the different pathological processes potentially underlying the ongoing neurodegenerative changes in diseases such as PD and AD have been performed. There has been shown to be evidence for a link between insulin resistance and alpha-synuclein (20) as well as tau (21) and beta-amyloid protein accumulation (22). This pathophysiological framework has prompted the hypothesis of an impaired insulin signaling to determine the brain's vulnerability toward protein precursor accumulation associated toxicity (23).

CLINICAL EVIDENCE

Following the promising preclinical findings for GLP-1RA in neurodegeneration and brain insulin resistance, first clinical trials have been conducted or are underway.

AD Clinical Evidence

In a 6-month double-blind randomized controlled trial, treatment of AD patients with GLP-1-RA liraglutide prevented the decline in brain glucose metabolism (measured with FDG-PET) but had no effect on plaque burden or memory function (24). To further elaborate a potential treatment effect of GLP-1-RA in AD, a 12-months multi-center study (ELAD - Evaluating the effects of the novel GLP-1 analogue liraglutide in Alzheimer's disease) was initiated but results are still pending (25).

PD Clinical Evidence

GLP-1RA Exenatide showed a positive effect on motor scores in PD in a proof of concept single-blind trial (26). These results have recently been confirmed in a double-blind RCT. Forty-eight weeks of treatment followed by 12 weeks off medication showed a sustained benefit on motor symptoms in patients suffering from PD, while it remained elusive to differentiate between a prolonged symptomatic or actual disease-modifying effect (27).

In summary, preclinical results for GLP-1-RA in the treatment of brain insulin sensitivity are showing great promise. Given the successful application of these compounds in the context of type 2 diabetes with positive clinical endpoints regarding cardiovascular events and a sufficient long-term safety profile, these drugs are well suited for the exploration of potential benefits regarding the treatment of brain insulin resistance. At this point, sufficient clinical data is lacking. The attempt for clinically relevant neuroprotection or disease modification in the context of PD and AD has been notoriously difficult albeit a plethora of promising preclinical results. To determine the role of GLP-1-RA in this context more RCTs are needed with a special focus on early initiation and sufficiently long follow-up periods.

INTRANASAL GLP-1 APPLICATION

To focus on the direct effects of GLP-1 on the brain potentially affecting central insulin signaling, intranasal GLP-1 application has been investigated. In patients with type 2 diabetes intranasal application of GLP-1-RA improved postprandial glycemic control by increasing insulin and inhibiting glucagon secretion (28). Further studies are needed to elaborate the exact mechanism of action for this route of application in order to further delineate between central and peripheral mediation. In mice, intranasal GLP-1 co-administered with cell-penetrating peptides showed a more widespread distribution throughout the central nervous system and was also able to improve cognitive dysfunction (29). Intranasal application of recombinant modified GLP-1 showed a pronounced and durable antihyperglycemic effect in mice (30).

While brain GLP-1 and brain insulin signaling might be heavily intertwined and connected reciprocally, they also might serve independent functions. Their response in the context of metabolic dysfunction for example might not necessarily be associated to one another (31). Given the widespread expression pattern of GLP-1 receptor and the multiple function domains of the system, it remains to be differentiated to what extent the pharmacological effects, dependent on brain insulin signaling, are orchestrated in close communication or function independently.

OTHER ANTI-DIABETIC DRUGS AND BRAIN INSULIN RESISTANCE

Insulin

Peripheral Insulin

The influence of insulin on brain activity was first investigated with hyperinsulinemic–euglycemic glucose clamps, and to date, abundant evidence has emerged that systemic insulin administration modulates cortical brain activity in humans (32–35). It is, however, difficult to discern the central effects of peripherally applied insulin from indirect signaling via other insulin-sensitive organs (36).

Of the peripheral insulins and insulin analogues used in the therapy of type 2 diabetes, most have not shown any effect on central insulin resistance, with the notable exception of insulin detemir. This insulin analogue was shown to increase cerebrocortical beta activity in obese subjects during an euglycemic clamp similar to lean subjects, an effect not found for the human insulin applied in the control group (37). Insulin detemir, applied peripherally, is characterized by a higher peak concentration in cerebrospinal fluid (CSF) and remained elevated for a longer period compared to NPH insulin, possibly on account of its lipophilic nature (38, 39). There is also data suggesting a reduced food intake/weight gain effect of insulin detemir (33, 40), further hinting towards a possible effect on central insulin signaling. While peripheral insulin effects on central insulin sensitivity should be further investigated, its therapeutic use is currently limited by the high doses needed to ensure adequate CSF levels and the resulting increased risk of hypoglycemia. Therefore, recent studies have concentrated on alternative application routes of insulin possibly capable of bypassing the blood-brain barrier (36, 41).

Intranasal Insulin

Intranasal administration of insulin has been proven effective in increasing CSF insulin levels with only a small fraction of the dose passing into systemic circulation, inducing central effects with a greatly reduced risk of hypoglycemia (42–44). There is also accumulating evidence for a positive effect of intranasal insulin application on central insulin sensitivity. In the 3xTg-AD mouse model of Alzheimer's disease, the application of intranasal insulin led to a nearly complete restoration of the damaged IR→IRS-1→PI3K signaling pathway as well as increased levels of synaptic proteins and reduced beta-amyloid levels (45). In a pilot clinical trial in 104 patients, intranasal insulin showed improved cognitive performance as well as restored brain insulin sensitivity as assessed by 2-fluoro-2-deoxy-d-glucose positron emission tomography (FDG PET/CT) (46). However, a large follow-up randomized controlled trial could not replicate the positive effects on cognition (43). Since brain glucose metabolism was not monitored and a midtrial change of the delivery system occurred, these results might be at least partially explained by a failure of intranasal insulin delivery to the brain. Considering the reduced hypoglycemia risk compared to peripherally administered insulin, further studies improving insulin delivery to the central nervous system and investigating its effects seem worthwhile.

Insulin Sensitizers

Metformin

The biguanide antihyperglycemic agent metformin is one of the cornerstones of type 2 diabetes therapy. Metformin has insulin-sensitizing effects and can cross the blood-brain barrier (albeit in a greatly reduced concentration), it is reasonable to assume that the drug might improve central insulin sensitivity (47). Recently, Gupta et al. demonstrated *in vitro* that metformin application induces normalization of glucose uptake as well as a complete remission of the increased tau phosphorylation in insulin-resistant neuronal cells (48). Clinical studies have confirmed metformin presence in CSF as well as its positive effects on cognitive impairment (49, 50). There is also emerging evidence for possible effects of metformin on mood disorders, specifically depression, in diabetics patients with diabetes and patients with polycystic ovary syndrome (51, 52). A recent RCT presented a similar effect in non-diabetic patients (53), although metabolic and cognitive changes were not addressed in this study, emphasizing the need for further prospective trials (54). Currently, no strong connection of these effects with brain insulin sensitivity could be established. However, metformin's profile as a safe and effective drug in clinical use as well as its emerging role as a neuroprotectant (55, 56) suggest possible therapeutic efficacy in neuronal insulin resistance, inferring a need for further comprehensive clinical evaluation.

Peroxisome Proliferator-Activated Receptor Agonists (PPARs)

PPARs have experienced a decline in clinical application in diabetes patients following severe cardiovascular side effects. While rosiglitazone had initially shown promising effects on brain insulin sensitivity as well as further neuroprotective effects (57, 58), a larger clinical trial found no significant effect on cognition (59). As the PPAR-gamma receptor targeted by most PPARs is primarily expressed peripherally, efforts are underway to investigate the effects of compounds combining agonism of PPAR gamma with effects on the more centrally abundant PPAR-delta. While there are promising initial results (60), there is a need for larger studies and further investigation of possible adverse events (61).

Incretin Analogues

Glucose-Dependent Insulinotropic Polypeptide (GIP) Analogues

GIP is an incretin hormone stimulating insulin secretion and reducing hypoglycemia similar to GLP-1. Outside of the pancreas, GIP receptors (GIPR) are expressed in neurons of hypothalamic nuclei linked to appetite control (62). Recent evidence shows improved glucose metabolism and reduced weight gain in CNS-specific humanized GIPR knockout mice, as indicated by enhanced glucose tolerance and decreased fasting levels of glucose and insulin (63). Conversely, GIP application centrally and peripherally showed similar results in the same study. However, when applied in CNS-specific GIPR knockout mice, the effects of GIP on body weight were lost, but improvements in glucose metabolism were replicated (63). While centrally applied GIP improved glycemic control in central GIPR knockout mice, these data suggest a more important role of peripheral/pancreatic GIP effects on glucose metabolism. GIP analogues have shown some promise in reducing oxidative stress and amyloid plaque load in cortex and hippocampus, indirectly improving brain insulin sensitivity by reverting amyloid mediated receptor desensitization (64), however relevant direct effects have not been demonstrated yet.

Dual/Triple Agonists

GLP-1 analogues have demonstrated promising effects on brain insulin sensitivity and GIP analogues have shown at least indirect effects. Thus, additional benefits of a novel dual GLP-1/GLP agonist seem possible, with several promising compounds being investigated (65). The concept of dual agonism is based on the activation of GLP-1 agonism and additionally GIP agonism which lead to the combined effect of increased glucose-dependent insulin secretion and additionally decreased glucagon production with improved effects on glucose homeostasis. Nonetheless, direct evidence for compound agonism to improve brain insulin sensitivity is currently lacking. However, enhanced blood-brain barrier transfer and potent neuroprotective effects of GLP-1/GIP dual receptor agonists (DA3–CH) (66) especially should encourage further studies. Additionally, novel concepts of triple agonism with long-acting GLP-1/GIP/Glucagon agonists have provided beneficial effects on peripheral glucose metabolism and their effects on neuronal signaling involved in glucose and energy homeostasis (67) open exciting treatment opportunities for metabolic diseases.

OTHER THERAPEUTIC AGENTS

Dipeptidyl-Peptidase 4 (DPP-4) Inhibitors

The DPP-4 substrate vildagliptin was shown to improve brain insulin sensitivity as measured via hippocampal insulin-induced long-term depression as well as the phosphorylation and expression of insulin receptors in testosterone-deprived and normal rats (68, 69). The competitive DPP-4 inhibitor sitagliptin on the other hand showed promising effects on recognition memory, hippocampal neurogenesis and specifically oxidative neuronal stress (70), which is inextricably linked with reduced brain insulin sensitivity (71). Although evidence in humans is lacking, a positive effect of DPP-4 inhibition on brain insulin sensitivity seems likely. However, it remains unclear if there is a specific effect beyond the improvements induced by elevated GLP-1 levels.

Sodium Dependent Glucose Co-Transporter 2 (SGLT-2) Inhibitors

Also well established in clinical therapy of type 2 diabetes, SGLT-2 inhibitors have been proposed to influence brain insulin sensitivity, possibly mediated by weight loss (36). There is evidence in rats showing improved brain insulin sensitivity after application of SGLT-2-inhibitors as well as improved brain mitochondrial function, apoptosis, and improved hippocampal synaptic plasticity (69). Clinical data are currently limited to one oral presentation suggesting a positive effect of SGLT-2 inhibitors on brain insulin sensitivity (72), and further studies will be needed to explore the mechanism of action behind this effect.

BRAIN INSULIN SENSITIVITY AND BARIATRIC SURGERY

Brain Insulin Sensitivity in Obesity

The action of insulin on peripheral tissues, specifically skeletal muscle and fat tissue, is ameliorated in obesity leading to consequences on glucose regulation and energy expenditure. This insulin resistance is the pathophysiological hallmark for a plethora of metabolic diseases such as type 2 diabetes bearing a high risk for cardiovascular disease and increased mortality. Insulin is also transported into the central nervous system (CNS) via receptor-mediated transcytosis in endothelial cells (73) and activates glial cells and neurons via insulin receptors that are abundantly expressed e.g. in the olfactory bulb, the cerebral cortex, the hippocampus, the hypothalamus, the septum, and the amygdala as has been shown in rats (74). In the CNS, insulin has numerous physiological effects depending on the region of action e.g.: homeostasis (hypothalamus), memory (hippocampus), reward (striatum), inhibition (lateral prefrontal cortex), attention (fusiform gyrus), sensory perception (insular cortex) and motivation (orbitofrontal cortex/anterior cingulate) (75). Obesity and metabolic disease such as type 2 diabetes are linked with structural alterations in grey matter density, cortical thickness, and white matter integrity (76–78). Besides structural brain characteristics in obesity, also multiple functional aspects of insulin action in the brain are affected. The effects of peripheral and central insulin action are partially diametrically opposed. I.e. by subcutaneous or intravenous insulin application, peripheral lipolysis is inhibited and the overall effects on muscle and fat tissue are anabolic (79, 80). However, using the route of intranasal application to selectively increase CNS insulin concentrations (beware those peripheral effects of intranasal insulin action may also occur) has shown anorexigenic effects in humans (81, 82). Also in animals central (invasive) administration of insulin has shown similar anorexigenic effects (83–85). Interestingly, the reduced food intake is not limited to homeostatic but also to hedonic aspects of food intake. In support of this, Hallschmidt et al. have previously shown that intranasal insulin application decreases post-prandial appetite and snack intake in healthy women (86). This interesting aspect of brain insulin action on satiety and food intake in humans implicates not only peripheral insulin resistance in obesity and type 2 diabetes but also central insulin resistance considering the hyperphagic aspects in the development of obesity and severe obesity. Indeed, cumulating evidence indicates that several aspects of insulin action in the brain are ameliorated in obese individuals. However, it is important to consider the various ways of assessing central insulin action. E.g. as opposed to intranasal insulin application one way to increase central (and peripheral) insulin concentrations but maintaining euglycemia is the euglycemic hyperinsulinemic glucose clamp. Here, insulin is administered at a constant rate via a peripheral vein and glucose is simultaneously infused, resulting in euglycemic blood glucose levels but high insulin concentrations. Applying the euglycemic hyperinsulinemic glucose clamp and magnetic encephalography (MEG) in normal weight and obese humans, Tschritter et al. have shown that the insulin-stimulated increase in cortical activity (beta and theta frequency) in normal weight humans was not observed in obese individuals (87). Accordingly, after intranasal insulin application evoked MEG-potentials by food stimuli in the fusiform gyrus were increased in normal-weight but not overweight individuals (88), and this can also be reproduced by functional MRI underscoring that the response to food objects during high insulin levels is reduced in the fusiform gyrus in overweight adults (89). Another way to assess central insulin sensitivity is 2-fluoro-2-deoxy-d-glucose positron emission tomography (FDG PET/CT) imaging to measure glucose uptake globally and in various brain regions. When separating peripheral insulin sensitive from resistant men, glucose uptake in the prefrontal cortex and the striatum (important network controlling appetite and reward) during a euglycemic hyperinsulinemic glucose clamp was reduced in the peripheral insulin-resistant group compared to insulin-sensitive individuals (90).

However, insulin levels during the clamp procedure in the insulin-resistant group were comparable to their fasting condition whereas three times higher compared to fasting in the insulin-sensitive group. Thus, reduced glucose metabolism is not necessarily reflecting brain insulin resistance. Several studies have shown that increasing insulin concentrations above fasting levels does not increase brain glucose metabolism in healthy individuals (91, 92). Hirvonen et al. also showed that insulin-mediated glucose uptake in the brain is not increased during the euglycemic hyperinsulinemic glucose clamp in healthy individuals with normal glucose regulation, however, in patients with impaired glucose regulation hyperinsulinemia during the clamp increased brain glucose metabolism (93). Others have also shown that glucose uptake in certain brain regions such as the caudate nucleus is increased during the clamp in obese but not lean individuals indicating abnormal food reward processing in obese individuals (94). Thus, increased insulin-mediated brain glucose metabolism is often viewed as part of cerebral insulin resistance. Accordingly, in an FDG-PET imaging study with more than 200 individuals, higher glucose metabolism in the left inferior temporal lobe, the right insula, anterior cingulate, as well as medial frontal and orbitofrontal regions was positively associated with body mass index (BMI) (95). Recently, it was shown that also basal cerebral glucose metabolism in the right insula and caudate nucleus was positively associated with future weight gain, as was glucose hypometabolism in the left cerebellum (96). Thus, insulin resistance in the brain potentially underlies regional distribution. Kullmann et al. have shown a reduction of cerebral blood flow in the prefrontal cortex only in lean healthy but not in overweight/-obese individuals in response to intranasal insulin application (97). Interestingly, the magnitude of the insulin-mediated response correlated with intraabdominal fat tissue volume. The strong link between impaired insulin signaling (and glucose metabolism) in the brain and obesity supports the notion that insulin resistance is not limited to peripheral tissues. E.g. in polycystic ovarian syndrome patients (a disease frequently linked with peripheral insulin resistance), prefrontal cortex and striatal inhibition after glucose ingestion were limited to insulin-sensitive women (98). One important question is whether insulin signaling in the CNS also affects peripheral insulin sensitivity and whether this is compromised in obesity? Interestingly, in normal weight individuals, intranasal insulin application during a euglycemic hyperinsulinemic glucose clamp necessitates higher glucose infusion rates to maintain euglycemia; however, in obese individuals, this peripheral insulin-sensitizing effect is absent (99, 100). Similarly, in patients with severe obesity in evaluation for bariatric surgery, insulin-stimulated brain glucose uptake during the euglycemic hyperinsulinemic clamp was positively associated with endogenous glucose production in the liver (101). Interestingly, baseline brain glucose uptake was linked with less improved fasting glucose levels after bariatric surgery implicating that brain insulin sensitivity is implicated in improved glucose metabolism after weight loss intervention.

BRAIN INSULIN SENSITIVITY, CONVENTIONAL WEIGHT LOSS AND BARIATRIC SURGERY

It is widely known that weight loss in humans improves peripheral insulin sensitivity and glucose regulation. However, does intended weight loss also affect brain insulin sensitivity? Using fMRI, previous studies have shown that blood oxygen level-dependent (BOLD) signal is decreased during an oral glucose challenge in healthy individuals but not in patients with type 2 diabetes (29, 30). However, only a very short-term but severe caloric restriction (450 kcal/day for 4 days) was necessary to normalize hypothalamic neuronal responsiveness to a glucose challenge (102). The observation that intact brain insulin sensitivity is linked with anorexigenic properties in humans and animals (81–85) imposes the conclusion that individuals with reduced brain insulin sensitivity might be prone to less successful weight loss during lifestyle interventions. Indeed, higher insulin-stimulated theta activity assessed by MEG during a euglycemic hyperinsulinemic clamp at baseline predicts lower body weight and lower visceral adipose tissue mass 9 months after lifestyle intervention (103). To date, we know that this effect may not be limited to short-term weight development, since follow-up analyses of the previous study have recently shown that individuals with higher brain insulin sensitivity at baseline not only show more pronounced weight loss during the lifestyle intervention phase but also showed significantly less regain of total body weight and visceral fat mass compared to individuals with brain insulin resistance (104).

Decreased brain insulin sensitivity in obesity is improved after short-term caloric restriction and predicts longitudinal weight trajectories with and without weight loss inducing lifestyle modifications. An invasive model addressing profound weight loss in the severely obese is bariatric surgery and current knowledge suggests that weight loss imposed by bariatric surgery may also affect brain insulin sensitivity. Indirect evidence comes from a study using fMRI, where obese women exhibited increased hypothalamic activity during visual stimulation with low vs. high caloric food pictures compared to normal weight controls and patients after Roux-en Y gastric bypass (RYGB) (105). More directly, Tuulari et al. have shown that increased insulin-stimulated brain glucose uptake during a euglycemic hyperinsulinemic clamp in obese individuals (which was not observed in normal weight controls) is reversed by bariatric surgery (sleeve gastrectomy and RYGB) (106). Similar findings have been shown recently for cerebral glucose metabolic rate 6 months after RYGB (107). Importantly, besides the restriction effects on food intake, RYGB (as compared to sleeve gastrectomy) also has malabsorptive effects, which usually result in higher remission rates of type 2 diabetes and better long-term glycemic control after surgery (108). Whether combined malabsorptive and restriction-based surgical procedures more strongly improve brain insulin sensitivity compared to solely restriction-based procedures such as sleeve gastrectomy or gastric banding remains unclear. So far, one study has addressed this question showing that RYGB patients exhibit lower brain activation (assessed by fMRI) to high caloric foods in brain areas of reward network systems like the orbitofrontal cortex, amygdala, caudate nucleus, nucleus accumbens, and hippocampus which was associated with lower palatability and appeal of food with high caloric density compared to patients who underwent gastric banding (109). Overall, it remains elusive whether brain insulin sensitivity significantly affects weight loss outcomes after bariatric surgery and/or vice versa and if these effects prevail over longer periods. In a heterogeneous study with 9 patients undergoing sleeve gastrectomy and 11 patients RYGB, brain glucose uptake (as a surrogate for brain insulin resistance) was reduced 6 months after bariatric surgery by approx. 5% but was still higher compared to control subjects (101). Indirect measures of functional connectivity by fMRI have been implicated to predict weight loss 12 months after sleeve gastrectomy (110) and after RYGB fMRI patterns in the hypothalamus were altered and more similar to lean individuals (111). In terms of brain insulin sensitivity, however, fMRI compared to FDG-PET imaging may reflect different aspects of brain insulin action. Recently, it was shown that compared to other studies, brain glucose uptake

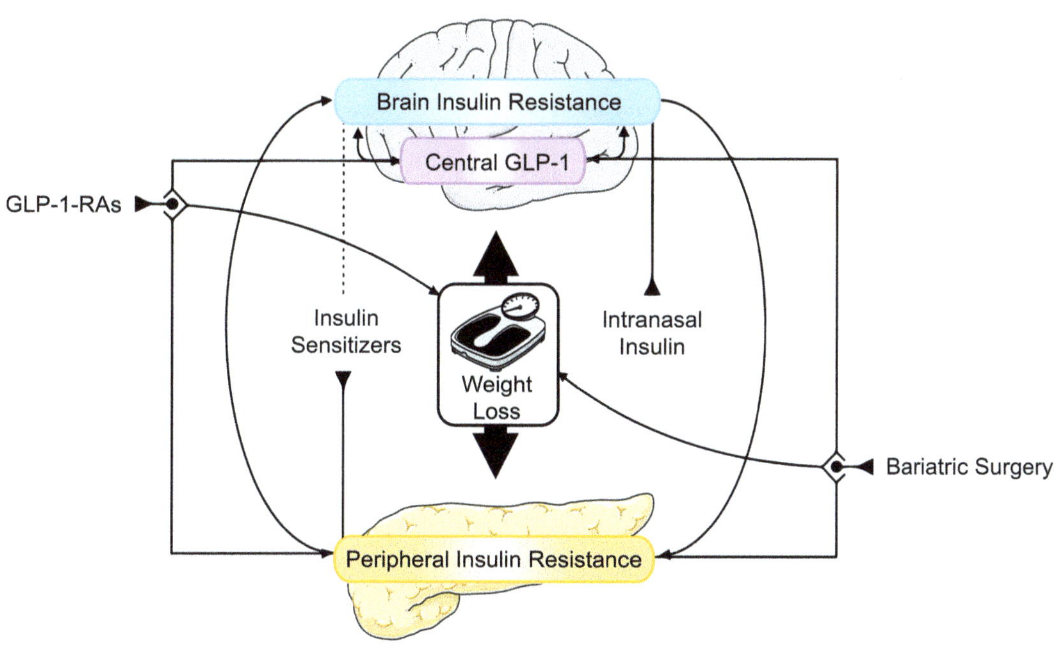

Figure 13.2 Intervention effects on brain insulin sensitivity.

(as a surrogate for central insulin resistance) is reduced during normoglycemia after RYGB, however cerebral blood flow assessed by fMRI is increased (112). Since hyperinsulinemic hypoglycemia is often observed after bariatric surgery, this study also applied a hyperinsulinemic hypoglycemic clamp (glucose concentration target 2.7 mmol/l) showing that after RYGB hypoglycemia induces increased activation in the thalamus and hypothalamus (regions implicated in glucose regulation) compared to the pre-surgical state. Additionally, considering that brain insulin sensitivity may also affect peripheral insulin action e.g. in the liver, whether the beneficial metabolic effect of bariatric surgery on peripheral insulin action is partially triggered by improved brain insulin sensitivity is not clear yet. In mice, it was recently shown that the beneficial effects of RYGB on resting metabolic rate and browning of visceral fat tissue are dependent on endocannabinoid signaling via the splanchnic nerve, which was not observed after sleeve gastrectomy indicating that CNS CB1 (endocannabinoid receptor 1) signaling is a key regulator for metabolic effects of bariatric surgery (at least in mice) (113). Another mechanism of how bariatric surgery may enhance brain insulin signaling may be via increased GLP-1 (glucagon-like peptide 1) signaling. Bariatric surgery is accompanied by increased GLP-1 levels which are known for its satiety and weight loss-inducing effects in animals and humans (114). Interestingly, Alzheimer's disease is commonly associated with type 2 diabetes and defective brain insulin signaling which goes along with serine phosphorylation of IRS-1 in both diseases. Exenatide (exendin-4), a GLP-1 analogue, is able to decrease levels of hippocampal IRS-1 serine phosphorylation and improve cognitive function in a mouse model of Alzheimer's disease (115). One could speculate, that the beneficial metabolic

TABLE 13.1
Intervention effects on brain insulin sensitivity

Intervention	Potential effects on brain insulin action	Potential mechanism
GLP-1 agonism	Ameliorate central insulin resistance	Modulating impaired IRS-1 signaling
		Reduce oxygen species formation
		Reduce protein precursor accumulation toxicity
Peripheral insulin detemir	Reduced food intake by enhanced insulin action	Increased cerebrocortical beta activity
Intranal insulin	Increased brain insulin sensitivity	Improved IR/IRS1/PI3K signaling pathway
Metformin	Improved central insulin sensitivity	Remission of the increased tau phosphorylation in insulin resistant neurons
Peroxisome proliferator activated receptor agonists	Improved central insulin sensitivity	Neuroprotective effects
Gastrointestinal polypeptide analogues	Improved central insulin sensitivity	Reverting amyloid mediated receptor desensitization
Dual agonists	Improved central insulin sensitivity	Neuroprotective effects
Dipeptidyl-peptidase 4 inhibitors	Improved central insulin sensitivity	Phosphorylation and expression of the insulin receptor
Sodium dependent glucose co-transporter	Improved central insulin sensitivity	Improved mitochondrial function and synaptic plasticity
Conventional weight loss/caloric restriction	Normalizes hypothalamic neuronal responsiveness to glucose	Unclear
Bariatric surgery	Reverses increased insulin-stimulated brain glucose uptake in obese humans	Unclear
	Improved central insulin sensitivity	Increased central GLP-1 signaling

effects of bariatric surgery are partially mediated by enhanced central GLP-1 signaling positively modulating brain insulin sensitivity. An overview of this working model is provided in Figure 13.2. An overview of intervention effects on brain insulin sensitivity is given in Table 13.1.

CONCLUDING REMARKS

Taken together, brain insulin sensitivity is linked with peripheral insulin action and is impaired in obesity resulting in higher insulin-mediated central glucose metabolism. Brain insulin sensitivity appears to be important in long-term weight regulation and outcomes of lifestyle interventions and bariatric surgery. Bariatric surgery itself most likely has beneficial effects on brain insulin sensitivity; however, the exact underlying mechanisms and whether this is obligatory for the long-term weight loss and metabolic effects of bariatric surgery remains to be shown.

LITERATURE

1. S. T. Ferreira, Brain insulin, insulin-like growth factor 1 and glucagon-like peptide 1 signalling in Alzheimer's disease. *J. Neuroendocrinol.* **33**, e12959 (2021).
2. S. M. de la Monte, M. Tong, N. Lester-Coll, M. Plater, J. R. Wands, Therapeutic rescue of neurodegeneration in experimental type 3 diabetes: relevance to Alzheimer's disease. *J. Alzheimers Dis. JAD.* **10**, 89–109 (2006).
3. H. Jinnouchi, S. Sugiyama, A. Yoshida, K. Hieshima, N. Kurinami, T. Suzuki, F. Miyamoto, K. Kajiwara, K. Matsui, T. Jinnouchi, Liraglutide, a glucagon-like peptide-1 analog, increased insulin sensitivity assessed by hyperinsulinemic-euglycemic clamp examination in patients with uncontrolled type 2 diabetes mellitus. *J. Diabetes Res.* **2015**, 706416 (2015).
4. K. Anthony, L. J. Reed, J. T. Dunn, E. Bingham, D. Hopkins, P. K. Marsden, S. A. Amiel, Attenuation of insulin-evoked responses in brain networks controlling appetite and reward in insulin resistance: the cerebral basis for impaired control of food intake in metabolic syndrome? *Diabetes.* **55**, 2986–2992 (2006).
5. F. Rodriquez de Fonseca, M. Navarro, E. Alvarez, I. Roncero, J. A. Chowen, O. Maestre, R. Gómez, R. M. Muñoz, J. Eng, E. Blázquez, Peripheral versus central effects of glucagon-like peptide-1 receptor agonists on satiety and body weight loss in Zucker obese rats. *Metabolism.* **49**, 709–717 (2000).
6. J. J. Holst, The physiology of glucagon-like peptide 1. *Physiol. Rev.* **87**, 1409–1439 (2007).
7. C. Knauf, P. D. Cani, D.-H. Kim, M. A. Iglesias, C. Chabo, A. Waget, A. Colom, S. Rastrelli, N. M. Delzenne, D. J. Drucker, R. J. Seeley, R. Burcelin, Role of central nervous system glucagon-like Peptide-1 receptors in enteric glucose sensing. *Diabetes.* **57**, 2603–2612 (2008).
8. K. Hunter, C. Hölscher, Drugs developed to treat diabetes, liraglutide and lixisenatide, cross the blood brain barrier and enhance neurogenesis. *BMC Neurosci.* **13**, 33 (2012).
9. S. C. Cork, J. E. Richards, M. K. Holt, F. M. Gribble, F. Reimann, S. Trapp, Distribution and characterisation of Glucagon-like peptide-1 receptor expressing cells in the mouse brain. *Mol. Metab.* **4**, 718–731 (2015).
10. T. R. Bomfim, L. Forny-Germano, L. B. Sathler, J. Brito-Moreira, J.-C. Houzel, H. Decker, M. A. Silverman, H. Kazi, H. M. Melo, P. L. McClean, C. Holscher, S. E. Arnold, K. Talbot, W. L. Klein, D. P. Munoz, S. T. Ferreira, F. G. De Felice, An anti-diabetes agent protects the mouse brain from defective insulin signaling caused by Alzheimer's disease- associated Aβ oligomers. *J. Clin. Invest.* **122**, 1339–1353 (2012).
11. K. Talbot, H.-Y. Wang, H. Kazi, L.-Y. Han, K. P. Bakshi, A. Stucky, R. L. Fuino, K. R. Kawaguchi, A. J. Samoyedny, R. S. Wilson, Z. Arvanitakis, J. A. Schneider, B. A. Wolf, D. A. Bennett, J. Q. Trojanowski, S. E. Arnold, Demonstrated brain insulin resistance in Alzheimer's disease patients is associated with IGF-1 resistance, IRS-1 dysregulation, and cognitive decline. *J. Clin. Invest.* **122**, 1316–1338 (2012).
12. C. Hölscher, Central effects of GLP-1: new opportunities for treatments of neurodegenerative diseases. *J. Endocrinol.* **221**, T31–41 (2014).
13. H. H. Hansen, K. Fabricius, P. Barkholt, M. L. Niehoff, J. E. Morley, J. Jelsing, C. Pyke, L. B. Knudsen, S. A. Farr, N. Vrang, The GLP-1 receptor agonist liraglutide improves memory function and increases hippocampal CA1 neuronal numbers in a senescence-accelerated mouse model of Alzheimer's disease. *J. Alzheimers Dis. JAD.* **46**, 877–888 (2015).
14. P. L. McClean, V. Parthsarathy, E. Faivre, C. Hölscher, The diabetes drug liraglutide prevents degenerative processes in a mouse model of

Alzheimer's disease. *J. Neurosci. Off. J. Soc. Neurosci.* **31**, 6587–6594 (2011).
15. A. F. Batista, L. Forny-Germano, J. R. Clarke, N. M. Lyra E Silva, J. Brito-Moreira, S. E. Boehnke, A. Winterborn, B. C. Coe, A. Lablans, J. F. Vital, S. A. Marques, A. M. Martinez, M. Gralle, C. Holscher, W. L. Klein, J.-C. Houzel, S. T. Ferreira, D. P. Munoz, F. G. De Felice, The diabetes drug liraglutide reverses cognitive impairment in mice and attenuates insulin receptor and synaptic pathology in a non-human primate model of Alzheimer's disease. *J. Pathol.* **245**, 85–100 (2018).
16. D. Jo, G. Yoon, J. Song, Role of exendin-4 in brain insulin resistance, mitochondrial function, and neurite outgrowth in neurons under palmitic acid-induced oxidative stress. *Antioxid. Basel Switz.* **10**, 78 (2021).
17. Y. Li, T. Perry, M. S. Kindy, B. K. Harvey, D. Tweedie, H. W. Holloway, K. Powers, H. Shen, J. M. Egan, K. Sambamurti, A. Brossi, D. K. Lahiri, M. P. Mattson, B. J. Hoffer, Y. Wang, N. H. Greig, GLP-1 receptor stimulation preserves primary cortical and dopaminergic neurons in cellular and rodent models of stroke and Parkinsonism. *Proc. Natl. Acad. Sci. U.S.A.* **106**, 1285–1290 (2009).
18. L.-Y. Zhang, Q.-Q. Jin, C. Hölscher, L. Li, Glucagon-like peptide-1/glucose-dependent insulinotropic polypeptide dual receptor agonist DA-CH5 is superior to exendin-4 in protecting neurons in the 6-hydroxydopamine rat Parkinson model. *Neural Regen. Res.* **16**, 1660–1670 (2021).
19. I. Aviles-Olmos, P. Limousin, A. Lees, T. Foltynie, Parkinson's disease, insulin resistance and novel agents of neuroprotection. *Brain J. Neurol.* **136**, 374–384 (2013).
20. G. Rodriguez-Araujo, H. Nakagami, Y. Takami, T. Katsuya, H. Akasaka, S. Saitoh, K. Shimamoto, R. Morishita, H. Rakugi, Y. Kaneda, Low alpha-synuclein levels in the blood are associated with insulin resistance. *Sci. Rep.* **5**, 12081 (2015).
21. E. Marciniak, A. Leboucher, E. Caron, T. Ahmed, A. Tailleux, J. Dumont, T. Issad, E. Gerhardt, P. Pagesy, M. Vileno, C. Bournonville, M. Hamdane, K. Bantubungi, S. Lancel, D. Demeyer, S. Eddarkaoui, E. Vallez, D. Vieau, S. Humez, E. Faivre, B. Grenier-Boley, T. F. Outeiro, B. Staels, P. Amouyel, D. Balschun, L. Buee, D. Blum, Tau deletion promotes brain insulin resistance. *J. Exp. Med.* **214**, 2257–2269 (2017).
22. S. Craft, Insulin resistance syndrome and Alzheimer's disease: age- and obesity-related effects on memory, amyloid, and inflammation. *Neurobiol. Aging.* **26 Suppl 1**, 65–69 (2005).
23. R. J. Mullins, T. C. Diehl, C. W. Chia, D. Kapogiannis, Insulin resistance as a link between amyloid-beta and tau pathologies in Alzheimer's disease. *Front. Aging Neurosci.* **9**, 118 (2017).
24. M. Gejl, A. Gjedde, L. Egefjord, A. Møller, S. B. Hansen, K. Vang, A. Rodell, H. Brændgaard, H. Gottrup, A. Schacht, N. Møller, B. Brock, J. Rungby, In Alzheimer's disease, 6-month treatment with GLP-1 analog prevents decline of brain glucose metabolism: randomized, placebo-controlled, double-blind clinical trial. *Front. Aging Neurosci.* **8**, 108 (2016).
25. G. D. Femminella, E. Frangou, S. B. Love, G. Busza, C. Holmes, C. Ritchie, R. Lawrence, B. McFarlane, G. Tadros, B. H. Ridha, C. Bannister, Z. Walker, H. Archer, E. Coulthard, B. R. Underwood, A. Prasanna, P. Koranteng, S. Karim, K. Junaid, B. McGuinness, R. Nilforooshan, A. Macharouthu, A. Donaldson, S. Thacker, G. Russell, N. Malik, V. Mate, L. Knight, S. Kshemendran, J. Harrison, C. Hölscher, D. J. Brooks, A. P. Passmore, C. Ballard, P. Edison, Evaluating the effects of the novel GLP-1 analogue liraglutide in Alzheimer's disease: study protocol for a randomised controlled trial (ELAD study). *Trials.* **20**, 191 (2019).
26. I. Aviles-Olmos, J. Dickson, Z. Kefalopoulou, A. Djamshidian, P. Ell, T. Soderlund, P. Whitton, R. Wyse, T. Isaacs, A. Lees, P. Limousin, T. Foltynie, Exenatide and the treatment of patients with Parkinson's disease. *J. Clin. Invest.* **123**, 2730–2736 (2013).
27. D. Athauda, K. Maclagan, S. S. Skene, M. Bajwa-Joseph, D. Letchford, K. Chowdhury, S. Hibbert, N. Budnik, L. Zampedri, J. Dickson, Y. Li, I. Aviles-Olmos, T. T. Warner, P. Limousin, A. J. Lees, N. H. Greig, S. Tebbs, T. Foltynie, Exenatide once weekly versus placebo in Parkinson's disease: a randomised, double-blind, placebo-controlled trial. *Lancet Lond. Engl.* **390**, 1664–1675 (2017).
28. H. Ueno, M. Mizuta, T. Shiiya, W. Tsuchimochi, K. Noma, N. Nakashima, M. Fujihara, M. Nakazato, Exploratory trial of intranasal administration of glucagon-like peptide-1 in Japanese patients with type 2 diabetes. *Diabetes Care.* **37**, 2024–2027 (2014).
29. N. Kamei, N. Okada, T. Ikeda, H. Choi, Y. Fujiwara, H. Okumura, M. Takeda-Morishita, Effective nose-to-brain delivery of exendin-4

via coadministration with cell-penetrating peptides for improving progressive cognitive dysfunction. *Sci. Rep.* **8**, 17641 (2018).
30. N. V. Rykalina, E. V. Askerova, N. V. Bulushova, D. G. Kozlov, Intranasal human recombinant modified glucagon-like peptide-1: high antihyperglycemic activity and duration of action in mice. *Bull. Exp. Biol. Med.* **169**, 53–56 (2020).
31. M. Heni, S. Kullmann, B. Gallwitz, H.-U. Häring, H. Preissl, A. Fritsche, Dissociation of GLP-1 and insulin association with food processing in the brain: GLP-1 sensitivity despite insulin resistance in obese humans. *Mol. Metab.* **4**, 971–976 (2015).
32. L. Benedict, C. A. Nelson, E. Schunk, K. Sullwold, E. R. Seaquist, Effect of insulin on the brain activity obtained during visual and memory tasks in healthy human subjects. *Neuroendocrinology.* **83**, 20–26 (2006).
33. M. Hallschmid, K. Jauch-Chara, O. Korn, M. Mölle, B. Rasch, J. Born, B. Schultes, W. Kern, Euglycemic infusion of insulin detemir compared with human insulin appears to increase direct current brain potential response and reduces food intake while inducing similar systemic effects. *Diabetes.* **59**, 1101–1107 (2010).
34. K. Porubská, O. Tschritter, A. Fritsche, H.-U. Häring, N. Birbaumer, H. Preissl, Processing of food-relevant visual stimuli measured by magnetoencephalography and its modulation by insulin. *Int. Congr. Ser.* **1300**, 516–519 (2007).
35. O. Tschritter, H. Preissl, A. M. Hennige, M. Stumvoll, K. Porubska, R. Frost, H. Marx, B. Klösel, W. Lutzenberger, N. Birbaumer, H.-U. Häring, A. Fritsche, The cerebrocortical response to hyperinsulinemia is reduced in overweight humans: a magnetoencephalographic study. *Proc. Natl. Acad. Sci.* **103**, 12103–12108 (2006).
36. M. Heni, S. Kullmann, H. Preissl, A. Fritsche, H.-U. Häring, Impaired insulin action in the human brain: causes and metabolic consequences. *Nat. Rev. Endocrinol.* **11**, 701–711 (2015).
37. O. Tschritter, A. M. Hennige, H. Preissl, K. Porubska, S. A. Schäfer, W. Lutzenberger, F. Machicao, N. Birbaumer, A. Fritsche, H.-U. Häring, Cerebrocortical beta activity in overweight humans responds to insulin detemir. *PloS One.* **2**, e1196 (2007).
38. D. P. Begg, A. A. May, J. D. Mul, M. Liu, D. A. D'Alessio, R. J. Seeley, S. C. Woods, Insulin detemir is transported from blood to cerebrospinal fluid and has prolonged central anorectic action relative to NPH insulin. *Diabetes.* **64**, 2457–2466 (2015).
39. A. M. Hennige, T. Sartorius, O. Tschritter, H. Preissl, A. Fritsche, P. Ruth, H.-U. Häring, Tissue selectivity of insulin detemir action in vivo. *Diabetologia.* **49**, 1274–1282 (2006).
40. B. M. Frier, D. Russell-Jones, T. Heise, A comparison of insulin detemir and neutral protamine Hagedorn (isophane) insulin in the treatment of diabetes: a systematic review. *Diabetes Obes. Metab.* **15**, 978–986 (2013).
41. M. Yarchoan, S. E. Arnold, Repurposing diabetes drugs for brain insulin resistance in Alzheimer disease. *Diabetes.* **63**, 2253–2261 (2014).
42. J. Born, T. Lange, W. Kern, G. P. McGregor, U. Bickel, H. L. Fehm, Sniffing neuropeptides: a transnasal approach to the human brain. *Nat. Neurosci.* **5**, 514–516 (2002).
43. S. Craft, R. Raman, T. W. Chow, M. S. Rafii, C.-K. Sun, R. A. Rissman, M. C. Donohue, J. B. Brewer, C. Jenkins, K. Harless, D. Gessert, P. S. Aisen, Safety, efficacy, and feasibility of intranasal insulin for the treatment of mild cognitive impairment and Alzheimer disease dementia: a randomized clinical trial. *JAMA Neurol.* **77**, 1099–1109 (2020).
44. V. Schmid, S. Kullmann, W. Gfrörer, V. Hund, M. Hallschmid, H.-P. Lipp, H.-U. Häring, H. Preissl, A. Fritsche, M. Heni, Safety of intranasal human insulin: a review. *Diabetes Obes. Metab.* **20**, 1563–1577 (2018).
45. Y. Chen, Y. Zhao, C. Dai, Z. Liang, X. Run, K. Iqbal, F. Liu, C.-X. Gong, Intranasal insulin restores insulin signaling, increases synaptic proteins, and reduces Aβ level and microglia activation in the brains of 3xTg-AD mice. *Exp. Neurol.* **261**, 610–619 (2014).
46. S. Craft, Intranasal insulin therapy for Alzheimer disease and amnestic mild cognitive impairment: a pilot clinical trial. *Arch. Neurol.* **69**, 29 (2012).
47. W. Lv, J. Wen, L. Li, R. Sun, J. Wang, Y. Xian, C. Cao, Y. Wang, Y. Gao, The effect of metformin on food intake and its potential role in hypothalamic regulation in obese diabetic rats. *Brain Res.* **1444**, 11–19 (2012).
48. A. Gupta, B. Bisht, C. S. Dey, Peripheral insulin-sensitizer drug metformin ameliorates neuronal insulin resistance and Alzheimer's-like changes. *Neuropharmacology.* **60**, 910–920 (2011).
49. A. M. Koenig, D. Mechanic-Hamilton, S. X. Xie, M. F. Combs, A. R. Cappola, L. Xie, J. A. Detre,

D. A. Wolk, S. E. Arnold, Effects of the insulin sensitizer metformin in Alzheimer's disease: pilot data from a randomized placebo-controlled crossover study. *Alzheimer Dis. Assoc. Disord.* **31**, 107–113 (2017).
50. J. A. Luchsinger, T. Perez, H. Chang, P. Mehta, J. Steffener, G. Pradabhan, M. Ichise, J. Manly, D. P. Devanand, E. Bagiella, Metformin in amnestic mild cognitive impairment: results of a pilot randomized placebo controlled clinical trial. *J. Alzheimers Dis. JAD.* **51**, 501–514 (2016).
51. M. Guo, J. Mi, Q.-M. Jiang, J.-M. Xu, Y.-Y. Tang, G. Tian, B. Wang, Metformin may produce antidepressant effects through improvement of cognitive function among depressed patients with diabetes mellitus. *Clin. Exp. Pharmacol. Physiol.* **41**, 650–656 (2014).
52. F. AlHussain, Y. AlRuthia, H. Al-Mandeel, A. Bellahwal, F. Alharbi, Y. Almogbel, O. Awwad, R. Dala'een, F. A. Alharbi, Metformin improves the depression symptoms of women with polycystic ovary syndrome in a lifestyle modification program. *Patient Prefer. Adherence.* **14**, 737–746 (2020).
53. M. S. Abdallah, E. M. Mosalam, A.-A. A. Zidan, K. S. Elattar, S. A. Zaki, A. N. Ramadan, A. M. Ebeid, The antidiabetic metformin as an adjunct to antidepressants in patients with major depressive disorder: a proof-of-concept, randomized, double-blind, placebo-controlled trial. *Neurother. J. Am. Soc. Exp. Neurother.* **17**, 1897–1906 (2020).
54. L. B. Martins, J. R. Braga Tibães, M. Berk, A. L. Teixeira, Diabetes and mood disorders: shared mechanisms and therapeutic opportunities. *Int. J. Psychiatry Clin. Pract.*, 1–13 (2021).
55. T. Jiang, J.-T. Yu, X.-C. Zhu, H.-F. Wang, M.-S. Tan, L. Cao, Q.-Q. Zhang, L. Gao, J.-Q. Shi, Y.-D. Zhang, L. Tan, Acute metformin preconditioning confers neuroprotection against focal cerebral ischaemia by pre-activation of AMPK-dependent autophagy. *Br. J. Pharmacol.* **171**, 3146–3157 (2014).
56. Y. N. Paudel, E. Angelopoulou, C. Piperi, Mohd. F. Shaikh, I. Othman, Emerging neuroprotective effect of metformin in Parkinson's disease: a molecular crosstalk. *Pharmacol. Res.* **152**, 104593 (2020).
57. N. Nicolakakis, T. Aboulkassim, B. Ongali, C. Lecrux, P. Fernandes, P. Rosa-Neto, X.-K. Tong, E. Hamel, Complete rescue of cerebrovascular function in aged Alzheimer's disease transgenic mice by antioxidants and pioglitazone, a peroxisome proliferator-activated receptor γ agonist. *J. Neurosci.* **28**, 9287–9296 (2008).
58. G. S. Watson, B. A. Cholerton, M. A. Reger, L. D. Baker, S. R. Plymate, S. Asthana, M. A. Fishel, J. J. Kulstad, P. S. Green, D. G. Cook, S. E. Kahn, M. L. Keeling, S. Craft, Preserved cognition in patients with early Alzheimer disease and amnestic mild cognitive impairment during treatment with rosiglitazone: a preliminary study. *Am. J. Geriatr. Psychiatry Off. J. Am. Assoc. Geriatr. Psychiatry.* **13**, 950–958 (2005).
59. C. Harrington, S. Sawchak, C. Chiang, J. Davies, C. Donovan, A. M. Saunders, M. Irizarry, B. Jeter, M. Zvartau-Hind, C. H. van Dyck, M. Gold, Rosiglitazone does not improve cognition or global function when used as adjunctive therapy to AChE inhibitors in mild-to-moderate Alzheimer's disease: two phase 3 studies. *Curr. Alzheimer Res.* **8**, 592–606 (2011).
60. S. Chamberlain, H. Gabriel, W. Strittmatter, J. Didsbury, An exploratory phase IIa study of the PPAR delta/gamma agonist T3D-959 assessing metabolic and cognitive function in subjects with mild to moderate Alzheimer's disease. *J. Alzheimers Dis.* **73**, 1085–1103 (2020).
61. S. M. de la Monte, Insulin resistance and neurodegeneration: progress towards the development of new therapeutics for Alzheimer's disease. *Drugs.* **77**, 47–65 (2017).
62. A. E. Adriaenssens, E. K. Biggs, T. Darwish, J. Tadross, T. Sukthankar, M. Girish, J. Polex-Wolf, B. Y. Lam, I. Zvetkova, W. Pan, D. Chiarugi, G. S. H. Yeo, C. Blouet, F. M. Gribble, F. Reimann, Glucose-dependent insulinotropic polypeptide receptor-expressing cells in the hypothalamus regulate food intake. *Cell Metab.* **30**, 987–996.e6 (2019).
63. Q. Zhang, C. T. Delessa, R. Augustin, M. Bakhti, G. Colldén, D. J. Drucker, A. Feuchtinger, C. G. Caceres, G. Grandl, A. Harger, S. Herzig, S. Hofmann, C. L. Holleman, M. Jastroch, S. Keipert, M. Kleinert, P. J. Knerr, K. Kulaj, B. Legutko, H. Lickert, X. Liu, G. Luippold, D. Lutter, E. Malogajski, M. T. Medina, S. A. Mowery, A. Blutke, D. Perez-Tilve, C. Salinno, L. Sehrer, R. D. DiMarchi, M. H. Tschöp, K. Stemmer, B. Finan, C. Wolfrum, T. D. Müller, The glucose-dependent insulinotropic polypeptide (GIP) regulates body weight and food intake via CNS-GIPR signaling. *Cell Metab.* **33**, 833–844.e5 (2021).
64. A. M. Duffy, C. Hölscher, The incretin analogue D-Ala2GIP reduces plaque load, astrogliosis and

oxidative stress in an APP/PS1 mouse model of Alzheimer's disease. *Neuroscience*. **228**, 294–300 (2013).
65. B. Finan, T. Ma, N. Ottaway, T. D. Müller, K. M. Habegger, K. M. Heppner, H. Kirchner, J. Holland, J. Hembree, C. Raver, S. H. Lockie, D. L. Smiley, V. Gelfanov, B. Yang, S. Hofmann, D. Bruemmer, D. J. Drucker, P. T. Pfluger, D. Perez-Tilve, J. Gidda, L. Vignati, L. Zhang, J. B. Hauptman, M. Lau, M. Brecheisen, S. Uhles, W. Riboulet, E. Hainaut, E. Sebokova, K. Conde-Knape, A. Konkar, R. D. DiMarchi, M. H. Tschöp, Unimolecular dual incretins maximize metabolic benefits in rodents, monkeys, and humans. *Sci. Transl. Med.* **5**, 209ra151 (2013).
66. C. Hölscher, Brain insulin resistance: role in neurodegenerative disease and potential for targeting. *Expert Opin. Investig. Drugs*. **29**, 333–348 (2020).
67. J. K. Kim, J. S. Lee, J. Choi, S. Y. Jung, S. H. Lee, I. Y. Choi, S. J. Kim, Novel combination of a long-acting GLP-1/GIP/glucagon triple agonist (HM15211) and once-weekly basal insulin (HM12460a) offers improved glucose lowering and weight loss in a diabetic animal model. *Diabetes*. **67**, 77-OR (2018).
68. H. Pintana, W. Pongkan, W. Pratchayasakul, N. Chattipakorn, S. C. Chattipakorn, Dipeptidyl peptidase 4 inhibitor improves brain insulin sensitivity, but fails to prevent cognitive impairment in orchiectomy obese rats. *J. Endocrinol.* **226**, M1–M11 (2015).
69. P. Sa-nguanmoo, P. Tanajak, S. Kerdphoo, T. Jaiwongkam, W. Pratchayasakul, N. Chattipakorn, S. C. Chattipakorn, SGLT2-inhibitor and DPP-4 inhibitor improve brain function via attenuating mitochondrial dysfunction, insulin resistance, inflammation, and apoptosis in HFD-induced obese rats. *Toxicol. Appl. Pharmacol.* **333**, 43–50 (2017).
70. V. A. Gault, R. Lennox, P. R. Flatt, Sitagliptin, a dipeptidyl peptidase-4 inhibitor, improves recognition memory, oxidative stress and hippocampal neurogenesis and upregulates key genes involved in cognitive decline. *Diabetes Obes. Metab.* **17**, 403–413 (2015).
71. M. Maciejczyk, E. Żebrowska, A. Chabowski, Insulin resistance and oxidative stress in the brain: what's new? *Int. J. Mol. Sci.* **20** (2019), doi:10.3390/ijms20040874.
72. M. Heni, S. Kullmann, R. Wagner, J. Hummel, C. Dannecker, A. Vosseler, L. Fritsche, K. Kantartzis, J. Machann, H.-U. Haering, A. Fritsche, H. Preissl, 148-OR: empagliflozin improves insulin sensitivity of the hypothalamus in humans with prediabetes. *Diabetes*. **69** (2020), doi:10.2337/db20-148-OR.
73. D. P. Begg, Insulin transport into the brain and cerebrospinal fluid. *Vitam. Horm.* **98**, 229–248 (2015).
74. J. M. Hill, M. A. Lesniak, C. B. Pert, J. Roth, Autoradiographic localization of insulin receptors in rat brain: prominence in olfactory and limbic areas. *Neuroscience*. **17**, 1127–1138 (1986).
75. S. Kullmann, M. Heni, M. Hallschmid, A. Fritsche, H. Preissl, H.-U. Häring, Brain insulin resistance at the crossroads of metabolic and cognitive disorders in humans. *Physiol. Rev.* **96**, 1169–1209 (2016).
76. M. Brundel, L. J. Kappelle, G. J. Biessels, Brain imaging in type 2 diabetes. *Eur. Neuropsychopharmacol. J. Eur. Coll. Neuropsychopharmacol.* **24**, 1967–1981 (2014).
77. S. Kullmann, F. Schweizer, R. Veit, A. Fritsche, H. Preissl, Compromised white matter integrity in obesity. *Obes. Rev. Off. J. Int. Assoc. Study Obes.* **16**, 273–281 (2015).
78. G. N. Bischof, D. C. Park, Obesity and aging: consequences for cognition, brain structure, and brain function. *Psychosom. Med.* **77**, 697–709 (2015).
79. D. Russell-Jones, R. Khan, Insulin-associated weight gain in diabetes--causes, effects and coping strategies. *Diabetes Obes. Metab.* **9**, 799–812 (2007).
80. S. I. McFarlane, Insulin therapy and type 2 diabetes: management of weight gain. *J. Clin. Hypertens. Greenwich Conn.* **11**, 601–607 (2009).
81. M. Hallschmid, C. Benedict, B. Schultes, H.-L. Fehm, J. Born, W. Kern, Intranasal insulin reduces body fat in men but not in women. *Diabetes*. **53**, 3024–3029 (2004).
82. C. Benedict, W. Kern, B. Schultes, J. Born, M. Hallschmid, Differential sensitivity of men and women to anorexigenic and memory-improving effects of intranasal insulin. *J. Clin. Endocrinol. Metab.* **93**, 1339–1344 (2008).
83. K. Honda, H. Kamisoyama, T. Saneyasu, K. Sugahara, S. Hasegawa, Central administration of insulin suppresses food intake in chicks. *Neurosci. Lett.* **423**, 153–157 (2007).
84. D. J. Clegg, C. A. Riedy, K. A. B. Smith, S. C. Benoit, S. C. Woods, Differential sensitivity to central leptin and insulin in male and female rats. *Diabetes*. **52**, 682–687 (2003).
85. L. M. Brown, D. J. Clegg, S. C. Benoit, S. C. Woods, Intraventricular insulin and leptin

reduce food intake and body weight in C57BL/6J mice. *Physiol. Behav.* **89**, 687–691 (2006).
86. M. Hallschmid, S. Higgs, M. Thienel, V. Ott, H. Lehnert, Postprandial administration of intranasal insulin intensifies satiety and reduces intake of palatable snacks in women. *Diabetes.* **61**, 782–789 (2012).
87. O. Tschritter, H. Preissl, A. M. Hennige, M. Stumvoll, K. Porubska, R. Frost, H. Marx, B. Klösel, W. Lutzenberger, N. Birbaumer, H.-U. Häring, A. Fritsche, The cerebrocortical response to hyperinsulinemia is reduced in overweight humans: a magnetoencephalographic study. *Proc. Natl. Acad. Sci. U.S.A.* **103**, 12103–12108 (2006).
88. M. Guthoff, K. T. Stingl, O. Tschritter, M. Rogic, M. Heni, K. Stingl, M. Hallschmid, H.-U. Häring, A. Fritsche, H. Preissl, A. M. Hennige, The insulin-mediated modulation of visually evoked magnetic fields is reduced in obese subjects. *PloS One.* **6**, e19482 (2011).
89. M. Heni, S. Kullmann, C. Ketterer, M. Guthoff, M. Bayer, H. Staiger, F. Machicao, H.-U. Häring, H. Preissl, R. Veit, A. Fritsche, Differential effect of glucose ingestion on the neural processing of food stimuli in lean and overweight adults. *Hum. Brain Mapp.* **35**, 918–928 (2014).
90. K. Anthony, L. J. Reed, J. T. Dunn, E. Bingham, D. Hopkins, P. K. Marsden, S. A. Amiel, Attenuation of insulin-evoked responses in brain networks controlling appetite and reward in insulin resistance: the cerebral basis for impaired control of food intake in metabolic syndrome? *Diabetes.* **55**, 2986–2992 (2006).
91. S. G. Hasselbalch, G. M. Knudsen, C. Videbaek, L. H. Pinborg, J. F. Schmidt, S. Holm, O. B. Paulson, No effect of insulin on glucose blood-brain barrier transport and cerebral metabolism in humans. *Diabetes.* **48**, 1915–1921 (1999).
92. Change in hexose distribution volume and fractional utilization of [18F]-2-deoxy-2-fluoro--D-glucose in brain during acute hypoglycemia in humans - PubMed, (available at https://-pubmed.ncbi.nlm.nih.gov/2227124/).
93. J. Hirvonen, K. A. Virtanen, L. Nummenmaa, J. C. Hannukainen, M.-J. Honka, M. Bucci, S. V. Nesterov, R. Parkkola, J. Rinne, P. Iozzo, P. Nuutila, Effects of insulin on brain glucose metabolism in impaired glucose tolerance. *Diabetes.* **60**, 443–447 (2011).
94. L. Nummenmaa, J. Hirvonen, J. C. Hannukainen, H. Immonen, M. M. Lindroos, P. Salminen, P. Nuutila, Dorsal striatum and its limbic connectivity mediate abnormal anticipatory reward processing in obesity. *PLOS ONE.* **7**, e31089 (2012).
95. J. Pegueroles, A. Pané, E. Vilaplana, V. Montal, A. Bejanin, L. Videla, M. Carmona-Iragui, I. Barroeta, A. Ibarzabal, A. Casajoana, D. Alcolea, S. Valldeneu, M. Altuna, A. de Hollanda, J. Vidal, E. Ortega, R. Osorio, A. Convit, R. Blesa, A. Lleó, J. Fortea, A. Jiménez, Obesity impacts brain metabolism and structure independently of amyloid and tau pathology in healthy elderly. *Alzheimers Dement. Diagn. Assess. Dis. Monit.* 12, e12052 (2020).
96. J. M. Kim, M. Jang, E. H. Kim, M. Kim, S. J. Choi, K. Kim, K. Pak, Y. K. Jeon, S. S. Kim, B. H. Kim, S.-J. Kim, I. J. Kim, Cerebral glucose metabolism differs according to future weight change. *Brain Imaging Behav.* **14**, 2295–2301 (2020).
97. S. Kullmann, M. Heni, R. Veit, K. Scheffler, J. Machann, H.-U. Häring, A. Fritsche, H. Preissl, Selective insulin resistance in homeostatic and cognitive control brain areas in overweight and obese adults. *Diabetes Care.* **38**, 1044–1050 (2015).
98. D. A. Van Vugt, A. Krzemien, H. Alsaadi, T. C. Frank, R. L. Reid, Glucose-induced inhibition of the appetitive brain response to visual food cues in polycystic ovary syndrome patients. *Brain Res.* **1558**, 44–56 (2014).
99. M. Hallschmid, C. Benedict, B. Schultes, J. Born, W. Kern, Obese men respond to cognitive but not to catabolic brain insulin signaling. *Int. J. Obes.* 2005. **32**, 275–282 (2008).
100. M. Heni, R. Wagner, S. Kullmann, R. Veit, H. Mat Husin, K. Linder, C. Benkendorff, A. Peter, N. Stefan, H.-U. Häring, H. Preissl, A. Fritsche, Central insulin administration improves whole-body insulin sensitivity via hypothalamus and parasympathetic outputs in men. *Diabetes.* **63**, 4083–4088 (2014).
101. E. Rebelos, H. Immonen, M. Bucci, J. C. Hannukainen, L. Nummenmaa, M.-J. Honka, M. Soinio, P. Salminen, E. Ferrannini, P. Iozzo, P. Nuutila, Brain glucose uptake is associated with endogenous glucose production in obese patients before and after bariatric surgery and predicts metabolic outcome at follow-up. *Diabetes Obes. Metab.* **21**, 218–226 (2019).
102. W. M. Teeuwisse, R. L. Widya, M. Paulides, H. J. Lamb, J. W. A. Smit, A. de Roos, M. A. van Buchem, H. Pijl, J. van der Grond, Short-term caloric restriction normalizes hypothalamic neuronal responsiveness to glucose ingestion in

patients with type 2 diabetes. *Diabetes*. **61**, 3255–3259 (2012).
103. O. Tschritter, H. Preissl, A. M. Hennige, T. Sartorius, K. T. Stingl, M. Heni, C. Ketterer, N. Stefan, J. Machann, E. Schleicher, A. Fritsche, H.-U. Häring, High cerebral insulin sensitivity is associated with loss of body fat during lifestyle intervention. *Diabetologia*. **55**, 175–182 (2012).
104. S. Kullmann, V. Valenta, R. Wagner, O. Tschritter, J. Machann, H.-U. Häring, H. Preissl, A. Fritsche, M. Heni, Brain insulin sensitivity is linked to adiposity and body fat distribution. *Nat. Commun.* **11** (2020), doi:10.1038/s41467-020-15686-y.
105. S. Frank, B. Wilms, R. Veit, B. Ernst, M. Thurnheer, S. Kullmann, A. Fritsche, N. Birbaumer, H. Preissl, B. Schultes, Altered brain activity in severely obese women may recover after Roux-en Y gastric bypass surgery. *Int. J. Obes.* 2005. **38**, 341–348 (2014).
106. J. J. Tuulari, H. K. Karlsson, J. Hirvonen, J. C. Hannukainen, M. Bucci, M. Helmiö, J. Ovaska, M. Soinio, P. Salminen, N. Savisto, L. Nummenmaa, P. Nuutila, Weight loss after bariatric surgery reverses insulin-induced increases in brain glucose metabolism of the morbidly obese. *Diabetes*. **62**, 2747–2751 (2013).
107. Metabolic surgery improves brain glucose metabolism, cognitive function and neuroplasticity in obese non-diabetic subjects - Virtual Meeting | EASD, (available at https://www.easd.org/virtualmeeting/home.html#!resources/-metabolic-surgery-improves-brain-glucose-metabolism-cognitive-function-and-neuroplasticity-in-obese-non-diabetic-subjects-7b4c2be8-f3ae-4c7d-a731-a0ec106ea05e).
108. K. M. McTigue, R. Wellman, E. Nauman, J. Anau, R. Y. Coley, A. Odor, J. Tice, K. J. Coleman, A. Courcoulas, R. E. Pardee, S. Toh, C. D. Janning, N. Williams, A. Cook, J. L. Sturtevant, C. Horgan, D. Arterburn, Comparing the 5-year diabetes outcomes of sleeve gastrectomy and gastric bypass. *JAMA Surg.* (2020), doi:10.1001/jamasurg.2020.0087.
109. S. Scholtz, A. D. Miras, N. Chhina, C. G. Prechtl, M. L. Sleeth, N. M. Daud, N. A. Ismail, G. Durighel, A. R. Ahmed, T. Olbers, R. P. Vincent, J. Alaghband-Zadeh, M. A. Ghatei, A. D. Waldman, G. S. Frost, J. D. Bell, C. W. le Roux, A. P. Goldstone, Obese patients after gastric bypass surgery have lower brain-hedonic responses to food than after gastric banding. *Gut*. **63**, 891–902 (2014).
110. H. Cerit, P. Davidson, T. Hye, P. Moondra, F. Haimovici, S. Sogg, S. Shikora, J. M. Goldstein, A. E. Evins, S. Whitfield-Gabrieli, L. E. Stoeckel, L. M. Holsen, Resting-state brain connectivity predicts weight loss and cognitive control of eating behavior after vertical sleeve gastrectomy. *Obes. Silver Spring Md.* **27**, 1846–1855 (2019).
111. S. van de Sande-Lee, F. R. S. Pereira, D. E. Cintra, P. T. Fernandes, A. R. Cardoso, C. R. Garlipp, E. A. Chaim, J. C. Pareja, B. Geloneze, L. M. Li, F. Cendes, L. A. Velloso, Partial reversibility of hypothalamic dysfunction and changes in brain activity after body mass reduction in obese subjects. *Diabetes*. **60**, 1699–1704 (2011).
112. K. E. Almby, M. H. Lundqvist, N. Abrahamsson, S. Kvernby, M. Fahlström, M. J. Pereira, M. Gingnell, F. A. Karlsson, G. Fanni, M. Sundbom, U. Wiklund, S. Haller, M. Lubberink, J. Wikström, J. W. Eriksson, Effects of gastric bypass surgery on the brain: simultaneous assessment of glucose uptake, blood flow, neural activity, and cognitive function during normo- and hypoglycemia. *Diabetes* (2021), doi:10.2337/db20-1172.
113. Y. Ye, M. A. E. Haija, D. A. Morgan, D. Guo, Y. Song, A. Frank, L. Tian, R. A. Riedl, C. M. L. Burnett, Z. Gao, Z. Zhu, S. K. Shahi, K. Zarei, A. Couvelard, N. Poté, L. Ribeiro-Parenti, A. Bado, L. Noureddine, A. Bellizzi, P. Kievit, A. K. Mangalam, L. V. Zingman, M. Le Gall, J. L. Grobe, L. M. Kaplan, D. Clegg, K. Rahmouni, M. Mokadem, Endocannabinoid receptor-1 and sympathetic nervous system mediate the beneficial metabolic effects of gastric bypass. *Cell Rep.* **33**, 108270 (2020).
114. C. R. Hutch, D. Sandoval, The role of GLP-1 in the metabolic success of bariatric surgery. *Endocrinology*. **158**, 4139–4151 (2017).
115. T. R. Bomfim, L. Forny-Germano, L. B. Sathler, J. Brito-Moreira, J.-C. Houzel, H. Decker, M. A. Silverman, H. Kazi, H. M. Melo, P. L. McClean, C. Holscher, S. E. Arnold, K. Talbot, W. L. Klein, D. P. Munoz, S. T. Ferreira, F. G. De Felice, An anti-diabetes agent protects the mouse brain from defective insulin signaling caused by Alzheimer's disease- associated Aβ oligomers. *J. Clin. Invest.* 122, 1339–1353 (2012).

INDEX

Note: **Bold** page numbers refer to tables; *italic* page numbers refer to figures.

For Product Safety Concerns and Information please contact our EU
representative GPSR@taylorandfrancis.com
Taylor & Francis Verlag GmbH, Kaufingerstraße 24, 80331 München, Germany

www.ingramcontent.com/pod-product-compliance
Ingram Content Group UK Ltd.
Pitfield, Milton Keynes, MK11 3LW, UK
UKHW052216190926
475702UK00006B/1038
9780367529482